MW00844409

Reinventing the Energy Value Chain:
Supply Chain Roadmaps for Digital Oilfields
through Hydrogen Fuel Cells

REINVENTING THE ENERGY VALUE CHAIN

SUPPLY CHAIN ROADMAPS FOR DIGITAL OILFIELDS THROUGH HYDROGEN FUEL CELLS

DAVID STEVEN JACOBY
&
ALOK RAJ GUPTA

PENNWELL BOOKS

Disclaimer

The recommendations, advice, descriptions, and the methods in this book are presented solely for educational purposes. The author and publisher assume no liability whatsoever for any loss or damage that results from the use of any of the material in this book. Use of the material in this book is solely at the risk of the user.

PennWell Books, LLC
10050 E 52nd Street
Tulsa, Oklahoma 74146 USA

866.777.1814
sales@pennwellbooks.com
www.pennwellbooks.com

Publisher: Matthew Dresher

Images: gettyimages

Library of Congress Cataloging-in-Publication Data

Names: Jacoby, David, author. | Gupta, Alok Raj, author.
Title: Reinventing the energy value chain : supply chain roadmaps for digital oilfields through hydrogen fuel cells / David Steven Jacoby and Alok Raj Gupta.
Description: Tulsa, Oklahoma : PennWell Books, [2021] | Includes bibliographical references and index.
Identifiers: LCCN 2021032022 | ISBN 9781955578004 (hardback) | ISBN 9781955578011 (ebook)
Subjects: LCSH: Energy industries—Management. | Renewable energy sources. | Business logistics.
Classification: LCC HD9502.A2 J35 2021 | DDC 333.79068—dc23
LC record available at https://lccn.loc.gov/2021032022

Printed in the United States of America

1 2 3 4 5 25 24 23 22 21

Contents

Figures and Tables

Dedication

This book is dedicated to my son Weston Fixler Jacoby for being more than I could ever be in so many ways, and for his inspirations on energy storage. I love you so much. And also, to my son Brent Fixler Jacoby and my daughter Camille Fixler Jacoby whose courage and wonderful radiant energy far exceed any quantitative units of measure used to describe the size of the largest energy giga-projects. Each of you has extraordinary talents and I can't wait for you to find, express, and exercise your awesome power and voice! And, of course, thank you Jessica for making it all possible and for making our family rich in so many ways.

Foreword

The energy transition away from conventional hydrocarbon-based energy towards cleaner sources has been an inevitable fact for approximately the past two decades. Growing and ever more clear evidence of climate change, together with changing consumer expectations and increasing intolerance of overt environmental pollution, have accelerated a shift in market behavior and governmental policy towards green and sustainable energy. For the past decade, we have also started seeing the decoupling of GDP growth throughout the world from oil price, a significant sign of change and for many something that should have been a primer for a renewed focus on a more diversified energy mix.

We have seen demand drop from 2015 due to developments in shale oil, directional drilling technologies, and enhanced exploration and production strategies and approaches. These developments had obviously driven the supply up to such sustained levels, which, coupled with the growth in renewables, natural gas, and increased efficiencies on the demand side, has meant a very different outlook for oil & gas. The drop in CAPEX and OPEX in the past few years, and especially during the COVID year of 2020, has added such pressure on the market with demand destruction (finished products and crude) that it has no doubt started creating a new normal to come.

But here I am reminded of the wisdom of the late Ahmed Zaki Yamani, the former Saudi oil minister, who once said that as the Stone Age didn't end because we ran out of stones, likewise the Oil Age will not end with the end of oil. Thus what we are seeing today is a true structural change in the energy market and a transformation from conventional (hydrocarbon) fuels driven by a marked increase in investments in research and development in renewables, natural gas, and hydrogen besides others.

Climate change has become a reality beyond doubt and no longer is it a localized problem on issues in certain parts of the world or with an impact on certain species or certain socio-economic and environmental consequences. Issues related to climate change including wildfires, abnormal and sudden weather changes with aggressive storms; unprecedented snowfall in certain parts of the world; heavy rains leading to flooding, hurricanes, cyclones, and other significantly damaging episodes—all these have become ever so much more frequent. The Paris accords and the SDGs have been a great step forward, unifying the approach and setting a direction globally with measured targets that many nations are working towards. Within this context, this book addresses aspects, and in some depth SDG 7: *Affordable and Clean Energy*; SDG 12 *Responsible Consumption and Production,* and SDG 13 on *Climate Action.*

For leaders who are using this book, what is important is the understanding of the investment costs in the R&D space required and the wastes that the renewables can also generate. So, while the book promotes greatly the diversification in energy it steers away from some of the sentimentalism surrounding renewables and is thus more realistic about them. So, we must start with the end in mind, and as such understand comprehensively the life-cycle analysis and make careful and scientific comparisons between the technologies and management of different energy types. Much of the foundational thinking on which this book is built concerns the differentiation between the value chain and the supply chain thinking and is thus suitable for those working in the supply and demand spaces and in every part of the supply chain including producers, suppliers, and distributors of energy. The authors have built many of the frameworks on the previous work, which is well referenced, and they have also included a guide at the end of the book for additional research and reading.

Further, this book is intended to provide a purposeful set of concepts and principles furnishing those investing in energy projects with a good overall guide to aid them in their decision-making. It also provides a very timely discussion on the optimization of supply chains and systems using modern digitalization technologies and details the opportunities and the threats; it also covers economic and operational strategies. One such major risk is cyber security, which has in some industries been identified as the single most significant threat. Also addressed and illustrated in this book are great examples from the oil & gas and power industries through the in-depth review of case studies covering 12 segment areas in the conventional, alternative, and renewable energy sectors. Built on what is clearly decades of significant research and pragmatic in-depth experiential understanding, this book provides an excellent resource for those wanting to understand current practices and thinking in this space. While reviewing authors' contributions, I was struck by the way the authors have been able to look at supply chain dynamics, which helps one in this energy space really start to understand the holistic challenges and opportunities. Not that it was not understood before, but clearly the COVID pandemic and the impact it had on stripping demand globally in such a short space of time demonstrated the fragility of the supply chains and the energy markets during such a crisis that would have generally only been imagined in a world war.

Academics, energy policy makers, energy economists, and investment advisors, as well as the current and future executives of the energy industry, could make good use of this book. I see it also as a sound reference for energy researchers in universities as well as energy and sustainability research institutes conducting research on energy markets and supply chains in this critical thought leadership space of the future.

It has been my great pleasure to review and provide this short foreword to this book. Today we are all more aware of the importance of a successful energy transition, against the backdrop of becoming a more sustainable world that is responsible towards humanity and the environment. Understanding the energy supply chain as a complete dynamic global value chain is fundamental to a more sustainable, prosperous, and stable future for all nations.

Dr. Waddah S. Ghanem Al Hashmi
BEng (Hons), MBA, MSc, AFIChemE, FEI, FIEMA, FBDIGCC, MIoD,
Senior Director—Logistics & Marine Assurance,
Emirates National Oil Company Limited (ENOC) LLC
Hon. Chairman—Energy Institute - Middle East & Senior Advisor,
Clean Energy Business Council (CEBC), Dubai, United Arab Emirates

Acknowledgments

This book is inspired by the innovators who are redefining energy and electrification value chains, including legends and legends-to-be such as Bill Gates, Elon Musk, Anders Dahl, Magnus Öhrman and Erik Hiensch of Vattenfall, Preston Roper of Enel X, Jo-Jo Hubbard at Electron, Rod Colwell of Controlled Thermal Resources, Rodolfo Ribeiro of CBMM, Robert Fenwick-Smith of Lightning Systems, Eric Byres of aDolus, David You of Trina Solar, Hans Smit of Ocean Minerals, Mohammed Almadhoun of PV Robot, Jason Steinberg of Scanifly, Meng Wang of Solar Earth Technologies, and Thomas Sisto at XL Batteries, just to name a few. You are marking a bright path for others to follow. Keep up the brilliant work! We thrive on your talent, energy, and creativity.

In parallel to these entrepreneurial endeavors, a number of distinguished leaders have been boldly improving the efficiency and performance of our oil, gas, and thermal power value chains, including entrepreneurs in the unconventional oil & gas exploration and production such as the late Aubrey McClendon of Chesapeake Energy and Mark Papa of EOG Resources, and so many others such as Harold Hamm of Continental Resources and Tino Ceniti of Calvert GTL who have expanded economically viable oil & gas reserves by challenging conventional wisdom and pioneering new forms of producing and transporting hydrocarbons at economical scale. This list of leaders includes, arguably more impactfully, *intrapreneurs* such as Alfred Kruijer from Shell, Stephen Turnipseed from Chevron, Nayef Al-Hajri and Khaled Baradi from Qatar Fuel, Hussain al Abbas from Saudi Aramco, Suresh Nair from Bharat Petroleum, Gary Wawak from Motiva, Dave Cox from GE Oil & Gas, Peter Krieger from Freudenberg Oil & Gas, and Åsmund Mandal from FMC Kongsberg, again just to name a few. You are continuously improving the industry's operational, business, and environmental profile, which is of strategic benefit at this juncture.

A number of esteemed and dear colleagues and friends provided helpful feedback on the draft manuscript, including Dr. Jaydeep Balakrishnan (University of Calgary), Adolfo Fehrmann Espinosa (Isagen), Peter Poggi (Tennessee Valley Authority), Roger Pelham (Chevron), Fritz Troller (EDF), Andrew Bender (Forbes), and Ann Lee (New York University). Thank you! The quality of the book has been enhanced thanks to your expert input. You are leaders in your fields. I always appreciate your experience and expertise, and value your advice.

I would also like to recognize the important leadership role that industry institutions such as API, COSO, DNV, IAEA, IEA, IEC, IHA, ISO, IMO, NERC, NERL, and so many others are providing in risk management, operations excellence, standardization, and best practices. In addition, several professional associations have developed extensive and useful bodies of knowledge in operations management and research; these include ASCM (the Association for Supply Chain Management), ISM (the Institute for Supply Management), CSCMP (Council of

Supply Chain Management Professionals), and their satellite chapters and affiliates. You are doing the hard work of moving the ball forward, despite daily technical and commercial hurdles, and the industry is better off for your efforts!

Finally, I am deeply grateful to PennWell Books and to Matthew Dresher in particular, for nurturing the development of the project from concept through to outline and full manuscript. Without your faith and patience, this book would not have come to fruition. While our initial discussion about this project related to writing a second edition of the book we published together in 2011, the world of energy changed so much in the interim that the ensuing project turned out to be much bigger and more important than that. Thank you for your patience, trust, and encouragement.

If you have any suggestions for improvements for future editions, kindly send them to david@bostonstrategies.com.

Acronyms and Abbreviations

AI	Artificial Intelligence
ANP	National Agency of Petroleum, Natural Gas, and Biofuels (Brazil)
APC	Advanced Process Control
AWEA	American Wind Energy Association
b	Billion
BBL	Billion Barrels
BOM	Bill of Materials
BOO	Build Own Operate
BOP (1)	Balance of Plant
BOP (2)	Blowout Preventers
BPD	Barrels per day
BSI (1)	Boston Strategies International
BSI (2)	British Standards Institute
BTM	Behind the Meter
BTO	Build/Built to Order
BTU	British Thermal Unit
C&I	Commercial and Industrial
CAES	Compressed Air Energy Storage
CapEx	Capital Expenditure
CCGT	Combined Cycle Gas Turbine
CCS	Carbon Capture and Storage (or Sequestration)
CEN	European Committee for Standardization
CFM	Cubic Feet per Minute
CHP	Combined Heat and Power
CNG	Compressed Natural Gas
CSP	Concentrated Solar Power
DBOO	Design, Build, Own, and Operate
DOE	Department of Energy
E&P	Exploration and Production
EIA	Energy Information Administration (United States)
EMEA	Europe Middle East and Asia
EOR	Enhanced Oil Recovery
EPC	Engineering, Procurement, and Construction
EPCI	Engineering, Procurement, Construction, and Installation
ESG	Environmental Social and Governance
ESS	Energy Storage Systems
ETO	Engineered to Order
FCC	Fluid Catalytic Cracking and Construction
FCCU	Fuel Cell Control Unit
FID	Final Investment Decision

FIFO	First In, First Out
FPSO	Floating Production Storage and Off-loading Vessel
FTM	Front of the Meter
GCC	Gulf Cooperation Council
GOSP	Gas-Oil Separation Plant/Process
GTC	Gas to Commodity
GTP	Gas to Power
GW	Gigawatt
HP	Horse Power
HPHT	High-Pressure, High-Temperature
HSE	Health, Safety, Environment
IAEA	International Atomic Energy Agency
IEA	International Energy Agency
IEC	International Electrotechnical Commission
IFC	International Finance Corporation
IHA	International Hydropower Association
IMO	International Maritime Organization
IOC	International Oil Company
IoT	Internet of Things
IIoT	Industrial Internet of Things
IPP	Independent Power Plant
IRENA	International Renewable Energy Agency
ISA	International Society of Automation
ISM	Institute for Supply Management
ISO	International Standards Organization
IT	Information Technology
k	Thousands
kg	Kilogram
KPI	Key Performance Indicator
KSF	Key Success Factors
kW	Kilowatt
LCCS	Low Cost Country Sourcing
LCOE	Levelized Cost of Electricity
LNG	Liquefied Natural Gas
LPG	Liquefied Petroleum Gas
LSTK	Lump Sum Turn Key
lwd	Logging While Drilling
m	Millions
M&A	Merger and Acquisition
Mbd	Millions of barrels per day
MEC	Main Electrical Contractor
MENA	Middle East and North Africa
ML	Machine Learning

MMBTU	Million Metric British Thermal Units
MMBTU	One Million BTUs
MRO	Maintenance, Repair, & Operations
MRP	Materials Requirements Planning
MT	Metric Ton
MTBR	Mean Time Between Repair
MTO	Make to Order
MTPA	Million Tonnes per Annum
MW	Megawatts
MWD	Measuring While Drilling
MWe	Mega Watt Electric
NIMBY	Not in My Backyard
NERC	North American Electric Reliability Corporation
NIST	National Institution for Standards and Technology
NM	Newton-meter
NOC (1)	National Oil Company
NOC (2)	Network Operations Center
NPV	Net Present Value
NREL	National Renewable Energy Laboratory
O&G	Oil & Gas
O&M	Operations and Maintenance
OCTG	Oil Country Tubular Goods
OECD	Organization for Economic Cooperation and Development
OFS	Oil Field Service
OEM	Original Equipment Manufacturer
OPEC	Organization of Petroleum Exporting Countries
OpEx	Operating Expenditure
OSHA	Occupational Safety and Health Administration
OT	Operational Technology
PBL	Performance-Based Logistics
PLC	Programmable Logic Controller
PPA	Power Purchase Agreement
ppm	Parts per Million
PHS (or PSH)	Pumped Hydropower Storage (or Pumped Storage - Hydropower)
psi	Pounds per Square Inch
RCRA	Resource Conservation and Recovery Act
REACH	Registration Evaluation and Authorization of Chemicals
RFI	Request for Information
RFID	Radio Frequency Identification
RFP	Request for Proposal
RFQ	Request for Quotation
RFx	Request for Information, Request for Proposal, or Request for Quotation

ROA	Return on Assets
ROI	Return on Investment
ROP	Rate of Penetration
RPA	Robotic Process Automation
RPM	Revolutions per Minute
RTU	Remote Terminal Unit
S&OP	Sales and Operations Planning
SCADA	Supervisory Control and Data Acquisition System
SCM	Supply Chain Management
SCOR	Supply Chain Operations Reference model
SME	Small and Medium Size Enterprise
t	Trillion
T&D	Transmission and Distribution
TCO	Total Cost of Ownership
Ton	Short Ton
Tonne	Metric Ton
TPA	Tonnes per Annum
TPM	Total Productive Maintenance
TPY	Tonnes per Year
TSCA	Toxic Substances Control Act
UAV	Unmanned Aerial Vehicle
ULCC	Ultra Large Crude Carrier
US (USA)	United States (United States of America)
VLCC	Very Large Crude Carrier
VMI	Vendor-Managed Inventory
VR (AR)	Virtual Reality (Augmented Reality)
WMS	Warehouse Management System
WTI	West Texas Intermediate
WTO	World Trade Organization

Preface

For most of the past two centuries operations management in the energy industry was not the talk of cocktail parties, to say the least. Boring, some might say.

In the oil business, drilling and production were concentrated in a few geographical areas, and refining was done at small scale refineries. In the 1930s, Saudi Arabian workers crushed minerals by hand to make drilling mud. The California-Arabian Standard Oil Company, now called Saudi Aramco, employed Bedouin tribesmen to guard their fields and supply lines, transitioning to government police as its operations expanded and then in turn to private security.

International trade of oil, gas, and coal grew steadily starting in the 1930s. Until the advent of domestic refining capacity and regional pipelines, Saudi Arabia exported crude by barge to Bahrain. By the 1940s, tanker trucks were transporting oil from the production site to refineries. Each one held 40–50 barrels, compared to 120–215 barrels for tanker trucks today. Pipelines were also used, but they were not the main mode of transportation. From the refinery, finished products were mostly shipped by rail. In the power industry, thermal power plants were supplied by coal via barge and rail. Aggregating trainload shipments of coal from mines to stock power plants was a significant advance. After World War II, the power industry saw a major transformation in production and supply logistics as more than 400 nuclear plants were built; nuclear plant construction largely ended after the Three Mile Island and Chernobyl accidents.[1]

As globalization, booming consumer demand, and deregulation drove unprecedented growth in the energy industry throughout the 2000–2015 period, there were major changes in oil, gas, and power supply chains and in supply chain management—the coordination of a defined set of activities involved in procuring a product and its associated services and solutions from the point of usage back to the ultimate supplier in a way that maximizes the business objective(s) such as profitability. Supply chain management became critical in meeting capital project schedules and making operations profitable. In addition, global investments increased the complexity and sophistication of supply chain management techniques and practices.

Capital expenditures within the oil, gas, and power generation industries grew at more than 15% per year between 2004 and 2008, before the global financial crisis and the BP Gulf of Mexico blowout that caused a temporary shutdown of deepwater drilling.[2] Saudi Arabia's power projects alone were estimated to total $130b of investment between 2010 and 2015.[3] The investment spurred technological advances. Oil & gas companies used information technology to incrementally improve operations, for example doing inline pipe inspection using "smart pigging," monitoring corrosion via satellite, and implementing supervisory control and data acquisition (SCADA) systems to manage the pipelines. Power companies invested heavily in process control systems and control rooms.

Then, in as much as the 2000–2015 period experienced unprecedented growth, the 2015–2021 period was like a roller coaster on the way down. Unconventional production, especially of shale oil, coupled with directional drilling, 3D seismic imaging, and other exploration and production technologies, drove supply to unprecedented levels. Meanwhile, global demand growth for crude oil shrank almost 10% annually between 2015 and 2019, down from 2.2 mbd in 2015 to 0.8 mbd in 2019, due to increasing adoption of renewable energy and natural gas and increased efficiency of internal combustion engines.[4] Strong demand in India and China offset declining demand elsewhere to some extent: non-OECD countries have represented the largest share of world oil demand since 2012, reaching 53% in 2018.[5]

Capital expenditure (CapEx) took a nosedive, as shown in Figure 1, forcing many oil, gas and power construction projects to be suspended, delayed, or cancelled. In 2020 and 2021, COVID-19 and historically low energy prices forced many energy companies to drastically cut capital and operating budgets. In oil & gas, most of the cuts in CapEx came from relatively new unconventional and offshore projects, as many shale and deepwater producers struggled to remain profitable below $40/barrel during 2015 through 2017, and could not stand a chance at turning a profit at oil prices that sank as low as $11/barrel in April 2020 during the COVID-19 lockdowns. Supply chain managers who were working on those assets were largely redeployed or let go.

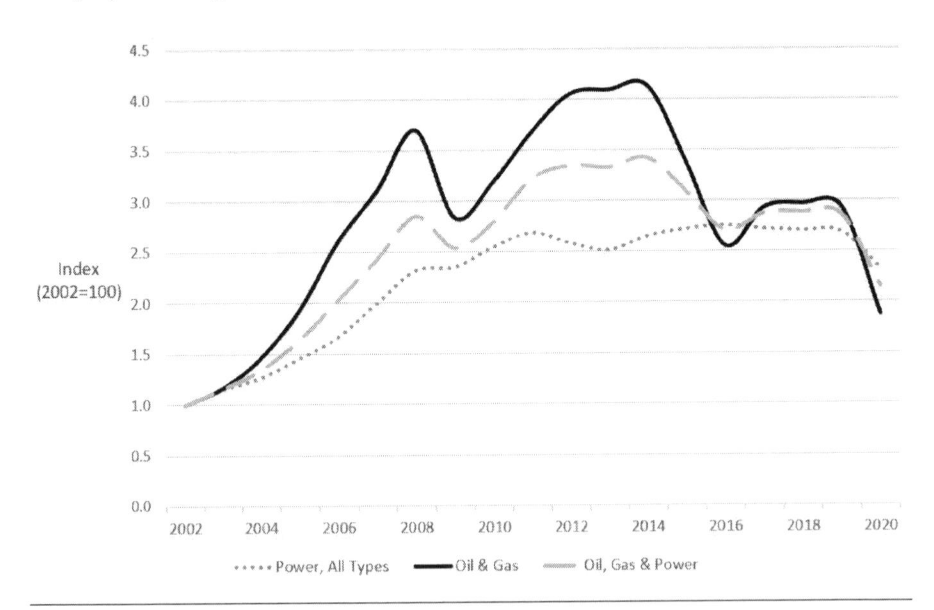

Figure 1. Evolution of Growth in Oil, Gas, and Power Markets
(*Source:* BSI Energy Ventures analysis of data from the International Energy Agency and IHS Markit)

Pundits are debating whether this drop in oil & gas is just another cyclical downturn or the beginning of a structural adjustment to a post-oil era. Some forecasts show upward cycles going forward[6] and others show oil declining in both absolute terms and relative to other energy technologies. Will renewables continue to outpace conventional energy or is this just another false start? In 1956 scientists predicted Peak Oil.[7] In 2007 a team at Boston Strategies International wrote a paper asking whether and when alternative energy technologies would replace oil.[8] By 2012, horizontal drilling and fracking had lowered the cost of oil production dramatically and suddenly made the US a global leader in oil & gas production, giving oil a new life.[9]

Renewables will continue to put increasing pressure on oil, gas, and conventional power because they have a *supply chain advantage*, especially solar, energy storage, and hydrogen for three reasons.

First, the ability to produce and use the energy in discrete, small, and modular quantities offers more opportunity to customize and innovate. There are four ways to create shareholder value through supply chain management: cutting costs, doing the same with fewer assets, proliferating products from the same production and distribution platform, and continuously developing and launching new products. A company's supply chain advantage is usually embedded in its business model and critical to its financial success. Because oil, gas, and power are produced in huge plants that are centralized and capital-intensive, the most obvious ways to generate financial benefit from their supply chains have been by cutting costs and managing asset productivity. By continuing to focus on cost and asset optimization they are playing on only half the ballfield. In contrast, solar panels, batteries for energy storage, and fuel cells are discretely manufactured and can be set up at small scale, including a single micro-module of PV called "pico PV" that is often sold in remote areas of developing countries to power a single electronic device. Batteries are assembled from cells into packs and from packs into batteries and from batteries into containers, so, again, they can be used at small scale. Fuel cells will also be able to be sized uniquely for each application and manufactured at varying production scales. This gives these companies a supply chain advantage that can translate to market growth opportunities. Of course, fossil fuel–based energy companies can and will customize and innovate, but it takes research, development, and out-of-the-box creative thinking to make a liter of oil uniquely valuable to each and every consumer, and most of the product differentiation is likely to occur downstream, close to the customer.

In addition to being able to operate at smaller scale and being in a better position to customize and innovate, some renewables are more likely to continually find *new* ways to generate shareholder value through customization and innovation than fossil fuel producers can find new ways to cut operating costs and optimize their asset bases. For example, hydrogen can be packaged for distribution in widely different and customizable units for different capacities of fuel cells, from utilities on an electric grid to minibikes. And PV solar, batteries, and energy storage are undergoing groundbreaking technological innovation in their basic chemistries that offers up new market growth opportunities, such as solar PV embedded in apparel and wearable devices. In other words, they will continue to push the edge of the frontier, which already has a competitive advantage compared to conventional oil, gas and power.

Finally, the supply chains for some renewable energy technologies can operate without the cost penalty associated with the bullwhip effect (see Appendix 1). To the extent that distributed power technologies supplant and replace the centralized ones, especially where rooftop solar replaces centralized power generation, and the material and components can become more standardized, less complex planning and scheduling would be required since those items and solutions could be available "Off the shelf." This could eventually smooth oil, gas, and power business cycles that exist today due to the long lead times and the associated lags and response times for price signals to flow from one end of the value chain to another. This volatility and business inefficiency has for decades amplified, in bullwhip fashion, price increases, capacity increases, and profit margin increases on the upswing of the cycle, and created deflation, layoffs, and losses on the downswing—experiences that everybody in the upstream oil & gas business is all too familiar with. For example, if everybody drove electric vehicles and charged them with the solar panels on their home and business rooftops, the economically costly bullwhip effect from long oil & gas value chains would disappear. Since this structural and behavioral inefficiency costs the industry about 10% of sales,[10] the cost differential between solar power options and oil & gas fueled vehicles could eventually fall by 10% more than the actual levelized cost differential between the two. The erosion of this destructive "bullwhip effect" could potentially represent a "tipping point" that accelerates the transition from fossil fuels to renewable power.

Renewable energy technologies such as wind and solar power are rapidly becoming mainstream and in many cases replacing the investment in oil and coal starting now, and natural gas starting in 2030. Forty percent of 2019 capital expenditure is in new energy technologies, as shown in Figure 2 below (which is based on 2019 data). These are ostensibly both clean and green, although there is debate over how green wind is if its intermittency is backed by gas-fired thermal plants, and how clean solar is if panels are thrown into landfills at the end of their useful lives.

Capital expenditure is already skewing sharply towards natural gas and renewable technologies. Oil, gas, and coal consumption has essentially already plateaued, as oil wells typically produce for 20 years or more.

As oil & gas CapEx flattens and eventually declines, projects involving other energy technologies will make up the difference.

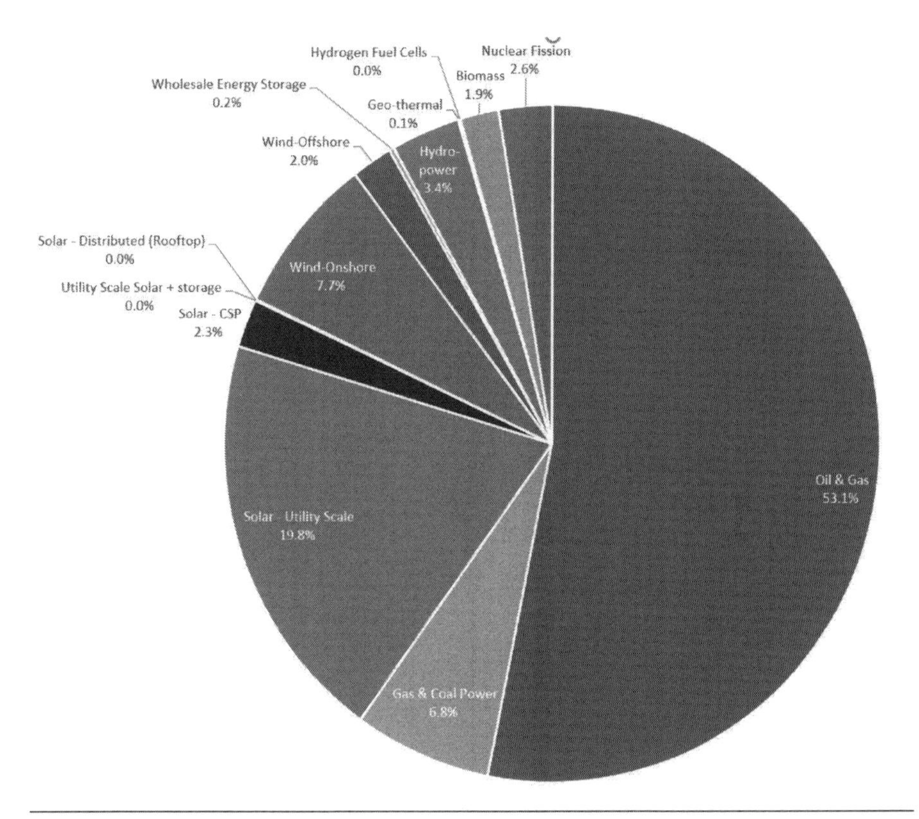

Figure 2. Annual Capital Expenditure by Energy Technology
(*Source:* BSI Energy Ventures)

Offshore wind, solar CSP, and hydropower projects involve CapEx of magnitudes similar to those of many oil & gas projects, which average $600 million of Total Invested Cost, and of gas-fired power plants, which average $935 million per project. Nuclear projects are far more capital-intensive, costing $13 billion on average, as shown in the table below. Offshore wind projects are also major endeavors, costing on average $1.7 billion. Solar CSP projects cost $830 million on average, and hydropower projects cost $532 million on average (see Figure 3).

Supply chain management for these projects involves many of the same planning, scheduling, sourcing, contracting, delivery, installation, and commissioning issues. The supply chains of these businesses have seven unique characteristics:

- Programs related to reliability, safety, asset management (risk, utilization, productivity), and life-cycle cost often dominate tactical logistics and inventory management, which is often the focus of supply chain management for fast-moving consumer goods.

Energy Type	Basis	Detail/Description	Example Project Size ($ millions)
Nuclear Fission	Advanced Nuclear (Brownfield)	2 x AP1000	$ 13,024
Wind-Offshore	Fixed-bottom Offshore Wind: Monopile Foundations	400 MW, 10 MW Wind Turbine Generator	$ 1,750
Thermal Power (Gas and Coal Fired)	Combined-Cycle, 1 combustion turbine generator, 1 steam turbine generator, 1 heat recovery steam generator (HRSG), Single Shaft, 90% Carbon Capture	H-Class	$ 935
Solar - CSP	Concentrated Solar Power Tower	Molten Salt Thermal Storage	$ 830
Oil & Gas	Gulf of Mexico	Deepwater Drilling Rig Cost	$ 600
Hydropower	Hydroelectric Power Plant	New Stream Reach Development	$ 532
Utility Scale Solar + Storage	Solar PV with Single Axis Tracking and Battery Storage	150 MWAC Solar 50 MW, 200 MWh Storage	$ 263
Wind-Onshore	Onshore Wind, Large Plant Footprint: Great Plains Region	200 MW, 2.82 MW WTG	$ 253
Biomass	50MW Biomass Plant	Bubbling Fluidized Bed	$ 205
Solar - Utility Scale	Solar PV with Single Axis Tracking	150 MWAC	$ 197
Geothermal	Hydrothermal	Binary Cycle	$ 126
Wholesale Energy Storage	Battery Energy Storage System	50 MW, 200 MWh	$ 69
Hydrogen Fuel Cells	Solid Oxide Fuel Cell	34 x 300 kW Gross	$ 67
Solar - Distributed (Rooftop)	7kw Home Rooftop	28 panels	$ 0

Figure 3. Illustrative Capital Project Sizes for Various Energy Technologies
Source: BSI Energy Ventures analysis of data from EIA[11]

- Highly engineered equipment operates in complex and customized systems, and the question of how much additional engineering and customization is worth is perennial. Technology and supplier selection is decided on the basis of life-cycle cost, and in the case of new technological platforms, there is little or no performance history on which to base accurate calculations.
- Long lead times constrain project schedules and make delays very costly if they bottleneck other activities or delay production.
- The prohibitive cost of downtime makes reliability and field responsiveness critical. Since many assets are remote and/or offshore, the question of how to accurately value the opportunity cost of lost production is a recurring one, as well as the issue of how to factor those costs into inventory parameters for items like capital spares.
- Long investment cycles mean decision-making must account for uncertainty. Construction decisions are made years in advance and actual construction projects usually last for years. Concessions are frequently awarded for decades, which increases the importance of risk management and analytical risk management tools.
- Remote, harsh, and frequently dangerous conditions, for example subsea and deepwater offshore, present large potential safety hazards that result in extraordinary price premia for safety and process reliability, raising another trade-off that is less common in other industries–how much is a

"safe" item worth above and beyond the cheap version that has a chance of failure? How much more should an explosion-proof light fixture cost than a regular one?

- Finally, the high public visibility of accidents and environmental problems elevates the importance of risk management, including supply chain risk. Even if the risk of serious problems is low, the possibility of having to face public scrutiny over a decision to save a small amount often makes managers wary of aggressive cost savings projects.

Onshore wind, biomass, geothermal, utility scale energy storage, and rooftop solar projects are often smaller in capital investment and more standardized in terms of equipment and services. Fuel cells also tend to involve comparatively smaller capital investments, but that may change as the transport applications that are prevalent today scale up to power utilities.

In addition, renewables *megaprojects* are already underway. In energy storage, LS Power's Gateway project in San Diego, California, defines 250 MW of storage for grid reliability and flexibility as "the new normal."[12] In onshore wind, the 20 GW Jiuquan farm in China, nearly ten times the size of most onshore wind megaprojects, is nearly complete (as of this writing).[13] Offshore, the 1.2 GW Hornsea project, currently under construction, is poised to be the largest offshore wind farm ever built.[14] In solar, the 1.5 GW Tengger Desert Solar Park in China became operational in 2017, and dozens of others of similar size are in construction or have been recently completed.[15]

As these and other projects move through their planning, construction, and operating phases, their complexity and novelty—many new value chains are being created afresh—necessitates a higher level of conceptual thinking and problem-solving in a less structured environment. As the highly mature supply chains that had evolved over a century in the oil, gas, and thermal power industries cede to a more dynamic technological and competitive set of existing and potential supply and channel relationships, this rethinking needs to consider the totality of value-added activities between the ultimate supplier and the ultimate customer.

Purpose, Scope, and Development of this Book

Supply chain strategies and management are critical to enabling new energy product and market opportunities, as well as digitalizing and modernizing oil & gas exploration, production, and distribution. Fuel cells, energy storage systems, and solar modules are typically manufactured in discrete or batch environments, which require different approaches to operations and supply chain management than do continuous flow operations such as oil & gas and thermal power generation. As new energy production technologies are added in order to satisfy growing world energy demand, energy supply chains will consist of many widely varying technologies, production modes, and distribution channels. Each energy technology presents a unique type of supply chain, as well as differences in supply chain types, profit margins, economies of scale, growth rates, and asset intensity. Applying supply chain skills from one energy segment to another segment could be a recipe for failure. Conversely, leveraging supply chain insights that pertain directly and correctly to a certain type of supply chain can open up new opportunities to lower costs, and in some cases, these supply chain insights can make business models feasible where they otherwise would have been deemed uneconomical or operationally infeasible.

This book provides a toolbox of techniques to successfully manage the range of complex trade-offs that are inherent in capital projects and operations & maintenance across energy technologies, and to apply best practice techniques to emerging energy industries—from the small to the large project, and from solar to nuclear and everything in between. While large economic decisions have sometimes historically been made with simple tools, today the range of available options is much wider, and the risks and rewards are commensurately higher, so they require rigorous logic, reliable data, and robust analysis methods.

It serves as an aid for making operational transformations that yield visible financial impacts at the enterprise level within energy companies. The decisions addressed include, for example:

- How much to commit before Financial Investment Decision ("pre-FID")
- Whether to bundle or unbundle complex products and services, such as combined purchase and operating/maintenance contracts
- How to calculate Total Cost of Ownership
- How to decide whether to make, buy, or rent/lease
- Whether to buy directly from sub-suppliers
- How to determine the optimal number of suppliers, and when to single source
- How to know the optimal contract term
- How to structure technology partnerships
- Whether or not to set up a joint venture or take an equity stake
- How to tender products under development but not yet commercialized
- How to organize to comply with local content laws
- How to achieve continuous cost reduction in brownfield investments

The scope addresses both value chains and supply chains. While it is not uncommon for industry professionals associate supply chain management with operations research and think of supply chains as tactical—determined, defined, and bounded by strategic decisions that are made at an executive level (not surprising given that supply chain management was historically an operational function based heavily on logistics and purchasing)—this book provides frameworks for unlocking a range of fresh new strategic options and possibilities based on creative, and more analytical, supply chain management. In this context, strategic supply chain management necessarily starts with an analysis of end-to-end value chains.

Figure 4 summarizes how we distinguish between value chains and supply chains in this book. Whereas *supply chain management* implies clearly defined actors, channels, and costs, and an ability to manage the intermediary processes using relatively scientific methods, today's leaders need to reason more broadly in terms of *value chains*. Whereas *supply chain management* typically aims primarily to create benefit for one entity, namely the one creating the demand, *value chain thinking* is usually to design strategies that will grow a market or technology and result in benefit to any or all of the parties along the path from the ultimate supplier to the ultimate customer. Whereas *supply chain management* typically takes place in the context of an already established procurement, logistics, or production initiative, *value chain management* may involve coordinating activities between parties to establish the basis for value creation and commercial relationships for the first time. Whereas *supply chain management* often refers to a specific segment of the value chain, *value chain* conceptualization usually considers the entire chain from raw material or source to end-of-life disposal at the ultimate customer. Hence, in this book we use the term *supply chain management* in the context of an established set of purchasing and operational interrelationships designed for the benefit of a single large producer, and the terms *value chain* and *value chain management* in the context of a more conceptual design of business relationships that support the development and growth of a market or a technology.

Aspect	Value Chain Design/Strategy/Management	Supply Chain Design/Strategy/Management
Objective	Grow or capture a market or technology, considering value addition potential across the chain	Create financial benefit for a specific company
Application	First-time or fresh look at business opportunities	Established actors and relationships, for optimization
Scope	End-to-end including all activities	The segment of the value chain that is relevant to a specific company

Figure 4. Value Chain versus Supply Chain

The book considers the needs of energy suppliers, producers, and distributors at every level (upstream, midstream and downstream). Although it covers to some degree conservation and methane leaks—and although several UN SDGs pertain to energy, especially Goal 13-climate action, the targets of those goals do not fall in the purview of most supply chain managers, so are not covered in this book. For environmental footprint of energy supply chains, refer to *The High Cost of Low Prices*[16] and *The Design and Analysis of Sustainable Supply Chains*.[17] This book is not a primer on supply chain management; for a primer on supply chain management the reader is advised to consult a more general book such as *The Guide to Supply Chain Management* by *The Economist.*[18]

The principles, frameworks, and models have been developed for this book, or in some cases come from previous presentations that David Steven Jacoby made to industry associations. The examples come from our client and project experience, our network of colleagues throughout the industry, and secondary research. All analysis is our own unless otherwise cited.

How the Book Is Organized

The Introduction provides a conceptual framework for value chain management in the energy sector. It lays out the objectives, key business processes, and performance metrics that provide useful guideposts. It offers first principles that should guide value chain initiatives in the energy industry and explains how to organize supply chain management activities.

After the Introduction, there are two types of chapters. The initial chapters on capital project and operations management explain overall tools and techniques that are relevant to energy supply chains, broadly speaking. The ensuing chapters show how these concepts apply to ten energy technologies: oil & gas, nuclear, solar, wind, gas and coal-fired power, hydropower, biomass, geothermal, energy storage, and hydrogen fuel cells.

Methods for CapEx Project Value Chain Risk Mitigation

This chapter lays out concepts and frameworks for making complex trade-offs involved in designing value chains, architecting supplier relationships, managing contract risk, and organizing engineering and constructing projects. It provides tools for managing the challenge of pre-FID (final investment decision) commitments and describes the theory and application of options as they relate to project size, project life, and technology/product mix. It is divided into the following sections:

1. Risk management—provides general approaches to managing risk, identifies commonly encountered value chain risks, and illustrates how to manage technology risk.
2. Structuring project work—addresses the complex issue of how to "define the buy"—whether to buy solutions or independent products and services, how to do a proper total cost of ownership analysis, which project governance mode to select, which activities to insource and which to outsource, and how to determine the optimal contract term.
3. Sustainability trade-offs—explores what kinds of initiatives and investments can offset carbon and other emissions and offers some ways of making decisions about these projects that are cost-effective as well as environmentally sound.

4. Materials and services unavailability—catalogues some frequently encountered materials and services shortages and provides some options for mitigating those risks.
5. Outsourcing—sheds light on which activities are sometimes outsourced, and what the risks and benefits of doing so are. An outsourcing decision-making process is provided.
6. Supplier partnering risks—shows how to structure partner relationships, how to structure alliances, how to determine the optimal number of suppliers, how to qualify suppliers once they are identified, and how to decide whether or not to establish a joint venture and whether to take equity in a partner.
7. Procurement bundling—explains the pros and cons of procuring from outside vendors in various configurations of bundling components, products, and services purchases.
8. Contract term risk—offers a way to determine the optimal term of contract commitment.

Methods for Operations and Maintenance Management Optimization

The chapter on trade-offs in operations and maintenance provides methods for decision-making trade-offs in operations. It covers the most impactful decision-making areas, namely:

1. AI/IoT—these two related sections cover the area of biggest opportunity for most O&M departments today, and concurrently the most vulnerable aspect of operations and maintenance.
2. Cybersecurity—with the rapid and pervasive expansion of AI/IoT applications, cybersecurity has become imperative.
3. Peak capacity strategies—this section on operating and maintaining complex systems provides frameworks and proven processes for managing overall equipment effectiveness, Total Productive Maintenance (TPM) and asset productivity, standardization, the cost of quality, throughput and debottlenecking, preventive and predictive maintenance, and equipment standardization, continuous cost reduction including Lean as it relates to inventory management, transportation management, outsourcing logistics and total supply chain activities (3PLs and 4PLs), and how to engage suppliers in performance improvement initiatives. It also deals with how to manage capital spares and other stochastic inventory and materials management dilemmas, and how to establish consignment and vendor-managed inventory programs.
4. Sourcing trade-offs—explains *Category Management* and provides frameworks for determining the optimal number of suppliers. It also explains how to prequalify suppliers, how to manage supplier relationships, and how to manage the tendering and contracting process.

5. Total Cost of Ownership—explains what TCO is, as well as some of the common leverage points for minimizing the total lifetime cost.
6. HSE considerations—this section identifies relevant risk management frameworks, policies and standards, enterprise risk management frameworks and methods, and operational risk mitigation processes and methods. It also clarifies supply chain's role in reducing carbon footprint. HSE practices for specific technologies can be drawn from these examples, which are subdivided into Oil & Gas Upstream/ Midstream/Downstream and Power (all technologies).

Technology-Specific Chapters

To update the initial version of this book (*Optimal Supply Chain Management in Oil, Gas & Power Generation* [PennWell, 2011]) we could have jammed all the renewable technologies into the Power chapter. However, this would have created a soup of many disparate and seemingly unrelated kinds of content. A second way would have been to organize it by the prevalence of each technology today. However, that would have unfairly penalized energy storage, hydrogen fuel cells, and other technologies that are almost commercially nonexistent today but may dominate the landscape in a matter of years. A third way would have been by their Levelized Cost of Electricity (LCOE). This would have captured both their prevalence today and their likely timing of more widespread adoption. However, this metric does not differentiate between the power technologies with regards to value chains or supply chain management.

To maximize the insight into value chain design, we decided to sequence the oil, gas, and power chapters according to their current supply chain positioning as described in Figure 10: Prevalent Supply Chain Strategy Positioning of Energy Companies. Capital-intensive energy technologies with centralized plant and equipment such as Oil & Gas (O&G), Thermal, Nuclear, Hydropower, Geothermal, and Biomass plants and projects are similar when it comes to supply chain management. They require extensive management of EPCs, and often extensive earth-working, drilling, or excavation. On the other hand, at least partially distributed technologies such as Solar, Wind, Energy Storage, and Fuel Cells (mostly for vehicles) require more customization and innovation, sometimes with customer-facing services such as installation. Since supply chain competency matures through four stages sequentially, namely *rationalization, synchronization, customization*, and *innovation*, the chapters are sequenced according to their dominant supply chain characteristics in that model, counterclockwise starting with those having the greatest inherent supply chain value generation potential:

1. Fuel Cells
2. Energy Storage
3. Wind
4. Solar

5. Biomass
6. Oil & Gas: Upstream
7. Oil & Gas: Midstream
8. Oil & Gas: Downstream
9. Geothermal
10. Gas and Coal-Fired Power
11. Hydropower
12. Nuclear

Each technology chapter follows a common structure. Each chapter begins with Chapter Highlights and an Introduction about that technology's relevance, and proceeds to review common themes of interest to the management of both CapEx and OpEx with respect to supply chain partners:

- Trade-offs and Management Techniques in Capital Project Management
  1. Technology Choice Risk Management
  2. Capital Project Risks and Mitigation
  3. Sustainability Trade-offs
  4. Supply Unavailability Risks and Mitigation
  5. Outsourcing Risks and Mitigation
  6. Supplier Partnering Risks and Mitigation
  7. Procurement Bundling Trade-offs
  8. Contract Term Risk and Mitigation
- Trade-offs and Management Techniques in Operations & Maintenance (note: HSE Considerations are considered only in the upfront chapter on Methods, and are not repeated for each energy technology)
  1. Internet of Things (IoT) and Artificial Intelligence (AI) Technology Choices
  2. Cybersecurity Risk Management
  3. Peak Capacity Strategies
  4. Sourcing Trade-offs
  5. TCO Trade-offs
- Roadmaps for unlocking latent value in supply chains at many companies, in our experience

The Conclusion reiterates key points and lays out key success factors for achieving the benefits articulated in the book.

The book has three appendices:

1. Bullwhip in the oil & gas supply chain—a study showing how the lack of effective supply chain coordination results in higher costs for upstream and downstream producers and their suppliers

2. Common Categories of Externally Purchased Equipment and Services for Energy Industry Capital Projects—a useful reference list for purchasing equipment and services
3. Registration, Evaluation, Authorisation, and Restriction of Chemicals (REACH)—a summary of this important and far-reaching legislation that affects many energy supply chains

A Glossary defines frequently used terms as they apply to oil, gas, and power generation companies. The definitions are intentionally focused on the applications described in this book, and thus are specific to the context of value chain management. They may therefore appear to differ from standard definitions found in more generic sources. They are also intended to be short, to the point, and practical, rather than comprehensive and academic.

Finally, a Bibliography provides a reading list for those who want to dive further into specific topics.

Introduction

Cleaner, Greener, and Smarter

The Big Shift

Two simultaneous tidal waves have hit the energy industry over the last ten years: strong pressure from renewable energy, and pervasive benefit and associated risks of technology, especially information technology.

Pressure for CO_2 reduction to combat climate change has made an indelible mark on every industry in the energy sector. Even without the regulatory and public pressure for clean energy, sharply decreasing costs of renewables, especially solar, wind, and energy storage, have made doing business the old way almost impossible. This has manifested itself in almost every energy subsector, for example:

- Nuclear power is continuously marching toward safer fuels and disposal regimes.
- Hydropower has become a more frequently evaluated option than before due to its carbon-neutral footprint.
- Gas and Coal-Fired Power plants are being scheduled for retirement.
- Oil & Gas refineries have been upgraded instead of adding new capacity.
- Solar has attracted unprecedented subsidies to accelerate its path toward low unit cost.
- Wind has been developed in conjunction with hydrogen to produce "green hydrogen" (when renewable energy is used to break water into hydrogen and oxygen, with the resulting hydrogen being used to power a fuel cell, all without any use of fossil fuels), and with energy storage to mitigate the problem of intermittent power from unpredictable wind patterns.
- Utility Scale Energy Storage has attracted unprecedented investment in a race for the most economically viable storage technology for applications at all scales.
- Fuel Cells have been undergoing pilot programs to demonstrate economic feasibility in contexts ranging from automotive to industrial, in addition to power utilities.

Technologies are galloping forward, especially driven by digitalization, and also including rapid advances in chemistry, materials science, and physics. Technology now affects every supply chain decision, from network design to procurement, installation, and logistics. To cite just a few examples:

- Robotic process automation (RPA) is automating workflows and elevating skills requirements.

- Artificial intelligence (AI) is making operations more effective and in some cases able to foresee solutions that remain obscure to most human analysis.
- Remote and robotic equipment are enabling safer and more efficient operations.
- Augmented and virtual reality (AR and VR) devices are allowing clearer view into operations that were formerly out of view.
- Cybersecurity is protecting Internet devices, and it is increasing in scope to include nearly every device.
- Applied chemistry is facilitating longer-term energy storage, as well as materials with higher tensile strength and properties that can be infinitely customized to specific industries and applications.
- Additive printing is allowing on-demand manufacturing of custom parts such as turbine blades, which can shorten repair times, cut costs, and increase uptime.

In order to understand the impact of this Big Shift as it relates to supply chains, it is necessary to understand the types of supply chains, and how to extract value from each type.

Supply Chain Types

Most people's understanding of supply chain management stems from the consumer products industry, in which thousands of stock-keeping units of fast-moving consumer goods flow through distribution centers and move on pallets onto retail shelves. While this characterizes *some* logistics flows that occur in oil, gas, and power supply chains, supply chain management in the oil, gas, and power industries is different in important ways.

There are six types of supply chains, as shown in Figure 5:

1. Extraction/Primary Raw Materials. Extraction-type supply chains involve physical plants that are large in scale and operate continuously. Shipping is typically by tanker or bulk rail in unit trains. Product values and inventory values are typically relatively low, while transportation costs are high as a percentage of delivered cost to the customer.
2. Process Manufacturing. Process manufacturers operate few, capital-intensive, and continuous production plants. The site location is usually determined by shipping costs. Reliability and predictability reduce costs, so more consistent transit times help synchronize the flow of transportation and inventory with the pace of production.
3. Discrete Manufacturing. Discrete manufacturers make and stock inventory, so inventory is a significant cost driver, and they have or

use a large vehicle fleet to move it around. Therefore, infrastructure projects that allow them to reduce inventory, transportation costs, or fleet assets will have a big impact. Discrete manufacturers are the most common type of manufacturer.

4. Design-to-Order Manufacturing. Design-to-order manufacturers do not ship product until it has been ordered, and usually ship directly to customers. They are usually engineering intensive and hold low inventory.
5. Distribution. Distributors buy finished product, usually add value to it, and resell it. Their profit depends on their ability to move product quickly and reliably. Unique transportation or logistics capabilities allow them to create supply chain advantages.
6. Reselling. Resellers, for example retailers and e-retailers, buy finished product and resell it in its identical state. They spend relatively large amounts on transportation, largely because their retail outlets and/or their customers are widely dispersed. Their success usually depends on excellent inventory management and close collaboration with the end customer.

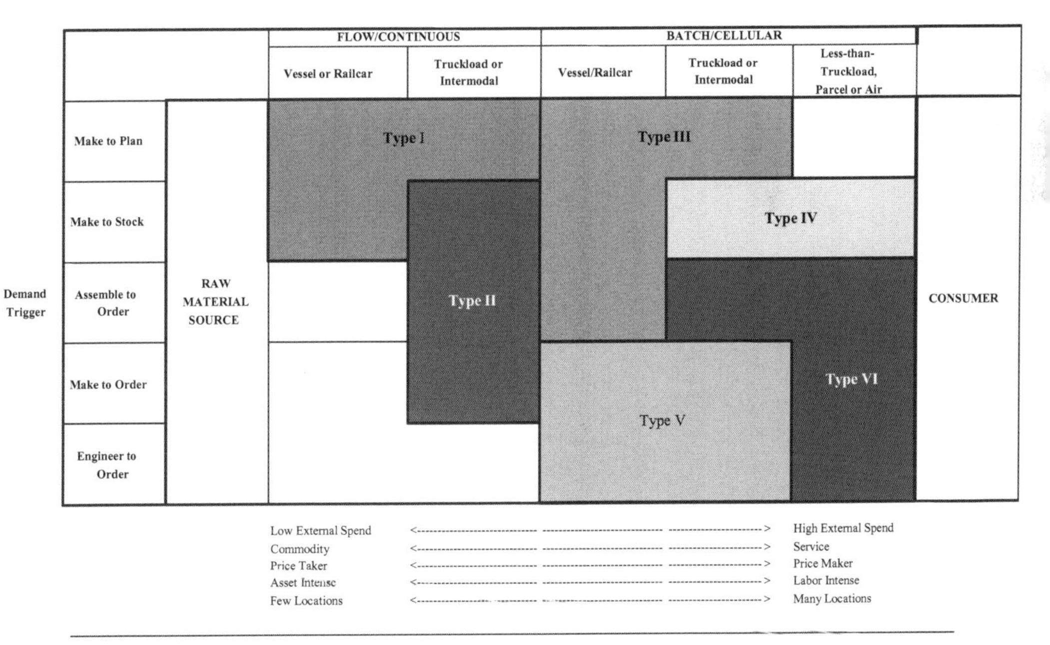

Figure 5. Supply Chain Types

(*Source:* Jacoby, David Steven, in *Guide to Quantifying the Economic Impacts of Federal Investments in Large-Scale Freight Transportation Projects.* Prepared for: Office of the Secretary of Transportation, US Department of Transportation, by: Cambridge Systematics, Inc., Economic Development Research Group, Inc., and Boston Logistics Group, Inc., August 2006. Page A-11)

Oil & gas, geothermal, gas and coal-fired power, utility-scale solar farms, wind, biomass, nuclear, hydropower companies, and the production of hydrogen fall into Type I supply chains (see figure 4). Energy storage and fuel cells involve Type III supply chains. Type I supply chains are Make-to-Plan or Make-to-Stock. They usually start with extraction and proceed to continuous production operations. Shipment is in bulk, usually by ship, rail, or truckload. For example, oil & gas are mined and produced in bulk, not in batches, and shipped in bulk tanker vessels. Type III supply chains involve some assembly, in which orders may be configured to the specifications or requirements of each customer. For example, fuel cells are built from discrete components, and electrolyzers have a general design but are usually customized to suit specific customer requirements.

Supply Chain Value Creation Strategies

There are four supply chain strategies: *Rationalization, Synchronization, Customization, and Innovation.*

1. *Rationalization* means managing operating costs through supply chain management to achieve cost leadership and greater profitability than those of competitors. *Rationalization* focuses on operating expense management rather than asset management, in particular for companies that are driven by quarterly earnings. *Rationalization* includes supply chain processes such as SKU rationalization, Kaizen, and value engineering, and procurement activities such as sourcing (performance management, supplier selection, etc.), production, and facilities management.
2. *Synchronization* means achieving reliable and flawless supply chain execution (right product at the right place at the right time) so as to be able to produce the same volume of output with less fixed assets (production capacity) and working capital (inventory) than those of competitors. *Synchronization* includes processes such as inventory management, maintenance, routing and scheduling, and demand planning, as well as improvement programs such as product life-cycle management, Six Sigma, design for assembly, standardization, and collaborative forecasting, planning, and replenishment.
3. *Customization* means differentially adapting products, services, and solutions to the needs of customer segments. Companies that pursue a *customization* supply chain strategy build a unique capability to use the supply chain to enhance customer relationships, which leads to higher gross margins. *Customization* embeds both responsiveness and flexibility. When *customization* strategy is used to target individuals, it is called Personalization.
4. *Innovation* means using supply chain activities to enable a stream of frequent and effective new product introductions, thus leading to an

acceleration of revenue that otherwise would occur over a longer time frame. Companies that are good at *innovation* are strong at research, development, and commercialization.

Henceforth in this text when the words Rationalization, Synchronization, Customization, and Innovation are italicized, they refer to the corresponding supply chain strategy. When they are not italicized, they are used in the conventional sense.

Asset-intensive companies have traditionally followed *rationalization* and *synchronization* strategies because cutting costs in a fixed asset environment directly increases Return on Investment (ROI), but asset-intensive companies can also follow *customization* and *innovation* strategies. For example, mining companies typically focus on asset utilization and economies of scale, which are *rationalization* strategies, based on a continuous production paradigm, but they could also crush and blend unique grades, and arrange delivery in dump trucks rather than in barges to optimize each customer's operations. Similarly, asset-light service companies often follow *customization* and *innovation* strategies because services are relatively easy to customize, but they can follow *rationalization* and *synchronization* supply chain strategies for a period of time. For example, a software company that is going through a blast of growth may decide to standardize everything about the product, its configuration, and its installation in order to capture the maximum amount of market share.

The most appropriate supply chain strategy depends on the industry and line of business. The part of the supply chain strategy that is "given" is dictated by the business's place in the value chain (see Figure 6).

		Rationalization	Synchronization	Customization	Innovation
Type of Business	Extraction	Positive	Negative	Negative	Negative
	Process (Continuous) Manufacturing	Positive	Negative	Negative	Negative
	Batch Manufacturing	Negative	Positive	Negative	Negative
	Make-to-Order Manufacturing	Negative	Positive	Positive	Positive
	Distribution	Negative	Negative	Positive	Positive
	Reselling	Negative	Negative	Negative	Positive

Figure 6. Correlation between Supply Chain Type and Supply Chain Strategy
(*Source:* David Steven Jacoby, *Guide to Supply Chain Management,* New York: Bloomberg Press, 2009, p. 53.)

Oil & gas, gas and coal-fired, nuclear, geothermal, hydropower, and nuclear operations have traditionally focused on achieving high return on assets due to the preponderance of capital investment EPC expenditure in their supply chain activities. Matching capacity with peak demand has always been a primary driver of both capital and operating decisions. Therefore, they have often pursued a *synchronization* strategy. Biomass-based power producers also need to minimize upfront capital investment, but since feedstock costs represent 40–50% of the total cost of electricity produced, and feedstock can come from many sources (see the Biomass chapter for details), they may influence financial success by decreasing the cost of feedstock and inbound transportation. The wide variety of burning and anaerobic processing technologies, combined with the potential to change feedstock, make biomass producers more likely to follow *rationalization* and *customization* strategies than producers with a more rigid fixed asset base.

New Strategic Supply Chain Options from Renewables

We plotted the supply chain strategy of each energy technology based on the supply chain strategy framework into a radar chart (see Figure 7), using financial data from bellwether companies in each category, such as Exelon, ExxonMobil, SunPower, Titan Solar Power, Acciona, Vestas, Hydro-Québec, Tesla, Pohjolan Voima, Vattenfall, EDF, Calpine, NextEra, Xcel Energy, and Ballard Power, in their respective categories. The data for each dimension was normalized to conform to a 10-point scale.[1]

- The *Rationalization-CapEx* Opportunity scale represents the Supply Chain Type (numbered 1-6, as in Figure 5) of the related Plant & Equipment Construction Capital Projects. The higher the number, the more capital cost take-out opportunity there usually is.
- *Rationalization-OpEx* Opportunity scale represents the Supply Chain Type for Product Shipments, using the same framework (refer to Figure 5). The higher the number, the more operating cost take-out opportunity there usually is.
- The *Synchronization Opportunity* scale shows the ratio of assets to sales revenue. More asset-intensive industries tend to be based on continuous, high-volume production, so improvements in capacity planning and asset utilization usually have disproportionately high financial benefits.
- The *Innovation Potential* scale shows the compound annual revenue growth rate, which is drawn from standards bodies such as DNVGL, industry associations such as the Solar Energy Industries Association, government agencies such as the US Energy Information Administration, think tanks, and the author's own estimates and adjustments.
- The *Customization Potential* scale represents the operating margin, which is a proxy for customizability since higher-margin and services businesses tend to be more customizable than commodity businesses.

In order to maximize the range of tools that optimal supply chain strategy and management can offer, companies' supply chain plots should score as high as possible on as many of the five attributes as possible. Supply chain management potential is maximized when a top score is reached on all five dimensions. Visually, this would be a full circle looking like a large cobweb. The higher the scores, the more

---

[1] The radar chart separates CapEx Rationalization Potential from OpEx Rationalization Potential to maximize strategic supply chain insights.

value can be derived from supply chain management tools and techniques. The best situation is to be able to constantly reduce costs, perfectly balance supply with demand, personalize energy delivery to each customer, and continuously develop new energy products that the market wants.

Conventional energy supply chains score high on *synchronization* and *rationalization,* as can be seen in Figure 7 below. Almost all the attributes point toward *Synchronization Potential* and *CapEx Rationalization Potential.* Upstream supply chain management typically focuses on a balance of cost containment and asset productivity, due to the relative importance of OpEx to CapEx. Electric and gas utilities have the most capital expenditure relative to OpEx (3.7 dollars of assets for each dollar of operating expense),[19] so they are biased towards asset productivity management (*synchronization*). Midstream and downstream activities have a greater proportion of operating expenses per dollar of assets (1.3 dollars of assets for each dollar of operating expense), so supply chain management in Midstream tends to focus more on operating cost reduction (*rationalization*) than on asset productivity (*synchronization*). E&P companies (upstream) fall in between (2.3 dollars of assets per dollar of operating expense), so supply chain management in E&P needs to a achieve a balance between CapEx reduction and operating cost containment.

Figure 7. Conventional Energy Supply Chain Strategy Profile

Renewable technologies are completely overturning value chain paradigms in the energy sector, and consequently the toolbox used to govern trading partner relationships. Because oil, gas, and power are produced in large capital-intensive plants, the most obvious ways to increase profit have been to cut costs and increase asset productivity. In contrast, solar modules, batteries, and fuel cells can be used at small scale as well as large, and their *modularity* gives them a huge supply chain advantage over conventional oil, gas, and power generation, both in tapping new markets and in lowering costs. Batteries start with cells, which are then assembled into packs, and the packs into batteries, and the batteries packed into containers, so, they can be deployed at any scale. Tiny solar micro-panels called "pico PVs" power single electronic devices like a mobile phone. Hydrogen fuel cells can be used to power mopeds. Battery material is being embedded in the clothes that you wear.

Most renewable technologies can be applied at both small and large scale, and the power plants are constructed from discretely manufactured components, more like a traditional manufacturing operation. This makes it more important to control variable costs, which leads to a *rationalization* strategy. Furthermore, they can be put together in different configurations, which open up opportunities for *customization* strategy. The addition of novel chemistries and other technological advancements makes the *innovation* strategy a potential, and sometimes an imperative, axis of competitive differentiation and a Key Success Factor.

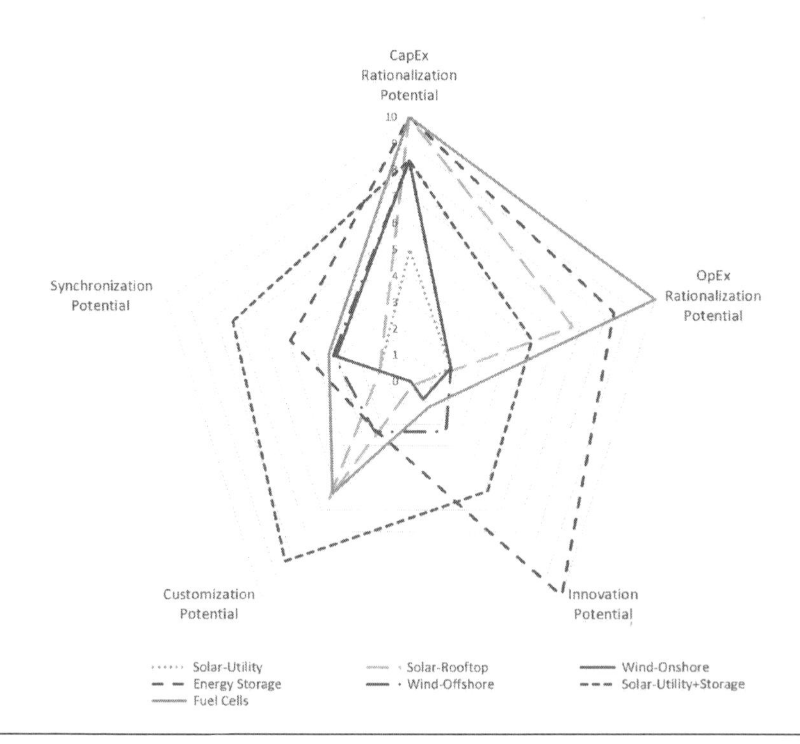

Figure 8. Renewable Energy Supply Chain Strategy Profile

For energy storage and hydrogen fuel cell manufacturers, *customization* and *innovation* supply chain strategies are not only opportunities but are also imperative for their operating and financial success. For example, flow battery and lithium-ion battery storage companies typically have one core technology but produce solutions that are specifically engineered for the size, flow rate, and storage duration of each power plant. Fuel cell manufacturers need to quickly and repeatedly establish new raw materials, stack design, and module configuration as significant innovations occur in Polymer Electrolyte Membranes (PEM), and possibly switch between PEM and other fuel cell designs as other advances in temperature, pressure, and materials change the optimal technology for specific applications.

Solar + Energy Storage is unique in being able to leverage the entire supply chain toolbox for competitive advantage, as shown in Figure 8 above. Solar power can be built at massive scale in concentrated solar plants involving billions of dollars of capital investment, while tiny solar panels are used by villagers in remote and poor areas of the world to power just a single mobile phone. Similarly, wind power can be harvested at huge scale, such as in the case of the planned 8.2 GW array off the shores of South Korea,[20] and at small scale with one turbine in a backyard. Small turbines and towers can be mass-produced with huge economies of scale, while massive turbines and towers are being custom engineered to capture maximum power including through smart tilting of turbine and blades and through artificial intelligence.

Renewable producers' access to the entire range of supply chain optimization techniques offers them a competitive advantage over centralized, asset-intensive power producers, which mostly need to focus on reducing capital costs. The breadth of application and the range of production methods, distribution channels, and cost structures make wind and solar uniquely able to find and exploit niche applications in a way that centralized, asset-intensive power producers cannot. This is game-changing. It means that even the large, centralized power producers need to find ways to Customize and Innovate, or else they will face commoditization and will eventually be outmaneuvered by renewables producers, who will be able to continuously find new ways of offering fresh and differentiated, and thus higher-value, solutions to their customers. The degree to which fossil-fuel producers think out of the box to be able to offer customized and innovative energy solutions will determine how long they can remain relevant.

Not only do renewables have a broader footprint across the strategies, but their improvement frontier is distributed across these strategies as well, which offers them more paths to profitability. The two juxtaposed charts in Figure 9 below show the supply chain value frontiers for Gas and Coal-Fired thermal power plants and for Utility-Scale Solar plants.

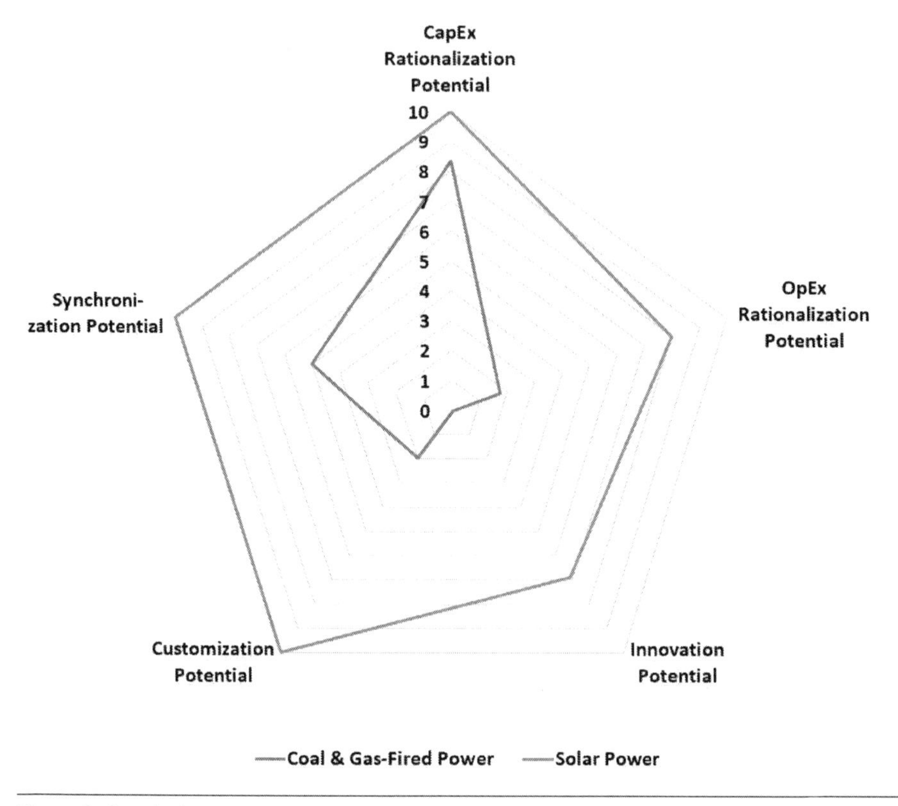

Figure 9. Supply Chain Value Frontiers for Coal and Gas-Fired versus Solar Power

Despite the strategic supply chain advantages of some renewable technologies, fossil fuel producers still have room for improvement. Oil & gas producers have been customizing both processes and products to extract oil & gas in different forms wherever it is found, for example through processes such as Enhanced Oil Recovery, the recovery and processing of stranded gas, Gas-to-Liquids, and sour gas processing solutions, and by more fully commercializing customized forms of hydrocarbons such as NGLs, heavy crude, and LNG. Nuclear power technology, which is usually considered non-renewable and thus is grouped with the fossil fuel producers in this context, has steadily engaged in an *innovation* strategy, putting out new generations of safer, more economical fuels and combustion processes—fission technology is now in its fourth generation. Russia in particular has taken a big step toward Customization in developing floating nuclear ships that can be dynamically moored offshore.

Roadmaps for Maximum Value Creation and Financial Impact

Each of the technology chapters identifies high-impact value chain development challenges and opportunities, and quantitatively rates the impact of addressing those challenges/opportunities, using the same scales as presented above. While the opportunities in conventional energy technologies tend to reinforce their *rationalization* and *synchronization* strategies by providing cost reduction and asset utilization improvements, the opportunities in the renewable technologies offer a broad array of potential benefits across *rationalization, synchronization, customization,* and *innovation* supply chain strategies.

Each chapter also contains a roadmap that defines key supply chain-related milestones based on these evaluations. As each energy technology evolves along the roadmaps presented in the following chapters, their strategic supply chain footprints will shift. The arrows in Figure 10 below show the direction of movement.

- Nuclear supply chains can become more customizable through mini-nuclear and floating nuclear, more rationalized by optimizing economies of scale, more innovative via pioneering fuels and waste technologies, and more synchronized by reducing the bullwhip effect of the capital investment cycles inherent in megaprojects and gigaprojects.
- Hydropower can become more synchronized (and reliable) through: 1) predictive maintenance with vibration sensing; 2) energy storage; and 3) financial hedging for decreased flow. They may also become more innovative through hybrid energy technologies including green hydrogen, and eventually potentially wave and current energy capture.
- Gas and coal-fired power supply chains can become more synchronized, customized, rationalized, and innovative. Predictive maintenance and reliability engineering can better align production with demand, and cybersecurity can assure stability of production. Co-production can better tailor combustion processes to the availability of local renewable resources while reducing CO_2 emissions. Performance-based agreements and life-cycle cost management can help to minimize capital and operating costs. And innovations in the uses for coal may sustain jobs that are in jeopardy as a consequence of the shift toward clean power.
- Geothermal power supply chains can become more cost-efficient (rationalized) by integrating hybrid energy technology formats such as district heating and Combined Heat and Power (CHP), and by adding binary plants to existing plants instead of building new ones. They can also be better aligned with demand (Synchronized) by adding energy storage.
- Upstream Oil & Gas supply chains can become more Innovative through digital oil fields and by converting product procurement to

services-as-a-product; more reliable by increasing implementation of cyber security protections; and more sustainable by using renewable energy to power operations.

- Midstream Oil & Gas supply chains are fertile ground for supply chain enhancements across all four supply chain strategies (*rationalization, synchronization, customization,* and *innovation*). Improvement opportunities include: 1) the expansion of LNG, 2) small-scale modular gas processing technologies, 3) drone monitoring, and 4) smarter process automation solutions. Seen through a supply chain lens, LNG is a supply chain *innovation,* gas processing technologies are supply chain *customizations,* drone monitoring is a *rationalizing* (cost-cutting) strategy, and smarter process automation solutions are *synchronization* (supply-demand matching) improvements.
- Downstream Oil & Gas supply chains can become more synchronized through: 1) more robust risk governance; 2) cybersecurity for IoT cevices and systems; 3) smarter predictive maintenance; and 4) low-sulfur fuel. Risk governance and cybersecurity assure production reliability.
- Biomass supply chains can become more Innovative through hybrid technologies, more Rationalized through Multi-Stage Process Routes and Pre-Treatments, and more sustainable by Regrowing Natural Resources.
- Solar supply chains can become more Rationalized and Innovative through: 1) IoT/AI, including smart tracking to optimally align the modules toward the sun; 2) securing the polysilicon supply chain; 3) integrating energy storage; 4) managing and minimizing life-cycle cost; 5) enabling distributed solar; 6) expanding the safe use of floating solar; 7) expanding the use of embedded solar; and 8) integrating rooftop solar with vehicle charging systems. Also, from an environmental and social point of view, social exploitation and CO_2-emitting production processes should be avoided, and end-of-life disposal of toxic PV waste should be anticipated.
- Wind supply chains can become more Rationalized by: 1) adopting construction standards to reduce capital cost; 2) developing and scaling Wind+Storage solutions; 3) ensuring cybersecurity; and 4) fully utilizing the power of Artificial Intelligence and machine learning to optimize equipment parameters in real-time.
- Energy storage supply chains are expected to create value by leveraging all four supply chain strategies, in particular by: 1) ensuring availability of critical minerals such as cobalt; 2) minimizing or eliminating the risk of overheating and fire; 3) proactively managing battery costs throughout the supply chain; 4) realizing multiple benefits simultaneously to increase the ROI; 5) reducing the cyber security attack surface; 6) deploying AI for battery operating system and load management; 7) integrating energy

storage with vehicle charging networks; and 8) developing economical battery recycling approaches.

- Hydrogen Fuel Cells can become safer and more Rationalized, Synchronized, and Innovative by: 1) reducing the potential for leaks and flammable conditions; 2) realizing economies of scale in the GDL and the fuel stack; 3) sourcing platinum to ensure adequate supply; 4) protecting the FCCU from cyberattack; 5) developing optimal chemistries for various power ranges and environments; and 6) creating a market for replacement membrane elements.

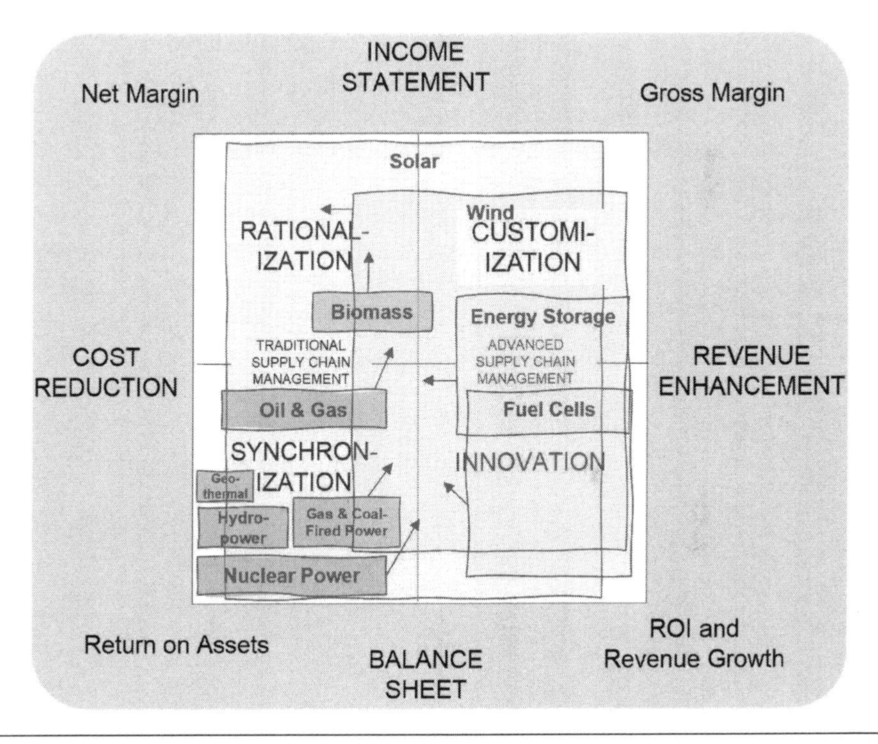

Figure 10. Prevalent Supply Chain Strategy Positioning of Energy Companies (*Source:* Boston Strategies International)

Supply Chain Cost

The costs associated with the supply chain management activities are typically embedded in financial metrics such as Cost of Goods Sold (for example, for maintenance and repair parts for an oil refinery) or capitalized into assets if they support a capital project, such as the delivery of wind turbine blades for a wind farm construction project. At the other extreme they are represented in detailed process metrics such as In-Transit Inventory Costs.

For comparability purposes in the context of the energy industry, we developed a metric that can be compared across energy companies: supply chain cost per million BTU of energy produced. It is built up from the costs at three energy processing stages, as follows:

- Upstream costs include the carrying cost of the capital investment, the annualized cash costs of installation if applicable, freight of materials and components to the construction site, and EPC management fees.
- Midstream costs include the carrying cost of the capital investment in infrastructure development (pipelines for O&G, and T&D for power), the carrying cost of the tanker and LNG fleet (for O&G)—these are proxies for the cost of fuel distribution, and the cash costs of line loss (for power).
- Downstream costs including the carrying cost of inventory (refined oil for O&G), and the costs of installation and servicing, often called "soft costs" or "balance of system" costs, associated with installation such as permitting, insurance, and installation (for distributed power).

Figure 11 shows this metric on a global basis for 2019. In many ways the supply chain cost correlates with the maturity of the industry. For example, oil & gas has a comparatively low supply chain cost, while fuel cells have a high supply chain cost. Over time, supply chain processes become standardized and systematically reduced. Nuclear power is an exception; most supply chain costs are high for nuclear projects due to the management costs associated with engineering, procurement and construction.

Technology	Trillions of BTU Produced	Energy Production Rank	Supply Chain Cost per Million BTU	Supply Chain Cost Rank
Oil & Gas	197,977	1	$ 1.75	1
Energy Storage	241	7	$ 1.96	2
Solar-Rooftop	6,490	3	$ 2.14	3
Geothermal	20	11	$ 4.74	4
Solar-Utility+Storage	16	12	$ 4.80	5
Hydropower	538	6	$ 6.67	6
Solar-Utility	7,373	2	$ 7.90	7
Wind-Onshore	1,793	4	$ 10.70	8
Combined Cycle	1,165	5	$ 17.16	9
Fuel Cells	3	13	$ 26.74	10
Wind-Offshore	179	9	$ 30.78	11
Biomass	200	8	$ 34.00	12
Nuclear	179	9	$ 48.31	13

Figure 11. Supply Chain Intensity of Energy Technologies
(*Source:* BSI Energy Ventures)

Supply Chain Performance Management—Metrics and Targets

Executive management at most companies expects the supply chain management function to reduce cost and increase net margin and return on assets, and to improve operations, especially to increase quality and availability of raw materials, intermediate services, and finished product. However, the same tools can in many cases be used to optimize around other objectives (for example, environment or local content).

Rationalization efforts have been demonstrated to create at least a 13% improvement in net margin.[II] *Synchronization* efforts have been shown to create a 10% improvement on Return on Net Assets (RONA). Supply chain initiatives in the *synchronization* phase often achieve the improvement in RONA by reducing forecast error, and thus achieving level production both within the enterprise and across trading partners, which decreases the need for inventory and fixed assets.

Most often, improvements are measured at the project level, where they are sought in targeted areas that contribute to a higher net margin and a higher return on assets such as:

- Reduced upfront purchase cost. Lower upfront cost is the most intuitive savings framework. However, it can be complicated by tiered pricing, promotions, discounts, and volume rebates.
- Reduced operating cost. Energy savings is a common way to reduce operating costs. Operating assumption variables can affect the savings, especially for large turbines and electrical distribution and control equipment.
- Increased throughput, or productivity. Improvements that increase the speed of a process such as drilling or refinery expansions can lead to higher production overall. Savings for these types of improvements can be calculated on the basis of either cost savings or profit enhancement. Cost savings might be estimated by, say, the reduction in the number of rig-days needed to drill a well. The savings per well then can be multiplied by the number of rigs in operation and the number of wells that need to be drilled over a period of time. If the increased productivity leads to reduced time to first production, the benefit may be improved profitability. To quantify this benefit, take the number of extra days of production and calculated additional profit based on a typical well, then apply an average output price per unit to get the profit margin, and multiply the resulting benefit by the number of units producing.
- Shorter lead times for the delivery of equipment or services. For capital items, one might determine that order lead times are constraining production that would otherwise be occurring. In this case, the decrease

[II] Based on a prototypical operation

in lead times can be multiplied by the weekly profit from greater production. For lower lead times on purchases that are held in inventory, inventory holding costs can be used as the basis for cost savings.

- Increased reliability or uptime. Longer time between failures means the same amount of production with less equipment, saving money by avoiding the need for equipment purchases. Benefits can be calculated according to how much less equipment is needed (in value, not in units).

Figure 12 shows a more comprehensive 11-point supply chain rationalization framework based on a "should-cost" methodology.

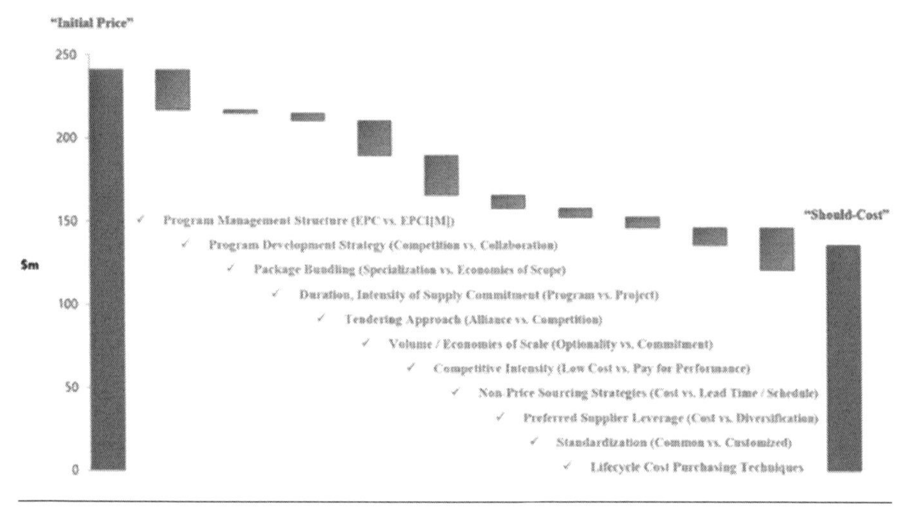

Figure 12. Supply Chain Rationalization Levers Based on "Should-Cost"
(*Source:* Boston Strategies International)

Metrics and measurements should be consistent with the chosen value chain strategy and the targeted benefits. The contribution of a well-designed value chain strategy and improvement program can be discretely measured by the improvement it generates in Economic Value Added, or EVA.[III] Efficiencies gained in traditional large-scale energy companies have improved performance measurably. In emerging energy technologies, success application can be transformational. Figure 13 shows the average impact on EVA of a cross-functional value chain creation initiative.

However, financial improvements don't account for the impact of added risk. Some improvement initiatives can be deployed at the expense of incurring risk. There are many complicated ways of measuring economic risk, most involving statistics and probability (and many more ways to measure operational risk—those will be covered in the next section). In general, executive managers discount

[III] EVA measures profit after tax less the true cost of capital employed. It is calculated as: Net Operating Profit After Taxes (NOPAT) - (Capital * Cost of Capital).

probabilistic financial analyses, preferring to trust their gut instinct around scenarios, of which one usually represents the most likely, or "baseline," scenario. The "tornado diagram" illustrated in Figure 14 is a relatively simple and intuitive way to measure and display the economic risk impact related to supply chain programs.[21]

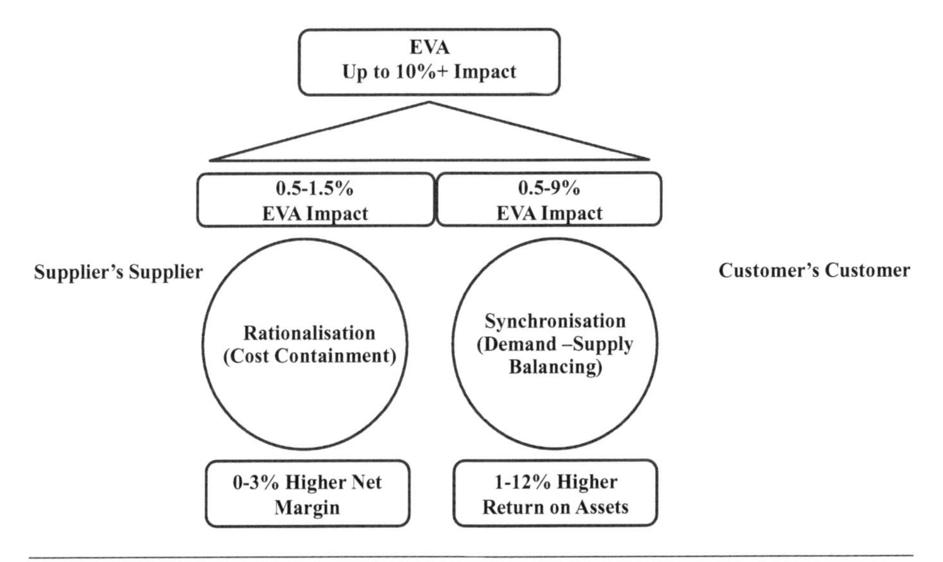

Figure 13. Quantification of Value Chain Strategy Benefits
(*Source:* Boston Strategies International)

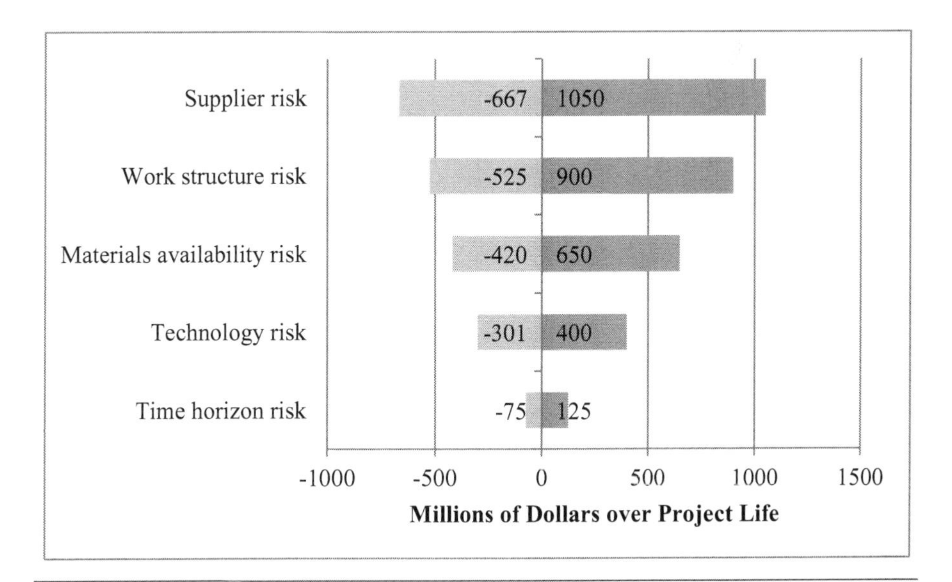

Figure 14. Quantification of Supply Chain Risks
(*Source:* Boston Strategies International)

Supply Chain Governance

Value chain decisions usually involve input from the C-Suite generally speaking, from Strategic Business Unit heads and from departments such as Research & Development and Marketing. Activities cover strategy, finance, mergers and acquisitions, and pricing, to name a few. Key decision-makers may include, for example:

- Chief Executive Officer (CEO)
- Chief Financial Officer (CFO)
- Chief Operating Officer (COO)
- Chief Technology or Information Officer (CTO)
- Chief Information / Data analytics / digitalization Officer (CIO)
- President of Business Unit
- Vice President or Senior Vice President of Marketing
- Vice President or Senior Vice President of Sales
- Vice President or Senior Vice President of Engineering
- Director, Vice President or Senior Vice President of Strategy
- Director, Vice President or Senior Vice President of Government & Regulatory Affairs
- Etc.

Operational supply chain management decisions are frequently made in departments such as Procurement, Production, Order Fulfillment, Materials Management, Demand Planning, Logistics, Customer Service, Production Control, and Transportation.[22] Many extractive energy companies assign most operational supply chain responsibilities to staff with formal titles like engineering, health & safety, drilling, and other departments that might not be recognizable as containing supply chain management content. For example, within the organization shown in Figure 15 below[23][24][25] are embedded supply chain managementactivities such as:

- Project risk mitigation
- Standardization and simplification of specifications (Engineering)
- Minimization of capital expenditure procurement (Procurement)
- Supply risk mitigation (Construction/Installation)
- Constraints management and throughput improvement, including Lean manufacturing (e.g., quality, TPM, etc.) = (Productivity Improvement)
- Upstream
    - Procurement
    - Project Design
    - Construction Management
    - Commissioning

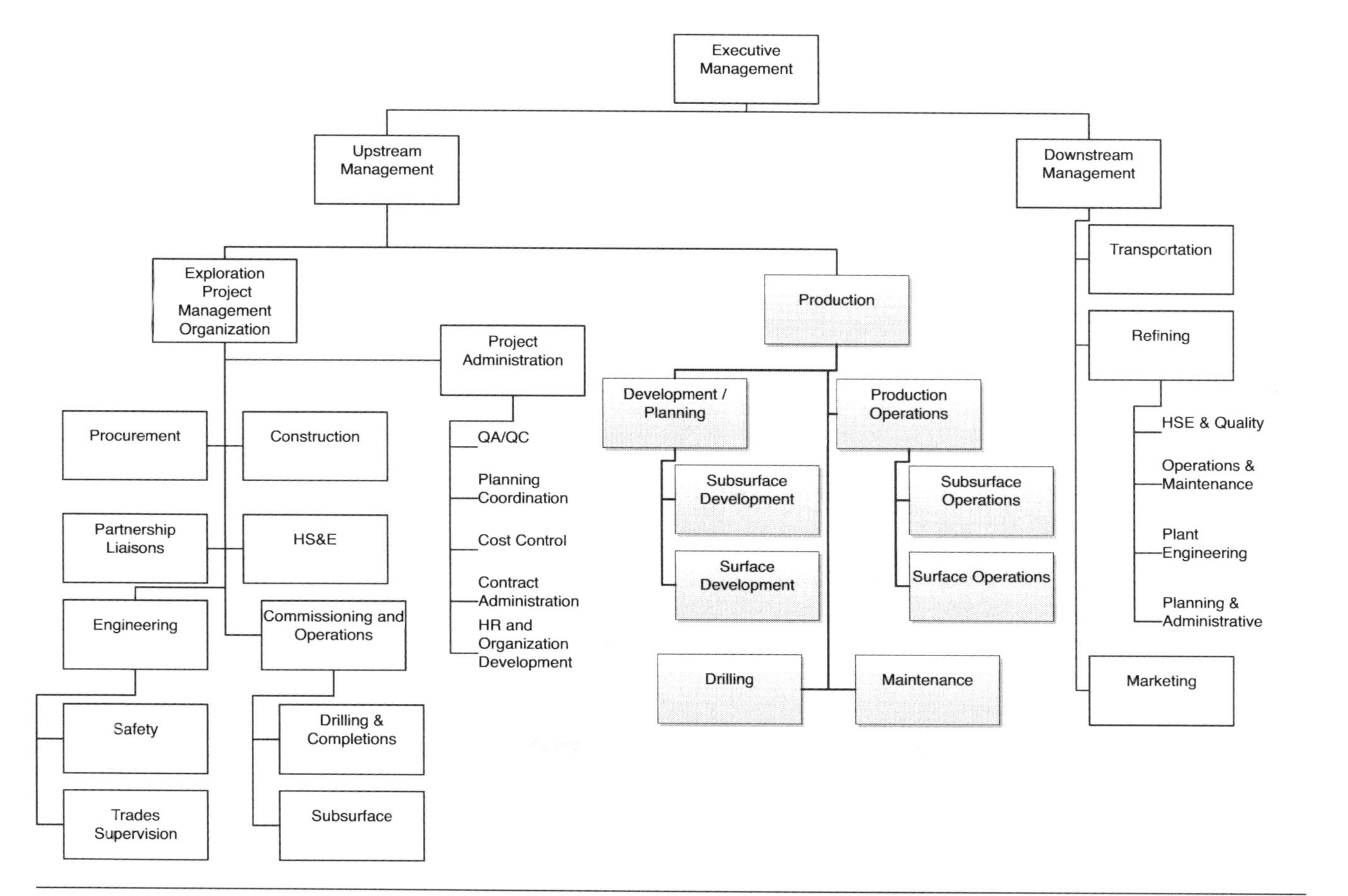

Figure 15. Illustrative Organization of Oil & Gas Company Operations Management Activities
(*Source:* Boston Strategies International)

  - Project Services
  - Drilling
  - Surface Operations
  - Maintenance
  - Planning
- Downstream
  - Export/Transportation
  - Operations & Maintenance
  - Plant Engineering
  - Marketing

Due to the preponderance of large capital projects, the expenditure per supply management employee is higher in the oil & gas industry and the power industry than in industrial manufacturing or financial services (see Figure 16). In 2010 oil & gas outspent power by 2:1; in 2020 power industry procurement personnel were responsible for 32% more external expenditure than their peers at oil & gas companies, a reflection of the decline in oil & gas capital expenditure since 2015 and of the emergence of large-scale renewable power projects.

Metric	Oil & Gas	Power	Aerospace	Manufacturing	Financial Services
Percent of Spend Influenced	80.5%	81.0%	85.60%	82.80%	81.70%
Cost Savings as % of Spend - Reduction	3.3%	2.7%	4.30%	3.40%	2.70%
Cost Savings as % of Spend - Avoidance	3.6%	2.1%	3.10%	0.80%	2.90%
Percent of Employees Strategic versus Tactical	30.0%	30.0%	29%	30%	41%
Spend Managed per Category Management Employee ($m)	124	164	222	52	86

Figure 16. Managed Spend per Supply Management Employee
(*Source:* The Metrics of Supply Management (Cross-Industry) Report, CAPS Research, 2018.[26])

Operations staff rely on associations like ASCM and ISM for continuous learning and training. Some companies have extensive internal training departments. For example, Saudi Aramco holds an annual Supply Chain Symposium in conjunction with King Fahd University of Petroleum and Minerals (KFUPM) to bring together key stakeholders and decision makers from suppliers, academia, and professional associations. By assembling local and world-renowned experts and exhibitors in the fields of supply chain and risk management, the initiative aims to develop a common language about supply chain management that can support extended supply chain management.[27]

Supply Chain Processes

As shown in Figure 17 below, typical industry supply chain management (SCM) processes include:

- Manufacturing and distribution site location (which in most industries is independent of proximity to natural resources and is assumed to be via fungible modes of transport like trucks, rather than specialized modes like VLCC crude tankers)
- Demand planning
- Strategic sourcing
- Purchasing and inbound logistics
- Production
- Inventory management and warehousing
- Distribution operations
- Performance measurement and management

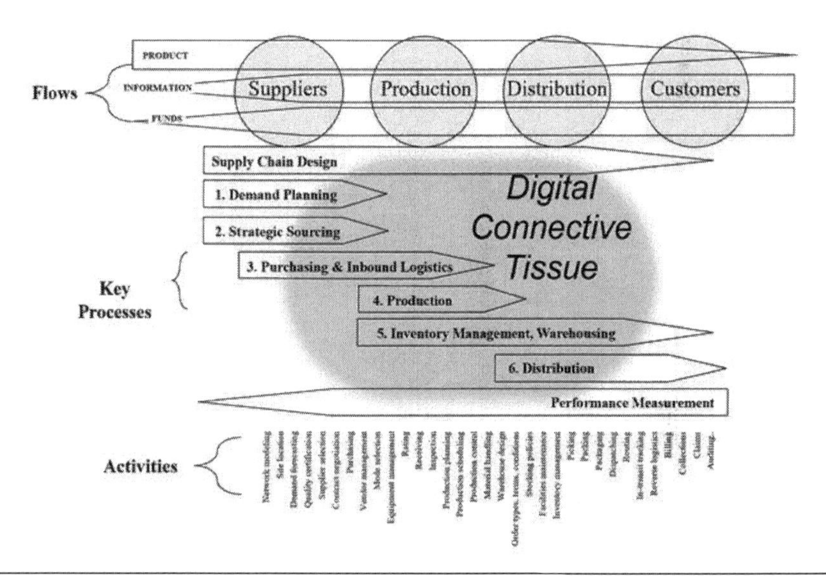

Figure 17. Typical Multi-Industry Supply Chain Management Processes
(*Source:* Boston Strategies International)

Oil, gas, and power companies are asset intensive, so many of their key business processes have to do with asset procurement, installation, deployment, operation and maintenance. A cross-section of upstream and downstream oil & gas business processes and power plant management processes looks quite different from analogous processes at consumer goods companies, following the pattern outlined below:[28]

- Negotiate or bid and acquire assets
- Model hydrocarbon resources/multi-plant or power grid network potential
- Acquire physical data (seismic, geologic, wind/wave)

- Decide project structure (BOO, BOOT, etc.)
- Develop a project execution plan
- Design generating systems
- Design facilities and installations
- Negotiate EPCI contracts
- Construct (rigs/platforms/vessels/pipelines/refineries/power plants)
- Drill or build
- Operate
- Maintain equipment
- Monitor and evaluate asset performance
- Abandon assets at the end of their useful life

Renewable power developers and operators need additional competencies related to operations planning and management, specifically managing operating expenditure (OpEx). Core skill sets include, for example:

- Constraints management and throughput improvement, including Lean manufacturing (including quality, TPM, etc.) to achieve continuous productivity improvements
- Operations and scheduling of plants and equipment
- Production planning
- Asset management
- Materials management
- Maintenance
- Monitoring and evaluation of asset performance
- Abandonment of assets at the end of their useful life

Following this integrated framework, each chapter in this book is divided into two major sections, and is further subdivided into topics that apply uniquely to upstream, midstream, downstream, and power. The two major subdivisions are:

- CapEx project supply chain risk mitigation
- Operating cost reduction

First Principles for Supply Chain Design and Improvement

A series of "first principles" can help to define the perimeter of potential solutions to the array of problems that are addressed in the CapEx and the OpEx sections. Three such overriding principles are:

- Consider the entire chain of actors and activities, from the ultimate source to the ultimate consumer.

- Minimize life-cycle cost by balancing capital and operating costs.
- Structure partner relationships to minimize contract, operating, and environmental risk.

Value chain design should consider the end-to-end chain. Alignment with suppliers of goods and services, and with customers, increases the likelihood that targeted benefits will be achieved. Suppliers' collaboration and adherence to a common and agreed set of practices and principles is essential to achieving cost reduction and risk minimization goals, and conversely divergence from such goals increases the chances of mishaps, as evidenced through BP's failed coordination with Transocean, Halliburton, and Cameron—which contributed to the Macondo spill. In addition, coordination of supply and demand forecasts and production and inventory levels can dampen the "bullwhip effect," an oscillation and amplification of demand throughout the value chain that adds 10% to the ultimate price of products in the oil & gas supply chain. As demand increases, oil price rises, which causes producers to expand capacity, forcing prices down and depressing demand for equipment, in a feedback loop.[29] Oil & gas drilling activity fluctuates about three times as much as production of refined product, indicating (among other things) that if refiners coordinated more with E&P there would be less volatility up and down the supply chain.[30] BSI further quantified this bullwhip effect in the upstream oil & gas supply chain, finding that it increases the cost of gasoline by 10% as oil producers, oil refiners, heavy equipment suppliers, and their component suppliers pass on the costs of inventory overages and shortages, poorly timed capacity investments, and inflationary prices (see Appendix 1). A similar problem occurs in a multi-tier electrical distribution system, where peak demand requires the use of "peaker" generating plants at the upstream end of the supply chain, which can be activated to cope with fluctuations in demand. Supply chain planners and operations researchers have designed hundreds of models to balance and optimize fluctuating demand through a system; this book will feature some of their applications, including the System Dynamics Model of a Four-Tiered Upstream Oil & Gas Supply Chain used by Boston Strategies International, a portion of which is shown in Figure 18.

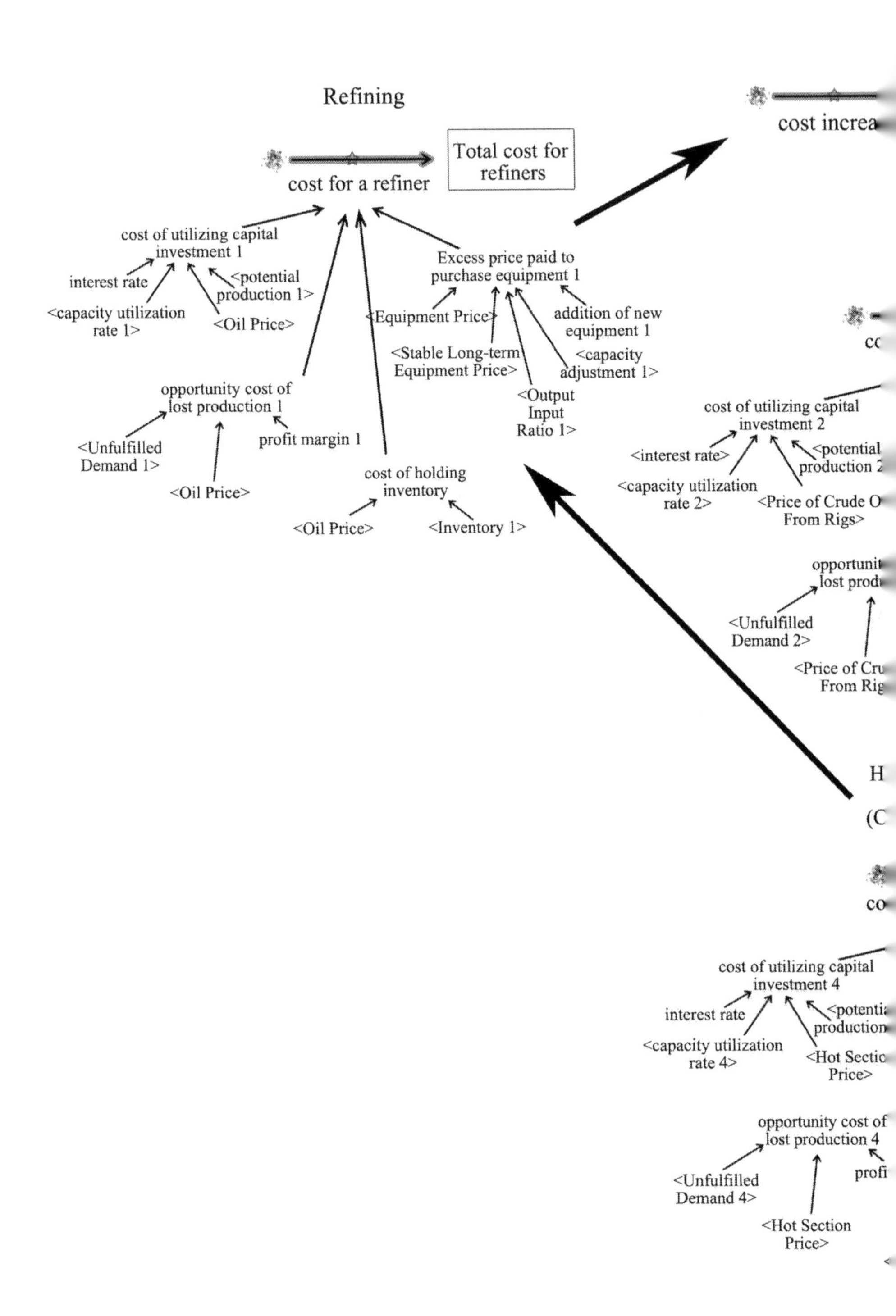

Figure 18. System Dynamics Model of a Four-Tiered Upstream Oil & Gas Supply Chain (*Source:* Boston Strategies International)

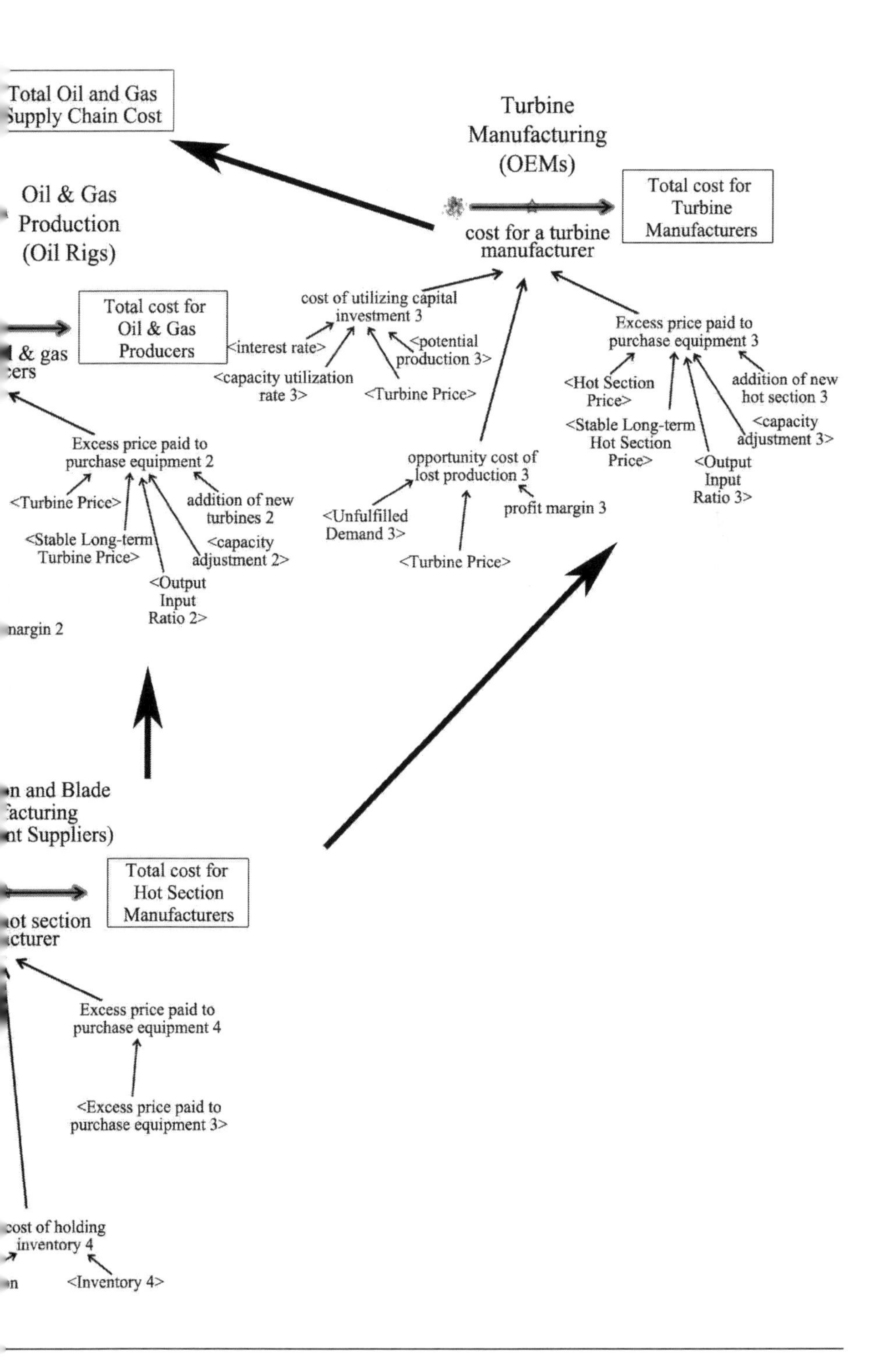
Total Oil and Gas
Supply Chain Cost
Turbine
Manufacturing
(OEMs)
Total cost for
Turbine
Manufacturers
Oil & Gas
Production
(Oil Rigs)
cost for a turbine
manufacturer
cost of utilizing capital
investment 3
Total cost for
Oil & Gas
Producers
<interest rate>
<potential
production 3>
<capacity utilization
rate 3>
<Turbine Price>
Excess price paid to
purchase equipment 3
<Hot Section
Price>
addition of new
hot section 3
<Stable Long-term
Hot Section
Price>
<capacity
adjustment 3>
<Output
Input
Ratio 3>
Excess price paid to
purchase equipment 2
<Turbine Price>
addition of new
turbines 2
<Stable Long-term
Turbine Price>
<capacity
adjustment 2>
<Output
Input
Ratio 2>
opportunity cost of
lost production 3
<Unfulfilled
Demand 3>
profit margin 3
<Turbine Price>
Total cost for
Hot Section
Manufacturers
Excess price paid to
purchase equipment 4
<Excess price paid to
purchase equipment 3>
inventory 4
<Inventory 4>

Supply chain choices must minimize life-cycle cost by balancing capital and operating costs over the relevant time horizon of the investment. Many decisions in these long investment horizons involve trade-offs between initial capital expenditure and ongoing operating expenditure. It is easy to save money on initial capital expenditure if one is willing to pay more in operating expenditure, and many suppliers would prefer to sell solutions that result in perpetual income for their companies. Although the idea is simple, execution of true life-cycle cost analysis is difficult for two reasons. First, with large and complex projects (see Figure 19 below), the life-cycle cost components are often so extensive that it is frequently hard to decide which are germane and which are not. Some may appear to be unrelated (for example, an investment in an extra module of an ERP system that needs to be in place to communicate with key suppliers), interrelated or contingent (investments may need to take place only if a milestone is reached), or sunk costs (for example, dredging to lay a cable, where the dredging may have had to be done anyway). The second challenge is getting reliable enough data to make good trade-offs. Frequently the data is simply unavailable—for example, reliability performance of a new technology platform for a gas turbine, or lifespan of wind turbine blades made in China versus comparable blades made in Europe.

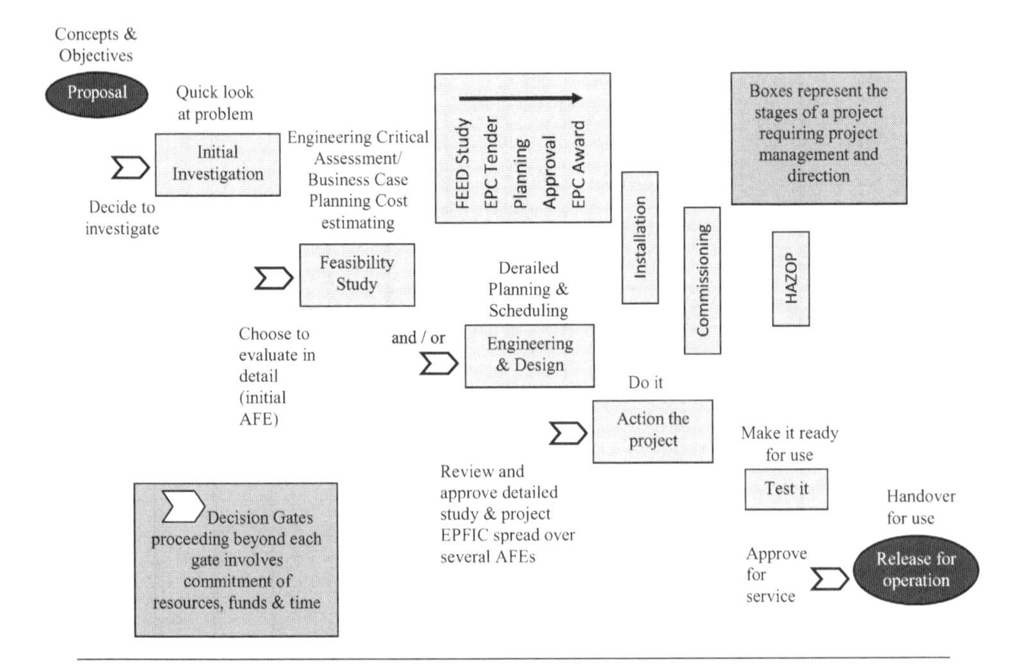

Figure 19. A Typical Oil & Gas Production Project Framework Divided into Stages & Gates (*Source:* Saeid Mokhatab, William A. Poe and James G. Speight, *Handbook of Natural Gas Transmission and Processing,* Burlington: Gulf Professional, 2006, p. 548.)

Third, supply chain management should minimize contract, operating, and environmental risk. Due to their long time frames, energy capital construction projects are inherently risky. A typical upstream or midstream project time frame could be seven years, as for an LNG export facility (see figure 20 below).[31] The time frame brings market risk, as costs and prices will change over the time horizon of construction, installation, and operation. The magnitude of investment typically brings supplier risk. The milestones bring go/no-go breaking points, which could halt the project, increasing or reducing the rate of return relative to initial forecasts. A particular challenge arises when suppliers require financial commitment before the final investment decision (FID) in order to deliver the product or service by the deadline. The role of supply chain management should be to minimize these risks. This may be at odds with management's leanings, especially if rapid implementation would accelerate substantial rewards, if savings could be had by reducing the investment of time or money, or by ignoring negative environmental externalities.

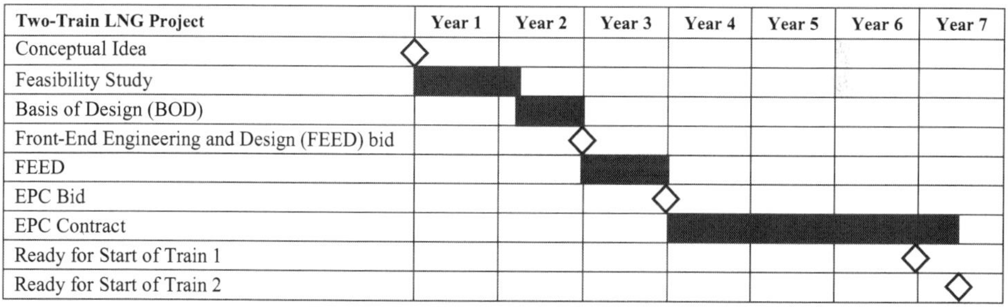

Two-Train LNG Project	Year 1	Year 2	Year 3	Year 4	Year 5	Year 6	Year 7
Conceptual Idea							
Feasibility Study							
Basis of Design (BOD)							
Front-End Engineering and Design (FEED) bid							
FEED							
EPC Bid							
EPC Contract							
Ready for Start of Train 1							
Ready for Start of Train 2							

Figure 20. Typical Project Time Frame for an LNG Export Facility
(*Source:* Michael D. Tusiani, *LNG: A Nontechnical Guide,* Oklahoma: PennWell, 2007, p. 125.)

Supply Chain Successes

The value chain management framework introduced earlier (in Figures 7 through 10) can be used to identify and create disruptive opportunities as well as to manage operations.

While disruptive value chain strategies can lead to exorbitant success, supply chain management programs are more predictable. By using the decision analysis tools in this book, fossil-fuel energy companies have achieved a 13% reduction in capital costs, a 9% increase in total annual sales revenue as a result of debottlenecking and increased throughput capacity, and a 1% reduction in total operating cost, while enhancing stewardship to safety and the environment. Renewable energy producers can do better because their business environments are at an earlier stage of maturity, which makes strategic value chain decisions much more impactful.

The implementation of best practices in traditional energy industries has yielded benefits such as (in descending order of their impact on overall corporate EVA):

- 13% reduction in capital costs, which would increase total corporate EVA by 0.8%, from 8.4% to 9.2%.

- 9.8% increase in total annual sales revenue due to debottlenecking and increased throughput capacity, which would increase total corporate EVA by 0.5%, from 8.4% to 8.9%. If capacity is constrained and all improvements resulted in output that could be sold immediately, total sales would increase by 45%, and EVA would increase by 1.9%.
- 1.0% reduction in total operating cost, which would increase total corporate EVA by 0.6%, from 8.4% to 9.0%.
- 1.6% reduction in inventory holding cost, which would increase corporate EVA by only 0.01%, from 8.4% to 8.41%, due to the asset-intensive nature of the business.
- On an EVA basis, the combined effect of these initiatives would raise corporate EVA from 8.4% to 10.7%. Furthermore, if capacity is constrained, the benefit could rise to 11.8%.

Rationalization Strategy Successes

A wide range of innovative companies in the new energy economy are offering products, services, and solutions that reduce costs for utilities and other energy providers. For example:

- Scanifly (USA) and other specialist drone operators use drones to reduce the soft costs of solar panel installation and maintenance.
- PV Robot (Egypt)[IV] reduces the operating costs of utility-scale solar arrays by automatically cleaning dust from the panel surfaces.
- Nikola, an American manufacturer of hydrogen-fueled tractors, is attempting to make trucking, which is a central supply chain link for almost every industry, cleaner and cheaper. If successful, the resulting technology could influence the direction of hydrogen fuel cell development for electric utilities and the oil & gas exploration, production, and refining industries.
- Baker Hughes underwent a supply chain transformation program, with cost reduction targets of $100m per year based on supplier categorization and rationalization, as well as strategic sourcing to shorten its value chain and implement lean manufacturing.[32]

Synchronization Strategy Successes

In the new economy, electric utilities and oil & gas companies are uniquely able to integrate, modularize, and bundle supply chain solutions and services in creative ways that allow them to redefine their value chains both upstream (at the wellhead or power plant) and downstream (through the interface with their customers). For example:

[IV] A joint venture with BSI Energy Ventures.

- Vattenfall, a Swedish utility that has pivoted from being a leading fossil fuels user to being a renewables technologies development leader by enabling wind power + storage through creative advances in technology through superior supplier and supply chain management.[33]
- Advanced Microgrid Solutions (AMS) is enabling Tilt Renewables' 100 MW Snowtown 1 Wind Farm with artificial intelligence–driven trading algorithms that optimize revenue by learning the trade when the farm's available production meets points of highest demand (see the Wind chapter).
- In the battery world, XL Batteries, is licensing a unique chemistry for long-life nonflammable flow batteries that enable intermittent solar and wind deployments. The company won top prize at New York Energy Week Disruptors Day.[34]
- Electra and Integrated Storage Technologies are American companies that make battery storage control systems that increase the efficiency of battery storage systems, which ultimately help to synchronize supply with demand.[35]

In the fossil-fueled energy economy, *synchronization* success can be measured and managed according to the metrics laid out in David Steven Jacoby's book *Guide to Supply Chain Management.*[36] According to that scorecard,

- Qatar Fuel excelled in 15 different supply chain metrics at one point, which netted it 58% return on capital employed and 46% return on net assets.[37] Here are some of the underlying *synchronization*-oriented metrics:
  - 6 sigma order and delivery cycle time reliability
  - >99% orders delivered by time customer requests
  - >99% of orders delivered by the time committed to
  - >99% stock accuracy
  - >97.6% uptime
  - >95% first pass yield
  - 5.6% cost of order fulfillment as % of order value
- Chevron (USA) successfully extended its internal Lean Six Sigma success to its suppliers. Its Lean Six Sigma program created financial benefit of $250–500 million per year. Hundreds of projects across the US, the UK, Angola, and Indonesia contributed to the achievement. An internal consulting group led by Stephen Turnipseed helps Chevron business units start and mature individual programs. As the programs mature, Chevron actively engages its suppliers and contractors, both to further existing improvement projects and to develop their own continuous improvement programs.[38]
- Bharat Petroleum Corporation achieved superior order accuracy, inventory management, and overall asset productivity, as did Nalco and BASF.[39]

- Nalco reduced its costs by $122m using Six Sigma. The company has a team of "blue belts" tasked with cutting operating costs by $100m per year.[40] BASF saved $600m from a supply chain improvement program called NEXT.[41]

Customization Strategy Successes

Software and control systems solutions providers are developing platforms that users, such as utilities and oil & gas companies, can configure to facilitate and secure their digitalization initiatives. This allows them to have more secure supply chains. For example:

- Tesla, an American integrated power and automotive company, is packaging battery technology in innovative and customizable energy storage solutions tailored for individual markets, from large power grids to individual homes and the electric vehicles that plug into their internal power networks. Tesla's model of integrated energy delivery at multiple levels could transform how much of the world accesses power and how much people pay for it.
- aDolus, a Canadian cybersecurity solutions provider, protects the energy industry from cyberattacks while it deploys essential Internet-based technologies for configurable smart devices and equipment.

Innovation Strategy Successes

As in the Gold Rush, many companies are attempting to develop the next big technological innovation in the energy industry. These companies, frequently core raw materials and clean fuels technology developers, are often seeking to disrupt the value chain from the inside out by making clean power more accessible or affordable. For example:

- Solar Earth Technologies, a Canadian technology company, is embedding PV cell technology in remote, industrial and off-grid power, significantly extending the range of surfaces that can generate solar power.[42]
- Ocean Minerals LLC is an American mining company that is alleviating the cobalt supply chain shortage by using rovers to harvest cobalt nodules from the sea floor. While the initial output could be an input to batteries for consumer devices and electric vehicles, it could equally end up being used in battery-supported energy storage networks in power grids, most likely in transmission and distribution.
- CBMM, a Brazilian mining company, is pivoting from being a supplier of ore to the steel industry, to being a technology and innovation leader in the fast-charging lithium-ion battery supply chain. The company's niobium chemistry also has application in high-performance metals designed to function under extremely high heat and pressure, which could accelerate the development of utility-scale fuel cells and nuclear fusion.

Conclusion to the Introduction

Supply chain management is about how to make management decisions that optimize the results of trade-offs between cost and other parameters in the value chain such as quality, risk, service level, throughput, safety, or environmental impact, to name just a few. When examining trade-offs there are basically two types: those related to capital project management, and those related to operations and maintenance (O&M). The next two chapters will provide specific frameworks, tools, and techniques to help supply chain practitioners make optimal trade-offs in capital project management and in operations and maintenance management.

Methods for CapEx Project Supply Chain Risk Mitigation

Chapter Highlights

1. Large capital projects are characterized by risk. There are five principal types of risk: time horizon risk, technology risk, materials availability risk, work structure risk, and supplier risk.
2. At a high level there are four general ways to manage risk: avoidance or pass-through, diversification or risk-sharing, hedging or offsetting, and minimization by locking in known conditions.
3. For capital projects, these risks can be adapted by expanding supplier risk to encompass partnering, bundling, and outsourcing, and updating the risks to include sustainability-related trade-off decisions. This yields eight types of capital project trade-off decisions:
   a. Technology choice risk: how to adopt the most efficient technology platform while minimizing the risks of infant mortality, maintenance cost, and obsolescence
   b. Capital project management risk mitigation: how to minimize the financial, liability and legal exposure of capital projects while also minimizing project management costs
   c. Sustainability trade-offs: how to maximize the use of clean power while maximizing profitability or minimizing cost
   d. Materials unavailability risk: how to procure reasonably priced critical raw materials with low risk of unavailability
   e. Outsourcing risks: how to get the work done at the lowest cost while retaining knowledge and core competences in-house
   f. Supplier partnering risks: how to form the most efficient working relationship while minimizing the risk of I.P. loss
   g. Procurement bundling: how to achieve the simplest category management buying approach with the least external expenditure
   h. Contract term risk: how to negotiate the longest-term contract with the least risk of a better option arising during the commitment period

4. To effectively manage technology choices, the cost and benefit implications of learning curve effects, infant mortality, economies of scale, and standardization benefits related to operating and maintenance should be considered.
5. To effectively manage capital project management risks, supply chain costs and risks are affected by the choice of project ownership structures, such as Design Build Operate (DBO), Build Own Operate (BOO), Build Own Operate Transfer (BOOT); whether EPC or EPCI is the better framework; and the trade-offs and contractual terms of alternative EPC and EPCI contract structures, including Time & Materials, Cost Plus, and LSTK (Lump Sum Turnkey).
6. To help optimize the impact of sustainability projects, managers can develop a prioritized list of potential energy efficiency projects ranked by CO_2 reduction and payback, benchmark costs for renewable energy resources including energy storage, solar PV, etc., and calculate the investment payback frameworks and financial ratios of potential sustainability projects using impact calculators and CO_2 conversion factors.
7. To ensure the availability of critical materials and services, managers can use substitutes, plan farther in advance, buy or control the source, and/or refurbish, recycle, or reuse to extend availability.
8. To avoid outsourcing mistakes, procurement and supply chain managers can use a formal process to evaluate the strategic, operational, and economic considerations of insourcing versus outsourcing before entertaining proposals from third party providers. Also consider bringing outsourced activities back in-house if conditions change.
9. To ensure alignment of interests with strategic suppliers, procurement management can adopt a clear stratification of suppliers that categorizes them in five tiers and modulate interactions with them according to a maturity matrix that defines tighter coordination with more strategic suppliers.
10. Procurement, outsourcing, and subcontracting initiatives should seek solutions rather than buying materials, products, services, or systems separately. Suppliers can often add more value by contributing their know-how to a specific application, and the pricing structure can encourage success thereby increasing operational performance.
11. Managers should use simulation tools to determine the optimal contract term and adopt multi-year contracts with optional extensions in order to maximize economies of scale.

Introduction

Large capital projects are characterized by risk-reward trade-offs such as market (price and volume) risk, materials supply risk, supplier risk, construction risk (sometimes offloaded to an EPC firm), and operational, supplier, technology, political, and regulatory risk.[43]

Capital project managers can seek market risk analysis from a number of specialist consulting firms; political and regulatory risk are special types that extend well beyond supply chain management. These are not covered here.

General Approaches to Managing Risk

Supply chain policies, processes, systems, and organizational structures can be used to avoid, diversify, minimize, or hedge risk (see Figure 21). While the bulk of this chapter will provide tools and techniques for managing each of the eight trade-offs cited above, a general framework for managing risk can guide and inform some of the more detailed tools and techniques.

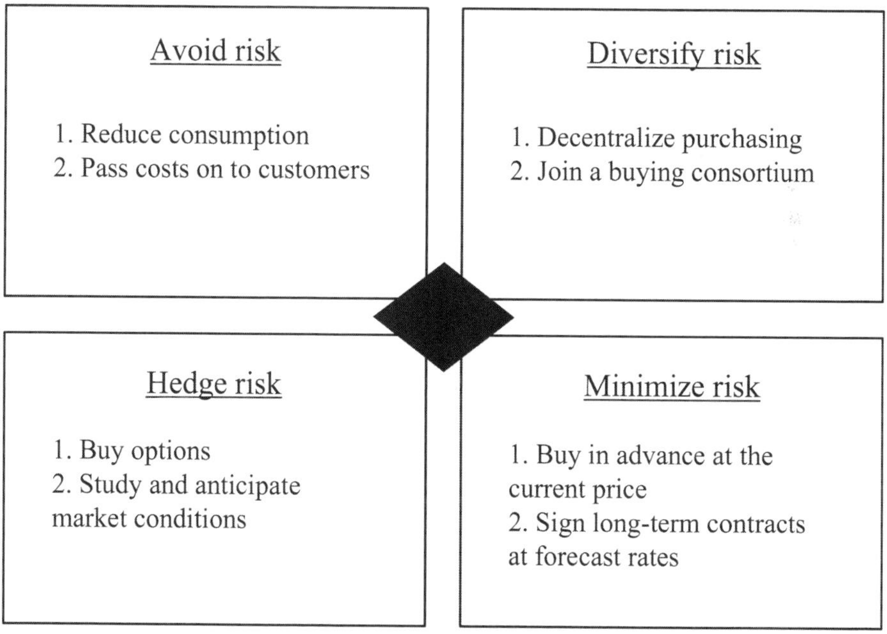

Avoid risk	Diversify risk
1. Reduce consumption 2. Pass costs on to customers	1. Decentralize purchasing 2. Join a buying consortium
Hedge risk	**Minimize risk**
1. Buy options 2. Study and anticipate market conditions	1. Buy in advance at the current price 2. Sign long-term contracts at forecast rates

Figure 21. General Strategies for Managing Risk
(*Source:* Boston Strategies International)

The easiest and in many cases the most effective risk management strategy is to avoid risk entirely by passing it through to customers.

- BSI studied how manufacturing companies dealt with the rising cost of energy.[44] Of eight risk mitigation strategies, the study found that customers allowed companies to institute price increases commensurate with the increased costs, and that this strategy largely mitigated the actual cost increases over the period studied. In this case there were two ways to pass through the price increases—through surcharges or by embedding them in the base sales price. Adding an energy surcharge resulted in a cost-neutral position. Cost increases were passed through when they occurred and retracted when they went away. An example of this would be the pricing formula for LNG in Europe, which passes through increases in exploration and production costs by linking LNG price to changes in the prices of fuels such as gas, oil, low sulfur fuel oil, and coal, and adjusting the price of LNG to match the change in the price of those fuels.[V] The other approach was to embed the cost increase in the base price. This proved to create a lucrative profit center, a strategy that was acknowledged by many interviewees. It not only shielded buyers from price increases but also resulted in windfall gains as energy prices subsequently declined but those increases were structurally embedded in the sales price.
- Another approach to avoid risk is to offload it to project managers (for example, EPC contractors) or use LSTK contracting parameters that shift any risk to the suppliers. EPC contracts typically delegate responsibility to a contractor for making sure the facility functions as intended. If for any reason the facility does not work as planned, the EPC rather than the operator is responsible for remedying the situation. This assures the buyer of a single point of contact, a guaranteed result, and, if the contract is for a fixed price, no cost risk. Under an LSTK contract (which may also be with an EPC firm), the lump sum aspect makes any overage the responsibility of the contractor. The turnkey aspect underlines the fact that the contractor must deliver a working facility, regardless of cost overrun status. About 6% of total petroleum industry expenditure runs through EPC contractors, according to ISM benchmarks.[45] The downside of most EPC and LSTK contracts is that they often embed either cost adders or buffers that protect the contractor against contingencies, and these buffers may exceed the average cost of the unexpected events. Also, EPC contractors may have an incentive to over-engineer since almost 60% of their contracts are on a time-and-materials basis.[46] For example, one operator complained

---

[V] $P_n = P_0 \text{ x } (W_1 \text{ x } F_1 / (F_1 \text{ at } t{=}0) + W_2 + F_2 / (F_2 \text{ at } t{=}0))$, where P_0 represents the original negotiated price (at time 0), W represents Weighting factors or percentages of alternate fuels, and F_1 and F_2 represent alternate fuels (often gas oil, low/high sulfur fuel oil, and coal). An inflation component may be added. Source: Gordon Shearer and Michael D. Tusiani, *LNG: A Nontechnical Guide*, Tulsa: PennWell, 2007, p. 329.

that EPC firms on a time-and-materials contract to build a water treatment plant would have an incentive to recommend an expensive bottom-mounted aeration unit instead of a cheaper and more efficient top-mounted system in order to boost their revenue.

A second strategy for managing risk is to diversify it, typically by sharing it between the buyer and supplier. Buyers prefer unlimited liability for the supplier while it is working on the buyer's site, while suppliers usually negotiate for liability caps at contract value or a percentage thereof. Negotiation of the liability caps can shift risk between the two parties. For example, on a $2 billion construction contract, the negotiated cap might end up being 20% of the value of the contract.

A third strategy is to minimize risk. Long-term agreements often minimize risk for both parties: the aforementioned BSI study showed that buyers realize (and suppliers concede) 3–8% savings, on average, by negotiating long-term agreements. This price reduction reflects the volatility that was removed from the supplier's future sales stream. However, a long-term contract can also accentuate some risk by locking buyers or suppliers into bearing the cost of underlying materials price increases. The more stable long-term contracts share or index these costs to ensure that one party won't suffer and eventually have to default under the weight of a cost increase.

The fourth strategy is to hedge risk. Contract price indexing can accomplish this. For example, a $1 billion specialty minerals processor indexed ten-year agreements to its costs, effectively negating any cost inflation and still allowing it to pass through an escalation of costs each year on top of that risk-hedged rate. Other companies buy products that contain extensive steel by specifying the tons of steel they contain and the price per ton at the time of the order, and then increasing or decreasing the price when delivered to the customer based on the change in steel prices during the order delivery lead time. Battery suppliers do the same thing with lead costs, and cable and some motor suppliers do this with copper prices. Polyvinyl pipe is sometimes indexed to resin prices. Other hedging techniques leverage financial options (traded on a mercantile exchange), futures contracts (traded on some stock exchanges), forward contracts (between two parties), and "real options" (a plan or contract that allows investments to be triggered, changed, or cancelled at certain milestone points at the discretion of the buyer).

A hybrid strategy is to anticipate risk. Some firms track and/or forecast the market, a practice known as supply market intelligence, in an attempt to select the most effective risk mitigation strategy for each circumstance. This is analogous to an intelligent demand forecasting module that evaluates a variety of forecasting methods each period and applies the one that would have had the least forecast error over the past several periods. One framework for supply market intelligence includes seven dimensions: capacity and scale, product and process technology, global opportunities and risks, costs, demand, channels and supply chain, and competitive dynamics (see Figure 22).

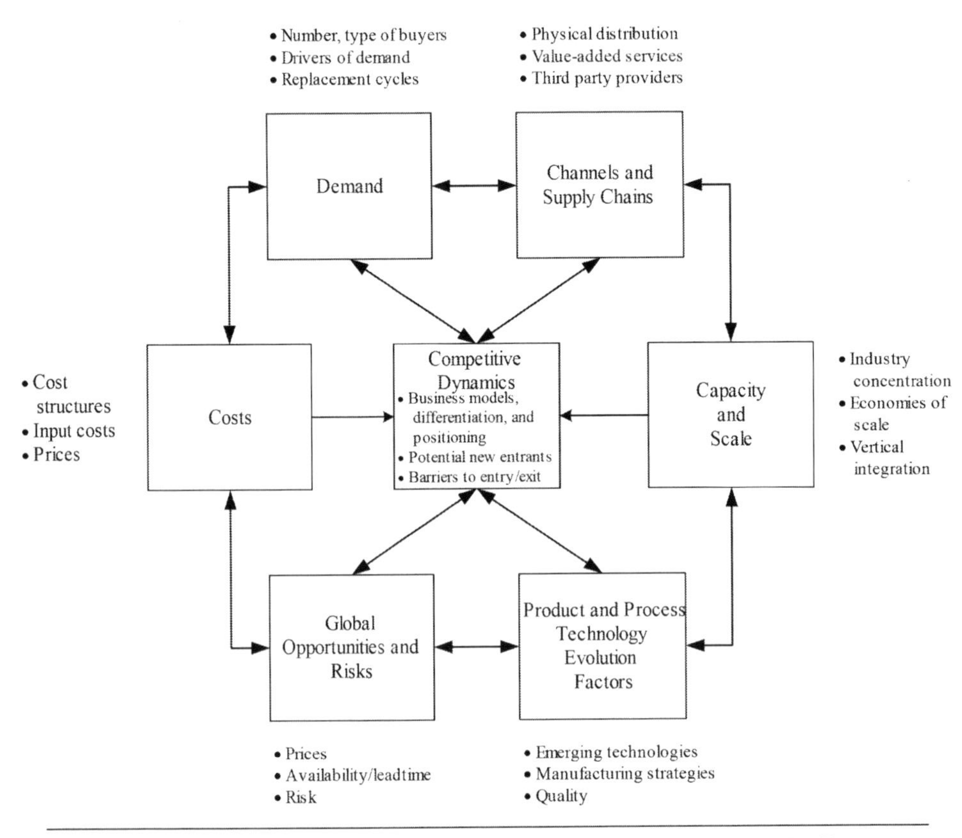

Figure 22. Boston Strategies International's Framework for Supply Chain Market Intelligence
Source: Boston Strategies International

Value Chain–Specific Risks and Trade-offs

Technology risk, work structure risk, materials availability risk, supplier risk, the size and duration of exposure to supply commitments, and contract term risk (financial exposure) can drive value chain economics and commercial success. To elucidate the risks and trade-offs pertinent to supply chain management, "supplier risk" may be expanded to encompass partnering, bundling, and outsourcing, and the list may be updated to include sustainability-related decisions that have become integral to management decisions as they may be required by law, regulatory guidance, or corporate strategy.

- Technology risk: it is substantial, and many companies minimize it by sticking with well-established technologies. However, making large, long-term capital investments on the basis of stale technology is all but assuring lower-than-market rates of return for potentially decades. Is there a methodology for quantifying this risk?

- Work structure risk: when working with large projects involving many systems and subsystems, operators can choose from a spectrum of solution bundling choices. They can opt to design, build, and operate in-house, or at the other extreme they can hire an EPC firm to engineer, procure, construct, install, and operate, effectively owning none of the value chain activities. If they keep it in-house, they can unbundle the whole supply chain down to the component level if they want. Which is the right balance of risk and cost-effectiveness?
- Sustainability trade-offs: a combination of laws, financial incentives and penalties, and corporate strategic objectives increasingly demand that managers are skilled at balancing the cost and effectiveness of projects that decrease carbon footprint, and at ensuring that suppliers are suitably screened and monitored for compliance to corporate and government environmental, social, and governance standards.
- Materials availability risk: embedded assumptions about materials availability, service prices, and skilled trade wages have caused many cases of *force majeure*. In one case, a manufacturer of expandable tubulars experienced a bottleneck in outsourced machining capacity that caused a bidding war for the machine time, thereby slowing production and raising costs. What is the best way to deal with supply availability unpredictability?
- Supplier risk: supplier failure can sometimes have a disastrous effect on a project, so many companies have extensive prequalification processes and require performance bonds. The choice of companies on the bid slate has a large impact on the competitiveness of bids, and the structure and intimacy of the ensuing relationship has a large impact on the cost efficiencies and performance levels achieved during the project. Furthermore, outsourcing decisions must balance cost savings with the need to maintain strategic core competences in-house and protect intellectual property. Partnering must balance or share all the risks identified here with strategic suppliers in a clear, comprehensive, and financially and legally judicious manner. Procurement bundling including category management must be done in a way that achieves maximum total cost savings considering both the "first costs" and the lifetime of operating and maintenance costs.
- Time horizon risk (exposure on commitment over more or less time): by choosing a time horizon for project investment, project managers are implicitly embedding a supply chain risk. If the time frame is too short, suppliers may not reap enough economies of scale or learning curve benefits to be able to lower costs enough to meet targets. If it is too long, the owner or operator may be transferring upside economies to suppliers and by committing to not re-source, the price or product technology

may become misaligned with the market. How can time horizon risks be measured, and what can be done about them?

Techniques for Capital Project Management Decisions

Capital project management is a very large part of supply chain management in the energy industry. Across all energy technologies, 84% of supply chain cost is incurred upstream, and most of that relates to construction of new plants. The framework presented here for managing capital projects explains how to manage supply chain activities to maximize asset productivity while minimizing the risk of unexpected future costs. Here are types of risk to manage (with trade-offs). Readers could construct a risk probability vs. severity matrix to prioritize them for their own situations.

1. Technology choice risk (how to adopt the most efficient technology platform while minimizing the risks of infant mortality, maintenance cost, and obsolescence)
2. Capital Project Management Risk Mitigation (how to minimize the financial, liability and legal exposure of capital projects while also minimizing project management costs)
3. Sustainability trade-offs (how to maximize the use of clean power while maximizing profit or minimizing cost)—especially for fossil-fuel based companies
4. Materials unavailability risk (how to procure reasonably priced critical raw materials with low risk of unavailability)
5. Outsourcing risks (how to get the work done at least cost while retaining knowledge and core competences in-house)
6. Supplier partnering risks (how to form the most efficient working relationship while minimizing the risk of I.P. loss)
7. Procurement bundling (how to achieve the simplest category management approach with the least external expenditure)
8. Contract term risk (how to negotiate the longest-term contract with the least risk of a better option arising during the commitment period)

This chapter considers these CapEx techniques in the order listed above based on a classification of their frequency versus their typical impact, with frequently used higher-impact tools topping the list, and infrequently used lower-impact tools at the bottom (see Figure 23).

Technology Choice Risk Management deals with how to adopt the newest and most efficient technology platform while minimizing the risks of infant mortality, maintenance cost, and obsolescence. For example, should you buy the next generation technology, in which case you have to buy sight unseen and pay more, and the

	Lower Likelihood or Prevalence	Higher Likelihood or Prevalence
Higher Impact (financial, operating, safety, legal, organizational)	• Supply Unavailability Risks and Mitigation • Outsourcing Risks and Mitigation • Strategic Partnering Risks and Mitigation	• Technology Choice Risk Management • Capital Project Risks and Mitigation • Sustainability Trade-offs
Lower Impact (financial, operating, safety, legal, organizational)	• Procurement Bundling Trade-offs	• Contract Term Risk and Mitigation

Figure 23. Trade-offs and Management Techniques in Capital Project Management

design bugs may not be worked out yet, or should you buy the previous generation tech which is ready off the shelf, cheaper, and proven? For a simple everyday life example, many people like me who upgraded from Windows 7 to Windows 8 years ago were very sorry—Windows 8 was a lemon. The people who decided to stick with Windows 7 saved a lot of hassle and money; they leapfrogged from Windows 7 to Windows 10 after Windows 10 was proven to be reliable and hassle free. In the energy industry, utilities putting up solar farms have to choose a PV technology—e.g., the proven and cheaper polycrystalline vs. the newer, more expensive, and "better" thin film. Thin film sounds better, but if thin film turns out to have problems (durability, fire hazards, etc.) they might wish they had stuck with polycrystalline. Every energy supply chain has analogous technology trade-offs, from wind turbines to battery chemistries to fuel cells to geothermal boiler types, and on and on.

Capital project risks and mitigation addresses how to minimize the financial, liability, and legal exposure of capital projects while also minimizing project management costs. Should you hire an EPC or manage projects in-house? If you engage an EPC, you pay a management fee to the EPC (the pricing mechanism varies, but one way or another they get paid), and you have less (or no) control. Are your project managers using risk management methods, frameworks, and procedures? This includes tools such as risk registers, heat maps, HAZID and HAZOPS studies, RACI matrices, and safety briefings. Are you buying equipment and services separately or bundling it together, and to what degree? In drilling you can buy downhole tools and bits and give them to an oilfield service (OFS) provider to drill with them, or you can hire the OFS to do everything and pay just one invoice. Furthermore, you can pay the OFS for his costs plus profit margin, or you can incentivize him to achieve higher productivity by compensating him by foot drilled. The choices have implications for cost, productivity, I.P., and efficiency. Finally, are you managing the project from beginning to end, or diversifying cost and legal risk at different stages? You can arrange for a project to be managed and handed off to the ultimate operator in one step or through multiple parties in multiple phases through five alternative project management structures, and the risks vary widely.

Sustainability trade-offs offer decision tools that help to maximize the use of clean power while minimizing cost. An example is using renewables for a power source. Are you using clean energy to power operations, and are you doing it at a loss in order to comply with regulations, or have you figured out a way to make clean energy pay for itself? For example, you may decide to use Carbon Capture and Storage (CCS) or not. If you do use CCS, you may use it simply to bury CO_2, or you may use that CO_2 in a productive operation. Have you found a way to redeploy the CO_2 to save money, or are you spending money to get rid of it? Is your supply chain organization tasked with making a profit from clean energy, or is clean energy a cost center for compliance purposes or PR benefit? Effective management of sustainability trade-offs would ensure access to affordable, reliable, sustainable and modern energy for all, the stated target of Sustainable Development Goal (SDG) 7 (clean and affordable energy), which aims to increase substantially the share of renewable energy in the global energy mix and double the global rate of improvement in energy efficiency by 2030.[47]

Supply Unavailability Risks and Mitigation is about how to procure reasonably priced critical raw materials with low risk of unavailability. Are you facing supply constraints on critical raw materials or components, and how are you mitigating those risks? It is illegal in many places to source cobalt from the Congo due to child labor laws, but there aren't many other affordable options. Sometimes trade barriers exacerbate materials sourcing strategies. Rare earths are used in magnets, but their availability and price are controlled mostly by China, so they are always expensive and sometimes unavailable. Barite, a drilling fluid, is in a similar predicament. The United States has erected tariffs on some types of solar PV; if you are sourcing PV into the US, sometimes supply managers have to work out alternate sourcing strategies.

Outsourcing Risks and Mitigation is about how to make trade-offs so the most work gets done at the least cost, while retaining knowledge and core competences in-house. Are you outsourcing too little or too much? How do you know how much is the optimal amount? For example, for a power company deciding to build a nuclear power plant without the help of an experienced EPC could be very costly and dangerous. On the other hand, for a component supplier, outsourcing manufacturing or R&D could save money short-term but gut the company of know-how in the long run.

Supplier partnering risks and mitigation is about how to form the most efficient working relationship while minimizing the risk of I.P. loss. How do you prequalify suppliers? When do you set up a joint venture with a supplier? Should you ever take an equity stake in a supplier's business? How do you manage the tendering process, and does that vary depending on the value and duration of the contract? How should you go to market for products under development but not yet commercialized?

Procurement bundling trade-offs is about how to achieve the simplest category management approach with the least external expenditure. How do you know whether to buy raw materials separately from components (if the raw material constitutes a very large proportion of the cost of the component, such as steel in a large fabrication)? Should you buy some components separately from other components (such as motors separately from pumps that they drive)? Is it better to buy products separately from the services that are associated with them (such as compressors separately from installation and commissioning of them)?

Contract term risk and mitigation is about how to negotiate the longest contract with the least risk of a better option arising during the commitment period. How can managers assess time horizon risks, and what can they do about them? If the contract is too short, suppliers may not reap enough economies of scale or learning curve benefits to be able to lower costs enough to meet targets. If it is too long, the owner or operator may be transferring upside economies to suppliers and by committing to not re-source, the price or product technology may become misaligned with the market.

Technology Choice Risk Management

Technology decisions are a perennial challenge in energy production and distribution management. While on the one hand new technology generally brings higher output, it almost always comes with a higher price tag and a risk that it will not deliver the anticipated benefits, or that it will have new product complications such as recalls or component failures. Since the investments are so large, it is hard to establish a verifiable pilot program that demonstrates the feasibility of the new technology at the scale at which it will actually be deployed. Furthermore, since the lead times are so long, suppliers cannot always invest in the extensive research and development needed to get to the next generation without pre-launch financial commitments from operators. Hence, a "chicken-and-egg" situation prevails, stymieing innovation and putting a drag on productivity improvements.

Whether it relates to subsea compressors, tension leg platforms, tripod foundations, or next-generation turbines, a proper evaluation of the technology should take into account current and projected acquisition costs; improvements in effective power ratings over time; learning curve effects; operating, repair, and maintenance costs; infant mortality rates; and supplier viability risk factors. Based on the total rolled-up cost over the time horizon of the project, and the risk factors for the most strategic suppliers, project management can use this information to decide whether to negotiate the desired outcome with target suppliers, or, if it is an option, to frequently re-source newer technologies as the market produces them.

Learning curve effects drive down the cost per unit of output for new technologies along a sometimes predictable curve (see Figure 24). For example, offshore wind turbines have become dramatically higher in power and lower in cost to the extent that various organizations have calculated, in a scientific way reminiscent of Metcalfe's Law, the time period during which the cost can be expected to be cut in half. In the wind industry, this is called the "progress ratio."

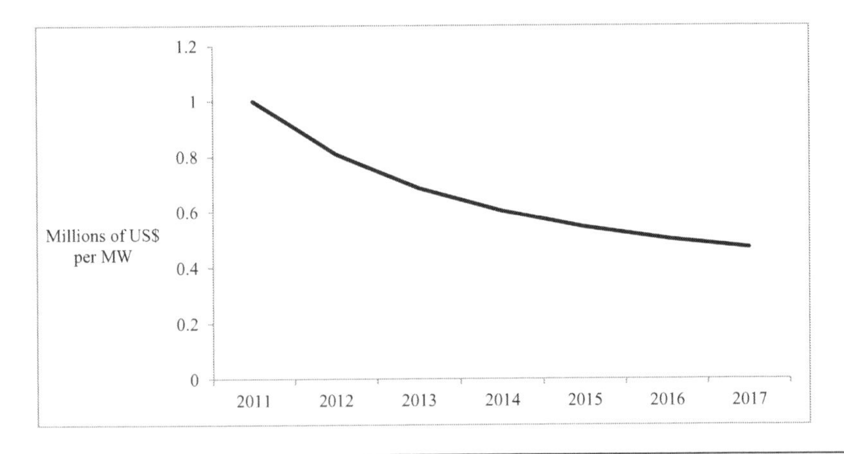

Figure 24. Illustrative Reduction in Cost per Unit over Time
(*Source:* Boston Strategies International)

Operating, repair, and maintenance costs typically decline as technology platforms and new models mature (see Figure 25). The initial period after model introduction has a higher repair and maintenance cost due to infant mortality. Statistically, the risk of unplanned failure decreases as the number of units produced and put into operation increases, due to learning in the production process, actual performance, and customer feedback. Sometimes the decrease in operating and maintenance costs can be very significant. For a long time, the cost per flight hour of Pratt & Whitney JT8 jet engines declined as the number of years since their launch increased, and GE's jet engines exhibited a similar phenomenon.

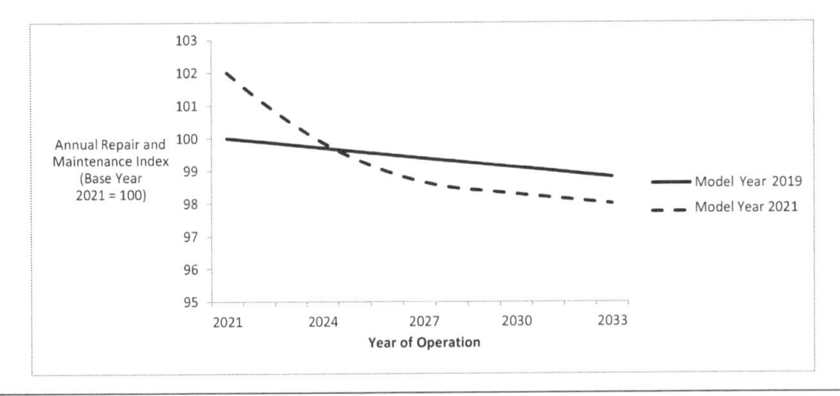

Figure 25. Repair and Maintenance Costs as a Function of New Model Introduction Year
(*Source:* Boston Strategies International)

If the slope of the total cost curve including economies of scale and all the other factors mentioned above (see Figure 26) is projected to be downward and steep, it may be hard to negotiate such a large decrease with suppliers. In this case, operators need to decide whether to bring facts to the negotiating table and hammer out a long-term agreement based on a sharp projected decrease in long-term costs, or whether re-sourcing on a periodic basis would be a more effective way of achieving the long-term cost target.

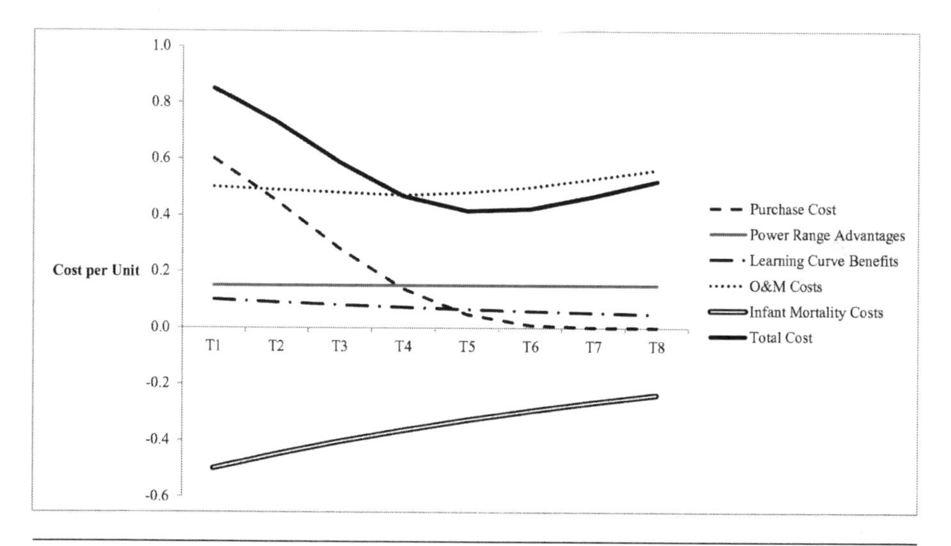

Figure 26. Illustrative Economic Impact of Technology Factors over Time (*Source:* Boston Strategies International)

All of this generally supports the conservatism of many oil, gas, and power companies when it comes to new technology. The important thing is that there is always a point at which the benefits of the new technology outweigh the risks and costs, and that point can be quantified with proper analysis.

When it comes to smaller capital expenditures, one way to hedge the technology curve is to rent rather than buy equipment. One major drilling tool supplier now earns about half of its revenue from rentals, as buyers are increasingly seeing the opportunity to lower costs and shift technology risk to the supplier.

Capital Project Risks and Mitigation

Build-Own-Operate Choices (BOO, BOOT, etc.)

At the highest level, the decision about what to do in-house and what to outsource drives most of the cost and risk in a project. Five alternative forms of project structure (see Figure 27) provide a spectrum of choices regarding how much cost and risk to bear, and how much to offload onto partners or suppliers.

In the following framework, a project's sponsor is the party that initiates the project. A sponsor can be an energy producer commissioning a capital project such as FPSO, an air separation unit, a refinery, or an energy storage-based "virtual power plant"; a utility building a power plant; or a government entity such as a power commission that offers concessions for facilities. The contractor is the firm that builds the project, which may or may not operate and maintain the facility once it is completed. The following configurations of ownership may be used to achieve the desired level of engineering, construction, operation, and maintenance risk (roughly in order of most to least in-house responsibility):

- DBO: Design, Build, Operate—the contractor retains full responsibility for all phases of the project; maintenance may be outsourced (if it is performed by the contractor then it is a DBOM contract).
- BOO: Build, Own, Operate—the contractor outsources the design phase, but retains control over the construction and operation aspects in-house; maintenance may be outsourced.
- BOOT: Build, Own, Operate, Transfer—the contractor takes full responsibility for the project, and then sells the resulting installation. This sets the company up to be a builder of whatever the product is, for example FPSOs or offshore wind farms.

Sponsors can also shift risk to third parties by hiring an Engineering, Procurement and Construction (EPC) or Engineering, Procurement, Construction and Installation (EPCI) firm to manage the project, and then by structuring the contract with the EPC to achieve the desired level of risk during the construction phase. Here are some options (in order of most to least in-house responsibility):

- EPC/Time & Materials
- EPC/Cost Plus
- EPC/LSTK (Lump Sum)
- EPCI-T&M (Time and Materials)
- EPCI/Cost Plus
- EPCI/LSTK (Lump Sum)

The complexity can be simplified to five basic types of agreement, as summarized in Figure 27 below.

	DBOM	BOO	BOOT	EPC(I)-LSTK	EPC(I)-Cost Plus or EPC(I)-T&M
Ownership of Site	Contractor	Contractor	Contractor/ Sponsor	Sponsor	Sponsor
Operator Control	Contractor	Contractor	Contractor/ Sponsor	Sponsor	Sponsor
Acquisition Cost Risk	Contractor	Contractor	Contractor	Contractor	Sponsor
Design Risk	Contractor	Sponsor	Sponsor	Sponsor	Sponsor
O&M Cost Risk	Contractor	Contractor	Contractor/ Sponsor	Contractor/ Sponsor	Sponsor
Prime Contractor Performance Risk	Contractor	Contractor	Contractor	Contractor	Sponsor
Change Orders	Contractor	Contractor	Contractor	Contractor	Sponsor
Project Management Resources	Contractor	Contractor	Contractor	Contractor	Sponsor
Contractor Risk Index					
Sponsor Risk Index					

Figure 27. Typical Assignment of Risks Assumed by EPC Contractors Under Five Types of Agreement (*Source:* Boston Strategies International)

Sustainability Trade-offs

The growing public environmental conscience is requiring tighter process integration with suppliers that have an eye for greener supply chains.

While sustainability in the oil & gas supply chain had been a concern for decades, the public focus on sustainability ramped up during the Deepwater Horizon drilling rig incident in 2010, and gained further steam over possible health and environmental risks due to hydraulic fracturing. The US EPA mandated public disclosure of certain formulations that would have previously been considered "trade secrets" or "proprietary ingredients" in fracturing fluids. Meanwhile, in Europe the Registration, Evaluation, Authorisation of Chemicals (REACH) protocol for the registration of chemical substances also kept the public mind on environmental impact (see Appendix 3 for details on REACH).

Over the last decade, growing concerns about global warming and CO_2 emissions have motivated oil & gas and fossil-fuel-fired power companies to search for ways to integrate renewable power into their own operations. One way to reduce environmental footprint and increase energy efficiency at the same time is by using renewable energy to power oil & gas drilling and production equipment such as pumps, compressors, and turbines. The converse, using fossil fuels to produce hydrogen, is called "blue hydrogen." The industry has seen many other types of creative hybrid applications such as floating solar and wind apparatuses to power offshore platforms. There are many ways to reduce carbon footprint and emissions, even while producing fossil fuels.

Tools available for managing the supply chain implications of sustainability initiatives vary widely depending on the scope of the initiative and the technology involved. These may include, for example:

- Standards such as ISO 20400 (Sustainable Procurement)
- Prioritized lists of potential energy efficiency projects, ranked by impact and payback
- Benchmark costs for renewable energy resources including energy storage, solar PV, etc.
- Investment payback frameworks and financial ratios
- Impact (e.g., carbon footprint) calculators and conversion factors

Materials and Services Unavailability Risks and Mitigation Strategies

Raw material resource scarcity has raised the cost and created shortages of critical raw materials, requiring a search for new technologies and substitute products. According to McKinsey, the 21st century needs a resource revolution akin to the labor revolution of the 20th century, wherein more people entered the workforce and labor productivity soared due to IT, in order to fulfill the rapidly growing needs of our global population.[48]

- Lithium and cobalt, both centrally important ingredients to the most popular battery storage solutions, come predominantly (51% and 62%, respectively) from China,[49] which puts the Chinese government in a position of extraordinary leverage over battery suppliers. In 2017 the US government announced an Executive Order (number 13817) to identify critical strategic minerals and build domestic supply chains to avoid being dependent on foreign countries, namely and especially China, for them.[50] This opens the door for potential active exploitation of lithium and cobalt mines in the United States, including the potential development of a "Lithium Valley" centered in the Salton Sea area in southern California.[51]
- Niobium, a relatively lesser known mineral in the transition metal category, has properties that harden and lighten the weight of certain alloys, which can be useful for construction and automobile manufacturing. However, 85% of the world's supply is located in Brazil,[52] which makes that supply potentially critical and strategic. Its strength-inducing qualities make it a potentially strategic metal for the next generation of energy technologies, such as fuel cells which at the utility scale suffer from durability limitations due to the unprecedented

levels of heat they sustain, and certain types of nuclear power which rely on atomic bombardment at high pressures. Executive Order 13817 may serve to accelerate exploitation of the Elk Creek niobium mine in Nebraska (USA).[53]

- Tungsten, a material used to harden drill bits and increase their durability, was in tight supply in 2008 when rig counts rose rapidly, creating both price inflation and bottlenecks. Similar to barite, the mineral is mostly mined in the interior of China and is vulnerable to supply disruptions from the country's ongoing mining sector reforms.
- Water resource depletion has impacted steel availability, which is used for huge offshore platforms in the oil & gas industry. Steel production depends crucially on the supply of iron ore, which in turn relies heavily on the water used to extract it. In its article "Dependencies and Regulatory Risks," McKinsey notes that almost 40% of iron ore mines are in areas with moderate to high water scarcity, and a lot of steel is produced in places where water is relatively scarce.
- Reserves of barite, which is used as a weighting agent for drilling fluid, have been rapidly depleted, driving drilling mud costs up dramatically. The world contains approximately 740m tons of identified barite reserves, according to the US Geological Survey in January. At demand growth rates forecast by Boston Strategies International, currently owned reserves could be completely depleted in less than 15 years. China is the world's largest barite exporter, but barite reserves are distributed in small pockets, and typically mined by small, independently owned mining companies. As a result, bad weather and poor transportation infrastructure disrupt supply, pushing prices up. Another supply risk is the stated intention of the Chinese government, in its latest five-year plan, to restrict the environmental impact of mining for minerals such as barite and to control their export more tightly, which has driven barite prices up by high double digits. In order to address the problem, major suppliers are scouring the globe to develop alternative barite sources.[54]

Raw material price and unavailability risk increases in periods of sharp economic growth. As just one example, metals prices both rose and fluctuated wildly during the first decade of this century, as shown in the chart below. As Daniel Yergin puts it in his book *The New Map: Energy, Climate, and the Clash of Nations*, "This BRIC era was characterized by what became known as the 'commodity supercycle'—high and rising prices for oil, copper, iron ore, and other commodities driven by strong economic growth in those countries."[55] Metals price fluctuations during the commodity supercycle are shown in Figure 28.

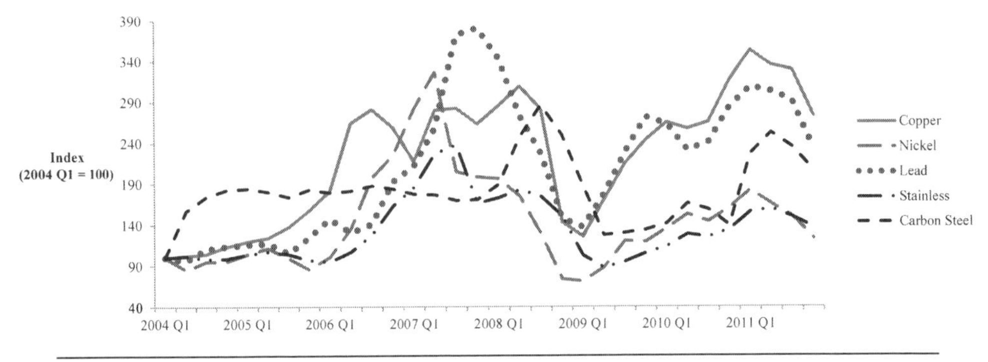

Figure 28. Fluctuation of Metals Prices During the Commodity Supercycle
(*Source:* Boston Strategies International)

In addition, technical services can also be in a state of shortage, especially during periods of disruption or rapid change, causing delays and bottlenecks. This can be especially acute in specialized requirements that require highly skilled, trained, and experienced engineers such as 3D seismic surveyors, fluid dynamic specialists experienced in deepwater pipeline transport, or FPSO megaprojects, or rotating equipment maintenance engineers knowledgeable in and willing to work in remote Arctic conditions.

Aside from purchasing futures contracts on a mercantile exchange, buyers have four options:

- Use substitute materials or services
- Plan farther in advance
- Buy or control the source
- Refurbish, recycle, or reuse the material or equipment, or retrain, reassign, or retain the resource to extend availability

Use substitute materials. As the automotive industry has transitioned from mostly steel to mostly plastic components to lower cost and weight, specialty chemicals buyers (catalyst manufacturers in particular) have replaced rare earths with other materials. Pump, turbine, and compressor manufacturers have replaced specialty alloys with other alloys that are less dependent on materials that are in short supply or in a state of continual depletion. Finding substitute materials takes time, so such a search is often the object of an ongoing initiative so a materials shortage does not interrupt production, and a price spike does not cause a budget variance crisis.

Plan farther in advance. Operators are often accused by suppliers of waiting until the last minute to order products and services. Some have made an attempt to plan farther ahead. Using master scheduling logic borrowed from MRP systems, planning for materials in shortage should take place in the Demand Planning Time

Fence, during which manufacturing assets can be redeployed. If planners wait until the Manufacturing Time Fence (line scheduling), there is a much greater chance of material or component unavailability or high prices. Stretching from a shorter to a longer time horizon allows more flexibility to pursue alternative strategies if there are any available.

Acquire the source. If material shortage appears to be chronic, it may make sense to acquire the source. Vertical integration makes sense when the cost of acquiring the materials through external sources exceeds the cost of procuring them internally.[56] Even if they could be procured at lower unit prices on the outside, the cost of searching for sources of supply and negotiating prices, and arranging logistics, transportation, and payment may be complicated in a tight market, and if sustained for long periods might justify vertical integration.

Refurbish, recycle, or renew. Over the last five years refurbishment has become a fairly popular alternative to buying new for a variety of equipment. Refurbished equipment is often less expensive and can have a shorter lead time, especially if the supplier builds refurbished equipment to stock. Recycling component materials has also become more prevalent, as evidenced by the introduction of new recovery processes for rare earth metals on the part of refinery catalyst manufacturers. For example, Grace Davison installed metal traps on its catalyst production lines, which recover about 2% of the total rare earth metal volume used to make the catalysts.[57] Using a larger analogy, enhanced oil recovery (EOR) is a large-scale analogy to the refurbishment concept: depletion has reached the point where recovery processes are widespread.

Outsourcing Risks and Mitigation

The decision about whether to insource or outsource an activity ("make vs. buy," as it is commonly called) drives cost in the same way as the ownership control decision discussed immediately above.

In addition, deciding to rent vs. buy can similarly change the risk/return profile of a work process by outsourcing part or all of it. Outsourcing affects not only risk profile, but also cost, effectiveness, and sustainability. Potential reasons for outsourcing may include:

- Lower cost and capital requirements. For instance, one production chemical supplier does not make any chemicals. It buys the base chemicals, mixes them, and resells them to the oil producer or service company. This allows the supplier to avoid tying up capital in facilities and gives it the flexibility to choose the best supplier for a given type of chemical without developing the chemistry itself.

- A more global network, including enhanced access to foreign markets. For example, one drilling and completion fluids supplier outsources production of most component chemicals to local suppliers to reduce shipping time and cost. Instead of investing in capacity in each region, it has developed a network of local suppliers. As most of the lead time for chemicals is related to shipping, outsourcing allows this supplier to deliver faster than competitors that produce in another region. In one instance, it arranged next-day delivery of a drilling fluid additive, sourced from a local partner, when the supplier with the next shortest lead time quoted a week.
- Prior investment in IT platforms that provide higher accuracy and reliability compared to in-house systems. Many companies outsource logistics primarily because the electronic manifests and other supply chain visibility systems cost too much to create and update in-house. Third party logistics providers continually invest in new information technology because they are serving a multitude of customers and can pass on a fraction of the cost to each one.
- Value-added services that would be challenging or expensive for the operator to do in-house. Often, the cost of in-house staff makes value-added logistics tasks more expensive than outsourced providers' equivalent services, for example, kitting, light assembly, mixing, blending, and similar customization operations that are performed on a finished or semi-finished product. Outsourcing these activities can often take advantage of the lower cost structure of an operation that is set up expressly with low-cost or specialized labor to perform them at a lower cost.

When considering whether to outsource an activity, one should consider the loss of skills and talent that will result. The loss of know-how could be permanent if the activity cannot easily be reconstructed in-house. One turbine user used the following criteria to determine if maintenance should be outsourced:

- Is it a core competence that should remain in-house even if there is a cost disadvantage?
- Is it less expensive to outsource it, including all in-house and outside life-cycle costs?
- If it is more expensive in-house, could the in-house process be engineered differently to reduce the cost enough to meet the outside cost?
- Would an investment be required to outsource the activity, and if so, would it pay off? Conversely, if outsourcing would allow equipment and tools to be liquidated, would any value be recaptured from selling the assets?
- Could the activity be brought back in-house in the future if the experience is deemed to be a failure?

- Does management have enough time to manage a transition from in-house to outside, or vice versa?
- Would renting equipment or asking the vendor to offer life-cycle solutions provide a viable alternative to outsourcing? If so, would this affect the pace of innovation or efficiency gains?

The outsourcing decision can be facilitated with a logical flowchart involving the above questions (see Figure 29). If the activity is critical to business operations, the first logical path is to ask a series of further questions about whether the activity is performed at world-class levels today or not, and if not, how to get it to that level. If it is not critical to business operations, then a different series of questions may be asked about outsourcing. Either logical pathway should consider the risks of taking that path, and, if risks are identified, then comparing the risk level and the mitigation strategies to the alternatives. A cost-saving offer by a third party can be very alluring, but if it ends up creating irrevocable dependency on the third party for mission-critical operations, the short-term gain is probably not worth it.

The same decision-making process should be followed in reverse if changing conditions could potentially invalidate or alter the original logic and math that justified outsourcing in the first place.

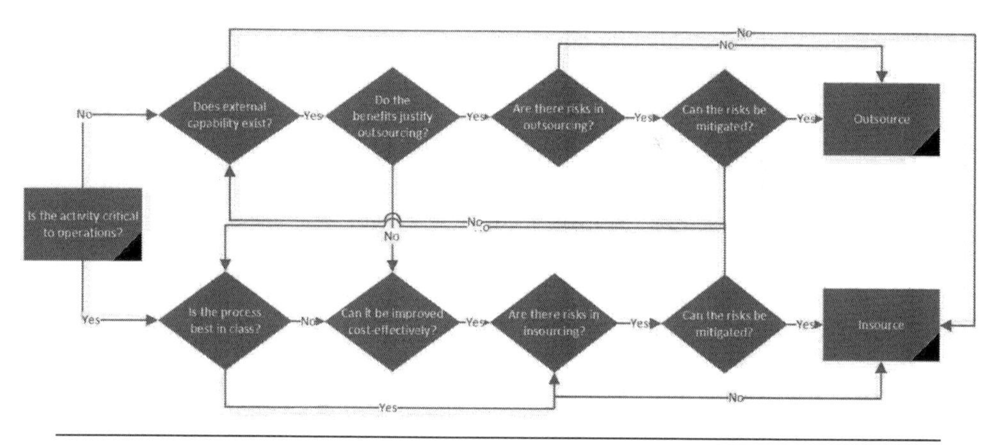

Figure 29. Outsourcing Decision Process

Supplier Partnering Risks and Mitigation

The oil, gas, and power industries have typically used a high degree of solution bundling due to project complexity, risk, task interrelatedness, and technology intensity. This often ensures compatibility of components within complex systems, and works well for suppliers because solution bundling usually allows the supplier to earn higher profit margins. Conversely, unbundling can often be a cost savings opportunity for operators. Operators regularly need to assess:

- Whether to buy raw materials separately from components (if the raw material constitutes a very large proportion of the cost of the component, such as steel in a large fabrication)
- Whether to buy some components separately from other components (such as motors separately from pumps that they drive)
- Whether to buy products separately from the services that are associated with them (such as compressors separately from their installation and commissioning)

In response to the commoditization of manufacturing, which had been catalyzed by aggressive strategic sourcing efforts, suppliers went through a period of adaptation to regain sustained losses in margins. As shown in Figure 30 below, while US GDP rose sevenfold between 1945 and 2005, manufacturing prices only rose threefold, squeezing manufacturing profit margins.

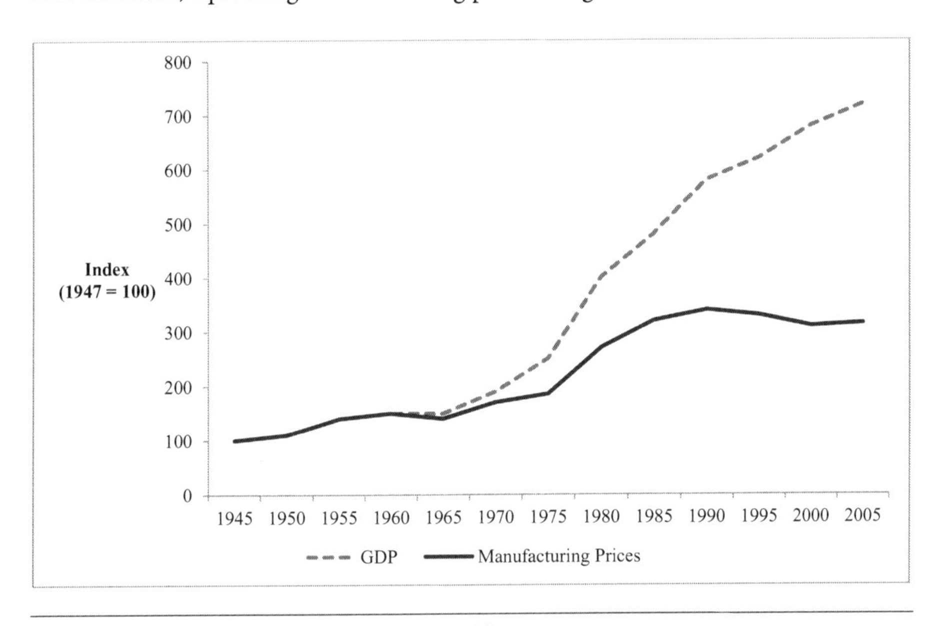

Figure 30. US Manufacturing Prices—The Long View
(*Source:* Author's analysis of data from the Bureau of Economic Analysis and the National Association of Manufacturers' report "The Facts about Modern Manufacturing," 7th edition.)

Bundling usually brings higher profit margins, so suppliers generally prefer to bundle. They can protect or enhance their margins most effectively by concurrently combining three strategies: layering on value-added services, building a technology edge, and positioning themselves as the premium provider (see Figure 31).[58]

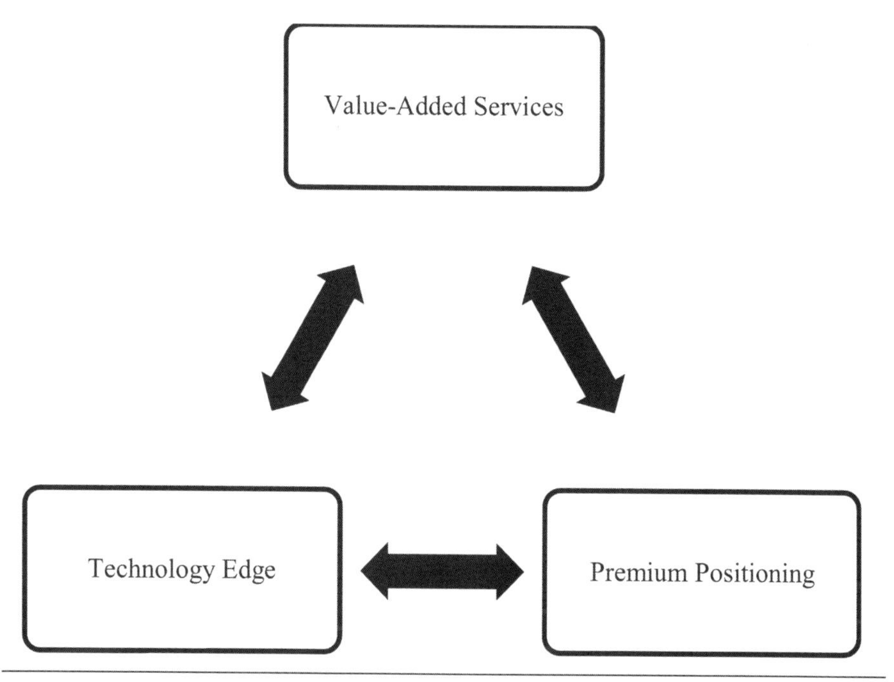

Figure 31. The Value Creation Triangle
(*Source:* Jacoby, David Steven. "How Will Western Manufacturers Survive?", 2008. p. 27.)

Structuring Strategic Partner Relationships

Once a supplier is selected, operators need to decide how to allocate resources to get the most out of the supplier. This starts with clarifying the degree of intimacy that will lower the cost of communication (and associated miscommunications and omissions) and increase the effectiveness of the combined resources by leveraging synergies.

Structuring Alliances

A comprehensive partner supplier program guides supplier relationships from the commodity stage through five levels of partnering toward strategic partnership (see Figure 32). The stages are transactional, preferred, value-added, alliance, and strategic.

- Transactional suppliers take orders as needed from the operator.
- Preferred suppliers have met some pre-specified conditions, such as a vendor or quality certification, that distinguish them from the transactional suppliers.
- Value-added suppliers have a framework agreement to provide a focused service.

- Alliance suppliers participate in joint development efforts with the buyer, such as oil contractor Halliburton has with many of the major oil companies in the development of oil fields.
- Strategic partners have established a common vision based on mutual needs and strategies. Companies that form true partnerships collaborate intimately with suppliers to achieve not only lower total cost, but also faster speed to market, more innovation, and better quality.

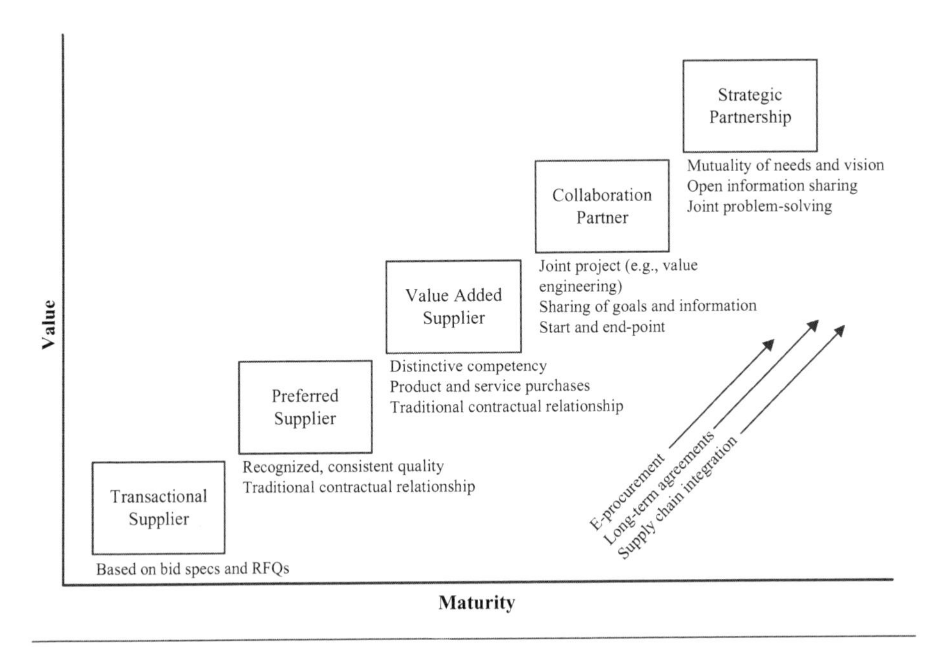

Figure 32. The Partnership Ladder
(*Source: Guide to Supply Chain Management* by David Steven Jacoby, further adapted from Michael Maccoby)

There are 12 dimensions to partnerships. On each of the 12 dimensions, the depth and nature of the partnership evolves as the relationship progresses through the five rungs of the partnership ladder, as shown in Figure 33.

Prequalification for a commodity supplier may consist of a simple, web-based application process. As the relationship progresses to a preferred supplier, site visits are required to understand the operating environment. When considering the same supplier for a value-added role, that supplier's past track record with the operator should be assessed. Alliance partners need to be approved at the division or business unit level, and strategic partners must be implicitly or explicitly blessed by executive management or board of directors.

Performance measurement escalates from being a one-time, project-based task to an ongoing and periodic assessment as the relationship becomes more symbiotic.

Level Number	1	2	3	4	5
Name	Transactional	Preferred	Value-Added	Alliance	Strategic Partner
Prequalification	Information exchange	Site visit	Joint track record	Divisional approval	Board vote
Performance measurement	Project-based	Periodic	Cumulative	Integrated	Perpetual
Certification		Basic	Silver	Bronze	Gold
Planning meetings		Project-based	Monthly	Weekly	Daily
Feedback mechanisms			Manager level	Multiple levels	All levels (360)
Framework agreement			MoU	Contract	Joint Charter
Joint long term planning			Tactical	Strategic	Strategic and Tactical
Dedicated liaisons			One	Many	Joint Teams
Joint investment				Case by case	Repetitive
Co-location				Project Office	Shared Resources
Joint R&D/I.P. Sharing				Limited Access	Co-development
Equity					JV

Figure 33. Partnership Maturity Model
(*Source:* Boston Strategies International)

Supplier certification or other recognition is often denoted by internally defined thresholds, and the level of certification can increase as suppliers become more strategic, creating "bands" of supplier capabilities; for example, one operator has defined bands of performance that it calls "bronze/silver/gold."

Communication frequency increases and management meetings become more frequent, often a natural byproduct of the size of the projects. This correlates with increased feedback and joint planning, and often requires the nomination of dedicated full-time liaisons from each organization.

Alliance and strategic partners co-locate facilities in order to increase communication and economize on facilities and overhead costs. Joint investments are considered and implemented, including joint research, development, and licensing of intellectual property. In a strategic alliance, the partners typically share equity in a joint venture organization, which is sometimes driven by local content obligations.

The opportunity presented by partnerships can pressure suppliers to "climb the ladder." While managing the supply base as it progresses through these stages, it is important for the buying organization to respect ethical norms. Practices for buyers to avoid include: issuing suppliers requests for quotes to benefit from free engineering while not intending to hire them (the ISM calls this "sharp practices"), sharing suppliers' prices with other suppliers, and accepting gifts or favors from suppliers.

Setting up joint ventures and taking equity stakes

Joint ventures should be considered only when the governors of both businesses determine that the strategies of both firms (perhaps within a certain market space) are sufficiently aligned that the two would be more powerful merging than competing. Large joint ventures can increase the business risk (the "beta," as it's referred to in equity analysis) completely independently of offering any supply chain benefits or synergies. So, when considering a joint venture, it is wise to first ask if the desired state could not be achieved through an arm's length transaction involving licensing or shareholding in each other's publicly traded stock. Often conventional contractual or trade-based relationships can be adapted to achieve the desired result without the "beta" risk that would accompany a business integration. In contrast, once a joint venture is formed, it may be costly and litigious to dissolve if things go awry.

Here are two examples of international licensing agreements between major suppliers and their international affiliates. Element 1 Corp. signed a licensing agreement with RIX Industries (RIX) for its proprietary methanol-based M-Series hydrogen generator technology for fuel cell mobility applications. Accordingly, the licensee was granted non-exclusive rights to manufacture and deploy the M-Series hydrogen generators in both North America and Europe.[59] Also, Cadenza Innovation, a US-based lithium-ion (Li-ion) energy storage solutions provider licensed its patented super-cell technology, design and all related know-how to Australian Li-ion battery manufacturer Energy Renaissance in 2019. This also included leveraging Cadenza's experience in designing manufacturing facilities, planning and deploying specific production lines for its battery technology, sourcing materials, and defining optimal value chains and workflows for battery manufacturing sites.[60]

Nonetheless, joint ventures are a popular form of supply chain integration due to cultural and legal factors in many countries. For example, Ballard Power Systems, a Canadian developer of fuel cells, entered into a joint venture with Weichai Power to form Weichai-Ballard Hy-Energy Technologies Co. The main objective of this JV is to manufacture fuel cell stacks and modules using Ballard's liquid cooled stack (LCS) technology for applications in commercial transportation in China.[61] Also, National Oil Varco (NOV) signed an agreement with Saudi Aramco to establish a joint venture in the Kingdom of Saudi Arabia (NOV owns 70%, and Saudi Aramco owns 30%). The JV aims to leverage NOV's manufacturing and fabrication facilities in the Kingdom and among NOV's facilities and is slated to provide high-specification land rigs and rig and drilling equipment, along with selected aftermarket services. The JV agreement also observed a commitment from the Saudi Aramco

Nabors Drilling Company to purchase 50 onshore drilling rigs for ten-year period.[62] Finally, in early 2011, Mitsubishi Heavy Industries (MHI) created a compressor manufacturing joint venture with Hangzhou Steam Turbine & Power Group Company, a Chinese firm that makes compressors for petrochemical plants. The JV gave MHI a stronger sales presence in mainland China and access to a lower-cost manufacturing base than its Japanese facilities. However, MHI supplies key components such as bearings from Japan to ensure quality and maintain control of the intellectual property.[63]

If management decides to form a joint venture, potential joint venture partners should discuss operating models while forming their mission and governance structures. Within joint ventures, the operating paradigm can favor domination by one of the partners, or alternatively it can be operated as a truly independent entity exercising control over its own destiny in the model of Well Dynamics, a smart completion supplier owned jointly by Halliburton and Shell. The degree to which each partner exercises control over operations in a joint venture should consider: 1) experience and success in operating similar businesses, especially among the senior management team; 2) which entity has assets that can drive the key success factors in the right direction; and 3) the degree of access that each partner may provide to the customer base of the other.

Procurement Bundling Trade-offs

However, while suppliers may universally (or nearly so) prefer solution bundling, buyers need to be discerning about when to bundle and when to unbundle. Three bundling configurations can optimize value and profit for an operator: selling complementary products or services, offering Product-as-a-Service, guaranteeing performance at the equipment level, and guaranteeing performance at the system level. Selling complementary products or services adds the least value; offering Product-as-a-Service adds more; and guaranteeing performance at the equipment or, better yet, the system level, can offer the most benefit to buyers. Buyers should determine which paradigm best suits their needs and embed the purchase paradigm in RFPs and generally in communications with suppliers.

Level 1—Selling complementary products or services. The most widespread method of bundling is to combine existing product or service offerings so that they complement each other, thereby reducing transactions costs (recall Coase's theory of the firm) for the buyer. Companies offer similar services as a "one-stop shop." The combination allows a buyer to reduce the number of suppliers (and thereby the cost of managing them), but more importantly it allows them to increase the quality and effectiveness of the solutions because multiple different products and services should be designed with integration and compatibility in mind. For example, these "complementary" bundling solutions should reduce overhead costs for operators.

In energy storage, Fluence offers a solution called Gridstack that combines the battery (the "cube"), the control system (the "OS"), and an analytics solution driven by artificial intelligence that provides "digital intelligence [that] improves system decision-making, asset performance, and operating costs with data-driven insights and dispatch algorithms."[64] In oil & gas, Schlumberger's acquisition of Smith International (2010) allows it to offer products to support the entire drillstring (from the bit up to the rotary table on the rig), adding bits to Schlumberger's portfolio and giving it full ownership of drilling fluids market leader M-I SWACO. Also, Baker Hughes' acquisition of BJ Services (in 2009) added pressure pumping services and equipment to its portfolio, allowing Baker to provide a broader range of services in more geographies than it could before the acquisition, as well as bundling different types of oil & gas production chemicals where either firm had a product with a strong reputation.

Level 2—Product-as-a-Service. In Product-as-a-Service, suppliers not only sell the initial equipment or services, but also continue to provide ancillary services throughout the life of the initial product or service. Examples are service contracts or "power by the hour." For example, in solar energy, Sunpower offers integrated project development, design, financing, construction, grid connection and operations. In oil & gas, Flowserve operates Quick Response Centers as part of its "Lifecycle Advantage" program. The program was created to deliver the lowest possible total cost of ownership by managing spare parts inventories and maintenance scheduling so as to reduce life cycle costs, increase mean time between repair and equipment life, and maximize uptime (reliability).[65]

Level 3—Guaranteeing performance at the equipment level. In solutioning, suppliers not only offer a product and a service, but guarantee a minimum level of performance for the equipment or service. Performance-based contracts and "value-based pricing" seek the maximum product or service performance, rather than the lowest cost. For example, one operator pays one of its turbine suppliers a bonus if the turbine's output exceeds the rated capacity. In a "performance-based logistics" (PBL) contract set up by the US Navy, parts availability was as low as 43% under the Navy's management, but under a PBL arrangement GE averaged 90–100% availability. The PBL also helped to eliminate 718 backorders and cut repair turn-around-time by 25%. The US Army instituted a similar program for GE's T700 engine, and GE, using lean management, cut engine turnaround times from 265 to 70 days.[66]

Level 4—Guaranteeing performance at the system level. Suppliers not only offer a product and a service but guarantee a minimum level of performance *within the process that the equipment supports*. The difference between Level 3 and Level 4 is that while Level 3 commitment relates to the equipment or service itself, Level 4 relates to the performance of the system in which the equipment is placed. PBL contracts in this configuration provide useful examples for assigning accountability and liability for overall system performance. The US Navy has set up various PBL arrangements for the Rolls Royce jet engines that it operates. The US Navy's C-17 Globemaster III Sustainment Partnership is a $1b per year program involving more than 1,400 Boeing personnel, consisting of mechanics at Boeing's depot, engineers in a second location, and technical representatives on every base. In another example involving the US Navy, Boeing coordinates with Pratt & Whitney, which makes the F117-PW-100 engine, and other C-17 subcontractors to maximize aircraft availability, the main Key Performance Indicator (KPI) of the program.[67]

Success may result in a large windfall to the supplier, especially if there is a multi-year contract based on a lump sum and if it obtains pre-negotiated financial incentives for higher performance and then hits the targets. In the oil & gas industry, a bearing manufacturer actually doubled the price of the equipment, and the operator was happy to pay the extra because the increase in uptime was worth far more than the higher equipment cost. Also, some drill bit manufacturers offer pricing on a cost-per-foot-drilled basis, instead of per bit, charging more on average per job but guaranteeing an increase in rate of penetration (ROP)—which justifies higher equipment spending due to the increase in revenue.[68]

Contract Term Risk and Mitigation

Determining the Optimal Term of Commitment

The contract term has as much impact on the total cost and risk profile of the project as the choice of whether to insource or outsource, although it is often decided by gut feel, without the help of analytical thinking. *A priori,* it is intuitive that commitments should be long-term because many of the projects' basic parameters are long-term:

- Long time to plan. Projects that take a long time to plan often require long forward views and extensive cost and economic estimates.
- Long time to build. If investments take a long time to build, then they require a long planning horizon, and the longer the time to build, the greater the risks that important planning assumptions will be or become wrong during the building time horizon.
- Deferred revenue streams. Along with large capital expenditures usually come deferred revenue streams and unpredictable future cost levels.
- High exit costs. To the extent that industries or projects are costly to exit, the planning horizon extends even further.

In fact, more than half of the oil & gas companies surveyed by Boston Strategies International said that a long-term stable pricing environment would aid in establishing steadier prices and operating profits that would help them minimize layoffs during downturns and re-hiring during upturns, thereby reducing long-term operating costs. A third of them said long-term stability would allow them to make more predictable R&D investments, which would result in higher exploration, refining, and distribution productivity due to faster and more consistent advances in oil & gas equipment technology.[69]

But how do you know which term is right? Ten, twenty, thirty, forty years? The term of commitment is sometimes decided arbitrarily. Oil, gas, power, water, and transport infrastructure providers frequently sign ten-year contracts, twenty-year operating agreements, and forty- or 50-year development and operating contracts. The term should be driven by the expected trend and volatility in the underlying cost and rate structures, and the economic dynamics that can be expected or forecast to change over the time frame of the agreement (economies of scale, scope, standardization, learning curve benefits, technology advances, and competition), which are illustrated in Figure 23.

One approach to optimizing contract term is to use an optimization model. BSI has developed a contract term model that uses the historical volatility in price movements and the likely savings that could be obtained from a long-term contract to test the expected net benefits at a spectrum of possible intervals on a long time horizon. The model uses previous engagement history to define probability distributions for the volatility and the savings and generates an optimal contract term through Monte Carlo simulations.

In several cases the simulation yielded optimization results of more than 20 years. One study[70] showed that it takes 22 years for an initial price shock to work its way through a capacity-price-cost-sales cycle, implying that long-term agreements in this context should be at least this long in order to completely eliminate market risk.

An alternative is to establish an initial agreement that includes structured options for expansion, contraction, modification, and cancellation as the project unfolds. This approach has the advantage of permitting more informed decisions and relying less on forecasts or bets. It also is economically more efficient, and results on average in higher NPVs, since it lowers the cost of capital investment by maintaining the upside opportunity while limiting the downside risk.

There are four relevant types of options, each of which corresponds to a type of analysis in "real options" theory. These are:

1. An option to expand (called a "call option")
2. An option to contract (called a "put option")
3. An option to terminate (called an "abandonment option"), and
4. An option to change the parameters (called a "switching option").

Option to expand. In the context of supply chain management and procurement contracts, options to expand are often called "capacity reservation agreements."

An example is an option to place or increase the number of orders or the size of orders, to increase the supplier's share of the expenditure. Suppliers will often grant a guaranteed capacity reservation on a subsequent order once an initial order has been confirmed. The contract typically specifies a fixed quantity of firm orders, and an option to obtain additional units at a future time for an agreed price. Once an agreement is signed, suppliers are often willing to flex the date when the option may be "called," so long as it has adequate capacity; however, there may be constraints on the specifications and ancillary terms (financing, insurance, warranty, etc.) that the supplier can accommodate, depending on production capacity and configuration at the time the "call" option is exercised. The value of options agreements is rarely publicly disclosed since they reveal executives' expectations about their companies' future expected spare capacity, which could impact pricing on other sales. Most deals seem to be perceived to clear at 5–10% of the normal purchase cost. In one example, Baker Hughes has optioned the rights to any barite found at Bravo Venture Group's new mine in central Nevada.[71] This also happens with aircraft, installation vessels, turbines, and a variety of other types of engineered equipment; additional examples are provided in the Project Risk Mitigation section of the Power chapter in Part 2.

Option to contract. The analogue to an option to expand is an option to reduce commitment. While these options can also be negotiated and their value quantified using real options methods, flexible capacity strategies on the part of the supplier can also increase downward flexibility. At the project planning stage, one way to make capacity flexible is to outsource peak load to a third party. This helps to keep capacity utilization steady and increase the reliability of service levels during troughs in demand.[72]

Option to terminate. Options to terminate can also be built into programs and contracts. However, rather than value and articulate the options to terminate, in some cases a "termination for convenience" clause may provide the same benefit. For example, KOC's LSTK contract stipulates that "the Company may, at any time, terminate any part of the Works or all remaining Works without giving any reason therefore, by giving written notice to the Contractor specifying the part of the Works to be terminated and the effective date of termination if different from the date of said written notice."[73]

Option to change. Similarly, almost all contracts have options for change orders at the discretion of the buyer. A change may be an order for a different product from the one originally ordered. Other frequently experienced change orders include changes to design or specifications, changes to delivery timing, alteration of subcontracting rules or approval procedures, and assignment of the contract.

Methods for Operations and Maintenance Management Optimization

Chapter Highlights

1. Six sets of operational risk and cost trade-offs should be judiciously managed: Internet of Things (IoT) and Artificial Intelligence (AI) Technology Choices, cybersecurity risks and strategies, capacity strategies, sourcing trade-offs, Total Cost of Ownership (TCO) trade-offs, and Environment Health & Safety (HSE) risks throughout the extended supply chain.
2. For cybersecurity risks, implementation of 13 best practices can contain the risk of malicious attacks; these include, for example, avoiding open source code, tightly managing cloud-based solutions, and applying standard contract language regarding cybersecurity measures in all vendor contracts especially for IoT solutions.
3. For AI trade-offs, category management should reflect product-services and systems solutions, and negotiate pricing on the basis of benefits and competitive benchmarks instead of unit costs, ensure proper data governance for IoT and AI-centric applications across the value chain, and initiate the use of AI in supply chain management activities such as smart supply-demand balancing, smart contracting, digital twins of production operations, use of robots or cobots in warehouses, and drone-based security surveillance of pipelines and transmission and distribution networks.
4. For capacity management, Overall Equipment Effectiveness (OEE) and Return on Net Assets (RONA) can be improved through Total Productive Maintenance (TPM); constraints management; debottlenecking; flexible capacity; standardization; continuous cost reduction; stochastic inventory management; vendor-managed inventory; JIT; and transportation and warehousing optimization.
5. For sourcing trade-offs, best practices include forming a category management organization; developing individual category strategies;

investing in prequalifying well-qualified suppliers; determining the optimal number of suppliers; prequalifying approved suppliers; managing the tendering process; and assuring local content management where applicable.

6. For Total Cost of Ownership, decisions should be based on the combined "first cost" and ongoing "life-cycle costs" of all aspects of the sourcing decision. Combined Purchase and Operating/Maintenance Agreements normally represent the synthesis of an analysis, a competitive process, and a negotiation reflecting the lowest total cost of ownership.
7. For HSE, accepted industry or international standards are effective tools for efficiently achieving supply chain safety goals. In addition, a Safety department, usually reporting to Quality or Operations, should conduct or contract for HAZID and HAZOP studies, maintain a risk register, and perform root cause analysis.

Introduction

Operating and maintenance costs, including energy, often constitute well over half of total life-cycle cost (with capital expenditure making up the rest) of equipment, and reliability can be greatly increased by the use of predictive maintenance using information technology and sensors. Yet planners sometimes estimate operating costs as a flat percent of capital costs and do not adequately leverage the power of best practices in operations and maintenance to drive down cost.

Trade-offs and Management Techniques in Operations & Maintenance

This chapter breaks tools for management OpEx into six categories:

1. Internet of Things (IoT) and Artificial Intelligence (AI) Technology Choices
2. Cybersecurity Risk Management
3. Capacity Strategies
4. Sourcing Trade-Offs
5. Total Cost of Ownership (TCO) Trade Offs
6. Health, Safety and Environmental (HSE) Considerations

As for CapEx, the chapters consider these CapEx techniques in the order listed above based on a classification of their frequency versus their typical impact, with frequently used higher-impact tools topping the list, and infrequently used lower-impact tools toward the bottom, as shown in Figure 34.

	Lower Likelihood or Prevalence	**Higher Likelihood or Prevalence**
Higher Impact (financial, operating, safety, legal, organizational)	• Sourcing Trade-offs	• Internet of Things (IoT) and Artificial Intelligence (AI) Technology Choices • Cybersecurity risk management • Peak capacity strategies
Lower Impact (financial, operating, safety, legal, organizational)	• HSE Considerations	• TCO trade-offs

Figure 34. Trade-offs and Management Techniques in Operations & Maintenance Management

Internet of Things (IoT) and Artificial Intelligence (AI) Technology Choices explains how to decide which technology investments will maximize throughput. Most companies know they need to digitalize their operations to remain competitive: how can you decide how fast you need to move and how much you should spend on digitalization, especially so-called digital twins, which are software images of physical manufacturing and distribution operations. For example, digital oilfields are an almost impossibly vast goal for oil & gas operators: there is no generally accepted end point. So how much fuel should a manager pour on the fire? Likewise for digital twins. Big 4 consultancies have established an ideal of complete digital replication of all physical operations, which is a financial bottomless pit. Where should one start, stop, and slow down with digital twins? Furthermore, robotics may be either dumb or smart. It's easy to figure out whether dumb robots are worth the investment or not based on their payback in efficiency. But how much are smart robots worth? They generate data that can be mined to make operations more efficient. The more data that is collected, the more data mining can be done, but we don't yet know the benefits of the data mining, so how are operations staff making decisions on smart robot investments?

Cybersecurity Risk Management addresses how to maximize IoT/Industry 4.0 operational benefits while minimizing risk of hacking and cyberattacks. All enterprises need to be taking advantage of the benefits of IoT technology, but running assets on Internet-enabled devices leaves operations exposed to cyberattacks. Predictive maintenance can reduce operating cost, but how do you control the security of the information flowing from field assets to the server? Similar challenges exist for robotic operations. And the more IoT, the more benefits, and the more risks. How can this conflict be managed in procurement and operations decisions? For example, handheld devices may allow for more efficient dispatch of field repair technicians, but it may also leave them vulnerable to attacks by malicious third parties hacking into their real-time location. What is the right choice in this situation? Also, real-time equipment monitoring typically increases asset utilization and reduces maintenance cost, but the use of cloud services likely exposes those assets to threats from malicious actors who could cause a lot more damage than the gain in asset productivity. How can this risk and reward be balanced?

Peak Capacity Strategies investigates how to satisfy peak capacity at the lowest fixed cost. Lean Manufacturing, dynamic pricing, stochastic inventory management, vendor-managed inventory (VMI), Just-in-Time (JIT), automation, and outsourcing are just a few of the tools that operators can use to mitigate the cost penalties associated with holding unused capacity if you own too much or encountering outages and shortages if you own too little. Each one brings some degree of control over the cost penalties of trying to satisfy peak demand with fixed assets and erratic withdrawals from Maintenance, Repair & Operations (MRO) inventory.

Responsible Sourcing Trade-offs explains how to use classic sourcing tools and techniques such as supplier prequalification, partnering, and bidding, while managing conflicts between lowest cost and environmental social and governance (ESG), including how to increase access to small- and medium-sized enterprises (SMEs) into procurement operations, and how to increase local content, which may in some cases tie into Sustainable Development Goal #9, sustainable infrastructure ("Build resilient infrastructure, promote sustainable industrialization, and foster innovation"). Specifically, SDG 9 aims to "[i]ncrease the access of small-scale industrial and other enterprises, in particular in developing countries ... and their integration into value chains and markets; upgrade infrastructure and retrofit industries to make them sustainable, with increased resource-use efficiency and greater adoption of clean and environmentally sound technologies and industrial processes; and support domestic technology development, research and innovation..."[74] Most companies have embedded social goals of procurement and operations in recent years, especially regarding local content and making it easier for small- and medium-sized businesses to win tenders. However, meeting these social goals often requires more time and energy than classic procurement: it would often be quicker, easier, and in many cases less expensive for buyers to award all business to large suppliers who could realize significant economies of scale. Instead, they establish and run training programs, skill upgrading programs, communication programs, and other outreach programs to offer these social benefits. How do they justify the additional budget for this? How do they decide how much of a bid price differential to allow for a local or a small or a minority supplier?

Total Cost of Ownership (TCO) Trade-offs looks at how to balance "first cost" with operating and maintenance costs to achieve the lowest lifetime cost. Tools and frameworks are presented to help standardize products, materials and processes, engineer value into products and processes, and measure and optimize quality levels.

Health, Safety & Environmental (HSE) Tools explains how to make trade-offs between maximizing output and respecting environmental, health, and safety goals. Although HSE is often a separate department, operational functions and departments often play a role in ensuring environmental health and safety, especially at suppliers and suppliers' suppliers, including complying with risk management standards, contributing to risk management registers, conducting PH&A Studies, and collaborating with HAZOP Studies, for example.

Internet of Things (IoT) and Artificial Intelligence (AI) Technology Choices

The Internet of Things (IoT) and 5G have created a wide-open opportunity for energy companies to make much of what they do "smart" instead of "dumb" or static. Every piece of equipment from diesel generators to pumps to motors to security systems is now a smart solution.

Artificial intelligence (AI) has been offering faster and better automation opportunities than had been thought possible even a few years ago. AI is even replacing "knowledge work" such as legal advising, which most people have thought of as impervious to automation. Data-oriented tasks are already mostly (62%) performed by computers. Complex and technical activities, which have been the domain of technical experts, are almost half (46%) done by computers today, up from 34% two years ago. Thirty percent of all managerial activities such as coordinating developing managing and advising, and communicating and interacting, are already performed by computers, up from 19% in 2018. Even basic reasoning and decision-making, which many have said could never be automated, are already 28% automated, according to the survey, which indicates that the rate of automation of even these "impenetrable" tasks is increasing by 4–5% per year.[75]

Operational managers and procurement staff are in the middle of this transition, as they need to design the operations that use them and procure the solutions. They should be equipped to:

1. Refine category management structures to accommodate solutions and product-services, in addition to or instead of traditional procurement designations such as "materials" and "services."
2. Conduct ongoing market intelligence on the state of smart devices, and in particular the ones that use machine learning and artificial intelligence.
3. Run or facilitate AI pilot programs to test emerging capabilities and embed those that add value and are ready for enterprise-wide rollout.
4. Establish data governance structures to secure the explosion of proprietary and shared digital data.

5. Evaluate the costs versus the benefits of rapidly emerging AI platforms in supply chain management activities such as Demand Planning, Procurement, Production Control, Warehouse Management, Fleet Management, and Customer Service. This should include, as applicable, the following:
   a. Supply-demand balancing that learns and rapidly runs pre-configured logistical scenarios that can be invoked in periods of disruption to find the optimal new supply chain design.
   b. Smart contracting, possibly with blockchain, that ensures complete traceability
   c. Digital replicas of physical production operations that allow for digital optimization
   d. Robots or cobots in warehouses
   e. Drone-based surveillance and security systems

Cybersecurity Risk Management

In the absence of adequate planning for cybersecurity, adversaries may operate undetected within active contracts, under subcontracts, or from outside, and companies may be inadvertently dealings with parties that are designated as Denied Persons by authorities. Hackers may use back doors, remote connections, or dormant or stolen credentials to access proprietary systems and files. Once access is obtained, they may insert malicious code that could bring systems down.

Energy companies can take steps to minimize the risk of cyberattack. For example:

1. Beware of open-source code in development environments, internally or via vendors
2. Conduct due diligence on cloud-based solution providers, and use dedicated servers
3. Use standard contract language about cybersecurity when negotiating with IT vendors
4. Integrate cybersecurity precautions into supplier qualification and onboarding procedures
5. Establish a policy that governs and limits development in adversarial environments
6. Monitor compliance against Denied Persons, disapproved vendors, and related lists and orders
7. Require internal and external vendors to validate the authenticity and origins of third party hardware and software

8. Require vendors to use strong authentication and cryptographic methods
9. Require vendors to manage credentials stringently, including periodic deprovisioning
10. Require vendors to deny communications with risky profiles and log denied access incidents
11. Use intelligence about active and potential threat sources to mitigate active threats
12. 1Require vendors to establish a documented patch process with safeguards against malicious actors
13. Verify patch authenticity via cryptography, hashes, certificates, or two-factor authentication

For further information on cybersecurity measures, readers may refer to the guidelines prepared by the Critical Infrastructure Protection Committee of the North American Electric Reliability Council, including provenance guidelines prepared by a working committee chaired by David Steven Jacoby.[76]

Peak Capacity Strategies

Overall Equipment Effectiveness (OEE) and Return on Net Assets (ROA)

The Overall Equipment Effectiveness (OEE) framework measures asset effectiveness by defining three types of capacity:

- Rated capacity (as determined by the original equipment manufacturer)
- Standard capacity (driven by equipment availability, which is based on scheduled uptime vs. total available time). Most often, this corresponds to the operator's normative, or expected output.
- Demonstrated capacity (actual production vs. the standard), which is affected by product quality or yield (good output vs. total output)

Actual capacity is equal to rated capacity times standard uptime times efficiency, or put another way, Time Available x Utilization x Efficiency, where Utilization = [Actual hours worked / Standard hours available] and Efficiency = [Standard hours produced / Actual hours worked].

The best performers have an OEE averaging 90%, whereas laggards have an OEE averaging 74%, according to a study by Aberdeen Group. The differences are due to:

- 2% unscheduled asset downtime vs. 12% unscheduled asset downtime
- 12% reduction in maintenance cost vs. 2% increase in maintenance cost
- 24% improvement in ROA vs. plan, compared to a 5% decrease in ROA vs. plan[77]

Another metric for measuring overall operational efficiency, as noted in the Introduction, is Return on Net Assets (RONA). RONA, which is calculated as Net Profit / (Fixed Assets + Net Working Capital).[78] This takes into account both throughput, as measured in OEE, and the efficiency of asset utilization.

Success in achieving either OEE or RONA depends on cultivating an organization and operational processes and systems that are focused on maintenance and asset management. Better-performing companies implement preventive and predictive maintenance processes, and measure return on assets at both the operational and the executive level. Consistent with total productive maintenance, the maintenance and production teams are aligned toward the same goals and contribute to joint brainstorming and problem-solving. Better-performing companies also spend more time and effort implementing maintenance management systems and asset management systems that tie into their corporate objectives.

Management should set a maintenance strategy either explicitly or implicitly, and clarify whether it applies to the whole company or whether it varies by plant, business, or facility. The strategy should correspond to the time horizon of expected active life of the plant. For plants and facilities that are expected to be productive for a long time, the strategy is generally to maintain the assets to maximize their production for the long run. Occasionally, if a major planned teardown, renovation, or major overhaul is impending, management decides to deliberately let assets deteriorate or operate them until they fail.

Even if a deliberate maintenance strategy is pursued, sometimes it makes sense to repair individual equipment only when the equipment breaks down (for example, when the cost of taking it offline for preventive maintenance is not worth the lost output during the downtime, and the cost of repairing it would be the same whether it breaks or not).

However, most often a preventive maintenance strategy best extends equipment life and results in the lowest long-term cost. There are three basic maintenance strategies, which are often combined into a hybrid solution, especially in large and complex facilities:

- Periodic, or time interval-based, is the simplest and oldest form of preventive maintenance.
- Usage-based maintenance schedules visits according to the number of operating hours or starts since the last maintenance, and often lowers maintenance cost if the proper interval can be determined accurately enough.
- Predictive maintenance checks for early warning signs of deterioration and takes corrective action before failure occurs. It differs from preventive maintenance because it is not periodic. Condition monitoring is a form of predictive maintenance. For example, the condition of bearings in large machinery can be monitored through electronic sensors and tracked via remote computer, and usually indicate problems

well before they fail, which facilitates purchasing replacement parts ahead of time. Remote condition monitoring also reduces onsite maintenance staff requirements and facilitates smoother maintenance workload planning.

Total productive maintenance involves employees in a long-term program of asset productivity optimization through workforce involvement (e.g., quality circles) and the use of problem-solving tools such as root cause analysis. Reliability-centered maintenance regroups an extensive body of knowledge about maintenance and focuses the enterprise's mission on improving performance toward an overall goal such as OEE.

Six Sigma, while not exclusively a maintenance concept, can be applied to maintenance. Six Sigma is the application of statistical process control to reduce variation in performance levels. It seeks conformity to a target level, which results in deviations only in the rarest of occasions. Six standard deviations from the norm represents only 3.4 defects per million items, which is equivalent to a result that is correct in 99.9999997% of occurrences.

Total Productive Maintenance and Related Concepts

Some people count quality costs as part of total cost of ownership. In many cases, however, since the cost of quality failure does not affect the equipment itself, it is usually accounted for separately so that the life-cycle costs can be used to compare and evaluate alternative equipment choices, while quality costs can be used to improve internal processes and systems.

Quality costs include internal failure costs, external failure costs, prevention costs, and appraisal costs.[79]

- Internal failure costs are most importantly the cost of re-jects, re-work, re-inspection, and re-placement. However, additional internal failure costs include corrective action, wasted material or other inputs, late charges, expediting costs, and early costs (opportunity costs).
- External failure costs include returns, warranty claims, liability costs (penalties/allowances), loss of customer goodwill, and lost sales.
- Prevention costs include identifying customer needs (market research), reviewing contracts, testing new products, certifying suppliers, and the cost of running a quality department and processes (e.g., statistical process control).
- Appraisal costs include source inspection, incoming inspection, the cost of measurement and lab equipment and services, and evaluation of field stock or service quality.

Zero defects[80] is a quality philosophy that emphasizes that ensuring good quality through conformance to requirements is cheaper than inspection and rework that are inherent when quality is bad. Cause and effect analysis[81] is a useful way to engage worker involvement in solving complex operational and maintenance-related problems.

Constraints Management, Debottlenecking, and Flexible Capacity

The capacity of a system is limited to the capacity of its bottleneck process. Therefore, in order to maximize the capacity of a value chain, capacity should be aligned at each step. To illustrate this, consider the case of LNG exporting that involves liquefaction, LNG tanker shipping, and regasification on the other side. If the heat exchanger used in the liquefaction process has a higher capacity than the tanker can intake during its time at berth, the capacity of the heat exchanger will be under-utilized. Similarly, if the capacity of the storage tanks or the seawater warming on the delivery side is lower than the capacity of the tanker or of the liquefaction process, the upstream process capacity will be under-utilized. Moreover, in the latter case there will be a bottleneck.

Debottlenecking is the process of successively identifying the binding constraint, eliminating that constraint, aligning other processes to the new throughput levels, and pursuing the next constraint.

Debottlenecking efforts must necessarily consider the entire value chain as a system. This is challenging because trading partners need to share information about their processes in order to identify bottlenecks, and many companies hesitate to share such information because it may cross a borderline into "confidential business information."

As demand levels change over time, capacity must flex up and down. Given that capacity takes a long time to build and remove in oil, gas, and power, there have historically been two basic strategies for matching supply to demand: chasing and leading. Chasing means adjusting capacity in response to changing demand, while leading means adjusting capacity in anticipation of a demand change. Both approaches assume some risk—in the case of chasing, the risk that capacity will not be available when needed, and in the case of leading, the risk that capacity will sit idle until demand catches up. There are ways to minimize these risks through "flexible capacity management." Flexible capacity management lowers the break-even point, thereby increasing profit potential without increasing the asset base. During ongoing operations (post-CapEx), five tactics can be used to make capacity flexible:[82]

- Increase temporary employees as a proportion of staff. This reduces the fixed labor cost base and therefore the break-even point. Temporary employees are easier to hire and let go when demand shifts, and if managed carefully can eliminate expensive overtime.
- Implement Lean concepts such as just-in-time, one-piece flow, level loading, cycle time compression, and make to order (MTO), in order to avoid building unnecessary inventory.

- Postpone disposition, dispatching, or finishing operations to as late as possible in the value chain in order to reduce capacity requirements and inventory obsolescence.
- Use demand planning techniques such as sales and operations planning (S&OP) and collaborative planning, forecasting, and replenishment to reduce overstocks and out-of-stocks.
- Dynamically price products to encourage customers to buy stock before it becomes obsolete, and conversely to earn extra margin on tight resources during peak periods. Once customers adapt their buying behavior to yield pricing, it has the effect of leveling capacity.

Standardization

Standardization of items and services—the reduction of the number of types or brands of equipment, or the number of component variations—can increase the purchased unit volume per item, thereby lowering the average purchase price as well as the life-cycle cost of spare parts, maintenance, repair, and training.

The case for standardization usually rests on the costs of *not* standardizing. For example, by operating multiple different types or brands of electrical equipment, an operator would need multiple spares pools, each possibly with its own service network. There would be overlapping and potentially conflicting documentation and training efforts, resulting in a duplication of effort and redundancy of resources. In turn, the redundancy would eventually cause some confusion, which would increase the likelihood of errors and deviations to operating procedures, which would reduce reliability and uptime.

In contrast, the potential benefits and savings from standardization are speed (e.g., *time to first oil* and faster to start up a new plant based on the use of familiar equipment), quicker installation, lower upfront acquisition cost based on volume discounts, lower facility costs through standardization of facility design and construction, lower training cost, lower operating and maintenance cost, and lower financial (e.g., technology and warranty) risk.

Standardization programs are often best directed at standardizing operating processes and systems, limiting the number of suppliers, technology platforms, or component types, rather than the actual configuration of equipment; the latter could reduce the suitability for diverse applications and operating environments, thereby reducing reliability.

Process standardization typically holds many opportunities, including for example work processes and standard operating procedures, work instruction cards (for maintenance), and design engineering practices. Most major oil companies have implemented standard process initiatives, for example:

- Shell has standardized design and engineering practices (DEP) which list the specifications and uses of various pieces of equipment. Combined with its Materials and Equipment Standards and Code (MESC) catalog, Shell has cut inventory levels in half for electric cable and achieved 30% price savings.[83]

- ExxonMobil actively participates in ISO and API to set standards, and it has performed case studies on specific standards in support of standardization.[84]
- Petrobras launched its PROPOÇO program to standardize well construction. The program established standards for projects, personnel, well design, and documentation. It encourages the sharing of information and best practices, and offers engineers detailed studies, calculations, and designs to optimize well construction.[85]
- Gazprom created a corporate standardization system in 2002, consisting of technical guidelines for product requirement specifications, and standardized processes in all aspects of the business. In addition to cost savings, the company expects standardization to increase conformance to governmental and internal regulations.[86]
- Petronas compiled pre-approved technical practices as a technical reference for engineers and a guideline for vendors and contractors.[87]

ERP platforms are a good place to begin standardizing systems. Many companies have multiple instances (versions, upgrades, modules, etc.) of the same ERP system in place, providing a high-level standardization opportunity. In addition, data definitions, data governance, and data formats and protocols are all rich targets for standardization initiatives. Petro-Canada has supported common standards in the industry. With a presentation at the International Standardization Workshop in Doha, Petro-Canada advanced common standards as a way to share best practices, delocalize procurement and investment, and reduce cost and delivery times.[88]

Standardization bodies, be they at the international level or at the national level, help facilitate standardization. Bodies include the International Organization for Standardization (ISO) and International Electrotechnical Commission (IEC), the American National Standards Institute (ANSI) and British Standards Institution (BSI), NORSOK (Standards Norway), OLF (The Norwegian Oil Industry Association), the International Renewable Energy Agency (IRENA), the American Wind Energy Association (AWEA), the Solar Energy Industries Association (SEIA), the International Atomic Energy Agency (IAEA), and regulatory bodies from countries such as Brazil and Japan, as well as various European countries. Organizations like CEN (Comité Européen de Normalisation) and CENELEC (CEN's electrotechnical division) bridge international and national standardization bodies. Various associations comprising contractors, suppliers, and operators from many industries complement the efforts of these standards bodies. For example, API (American Petroleum Institute) and ASME (American Society of Mechanical Engineers) write standards that are reviewed and accredited by ANSI, whereas EEMUA (Engineering Equipment and Materials Users Association) and IP (The Institute of Petroleum) are British organizations that write standards administered by BSI. In these cases, the official standards bodies audit the standard-writing procedures of the industry groups and recertify them periodically.

Over time, the standards of the different regional, industry, segment, and engineering disciplines are merging. For example, the Norwegian oil industry has developed 80 harmonized specifications, called NORSOK standards, that are freely available on the Internet. The effort started in 1993 to make drilling on the Norwegian Continental Shelf more cost-competitive by replacing company specifications and detailed regulatory requirements to cut costs by up to 50% and project completion time by up to 25%. The standardization has saved money by shortening project schedules, eliminating redundant engineering, and reducing both capital and operating costs.[89] NORSOK had a goal to replace NORSOK standards with international standards as soon as possible. This effort has resulted in fewer company specifications, and the Norwegian Petroleum Directorate (NPD) has reduced the volume of its regulations from over 1,200 pages to just 300 pages through the use of referenced standards, including both international and global standards such as API, IEC, ISO, and NACE.

Achieving Continuous Cost Reduction

Continuous cost reduction can drive OpEx costs down through initiatives such as flexible demand planning, proper management of erratic inventory consumption, consignment and vendor-managed inventory programs, logistics outsourcing (3PLs and 4PLs), and lean distribution, including just-in-time. For information on continuous improvement and continuous cost reduction, readers are advised to consult the extensive body of resources in the field of Lean Manufacturing, Lean Distribution, Quality Management, and the like, as expanding on the scope of these programs would exceed the page count allowed in this book.

Stochastic Inventory Management

For basic inventory management requirements, oil, gas, and power inventory management uses the same tools as other industries—demand forecasting methods, MRP and reordering or replenishment methods, stockkeeping rules, etc. Inventory serves the basic purpose of satisfying demand during the order lead time (this is called cycle stock), which is normally constant across multiple cycles of production and ordering.

Flexible inventory planning mechanisms allow a flexible response to changes in demand and outlook. Flexible inventory management methods include, through dynamic inventory management, monitoring downstream demand and increasing the frequency of demand.[90]

- Managing each class and volatility combination uniquely and dynamically. If fast-moving inventory is Class A and slow-moving inventory is Class B, and volatile demand items are Type 1 and stable demand items are Type 2, then for A1 inventory replenish on a JIT basis and engage in financial hedging, and for B1 inventory add an extra layer of safety stock.
- Tracking customers' customers' demand rather than just immediate customers' demand. The added "anticipation lead time" increases the chances of having adequate time for replenishment or reducing inventory rapidly in response to a sharp drop in demand.
- Increasing the frequency of demand (production, sales, and inventory) planning so that fewer items are unavailable due to being caught between replenishment cycles.

Independent demand items (spares are independent demand items because they are drawn directly according to an end-user requirement rather than being a component in another item that is being manufactured) are most frequently replenished using reorder points (while dependent demand items are replenished using bill of material (BOM) explosion and offsetting techniques in MRP systems). The reorder point is equal to:

Demand During Lead Time + Safety Stock

where

Demand During Lead Time = # Weeks x Units/Week

and

Safety Stock = Standard Deviation of Demand times a Safety Factor

Reorder quantities are often determined by an Economic Order Quantity (EOQ) calculation.[VI]

The unique challenge that the oil, gas, and power industries face—as do all process industries, from paper milling to aluminum, glass, and cement production—is that some items, particularly capital spares, have an irregular demand pattern that has traditionally been hard to forecast, and a single out-of-stock incident of even a small item could shut production down. Intermittent demand makes inventory planning difficult and forecasts unreliable. As a result, materials managers often keep a buffer of just-in-case inventory, which inevitably doesn't prevent all stock-outs. Sooner or later, financial controllers blame inventory managers for what looks like excess inventory in some items and simultaneous high-visibility stock-outs—surely somebody must not be doing their job!

Spares are different from other inventory problems in five ways, adapted here to oil, gas, and power applications from an insightful book by Philip Slater:[91]

[VI] Other replenishment methods include min-max, total cost, and grid search.

- They have unpredictable demand that is stochastic and intermittent. Because of this, the replenishment rule is usually “lot-for-lot” or “one-for-one” due to the high value and customized applications of each unit.
- The target stock level is often zero (order on demand) or one (one-for-one reorder), which increases the chance of a stock-out considerably compared to a high-volume, small-value item which is stocked by the hundreds. Quite frequently, supplier lot sizes are larger than EOQ order quantities, so the reorder quantities are larger than the reorder point, which means that there is “overstock” as soon as each replenishment order is received.
- MRP systems, if set to replenish according to the classic algorithms, draw the wrong conclusion from an item consumed at irregular frequencies. They assign it volatility and thereby replenish more than required. Over time the replenishment ends up being driven more by exaggerated restocking patterns than by actual consumption. Spares also differ from typical inventory management paradigms because the value of replacement parts is high and also widely variable (inventory carrying costs for capital spares make up 25–35% of total inventory value, according to Slater), and because stock-out costs can be disproportionate to the cost of acquiring or stocking the spare.
- Stock-out costs are hard to calculate and are disputable if they are calculated. If a one-for-one replenishment of a high torque drive (HTD) belt for a draw-works on an oil rig needs to be replaced, and the rig will be shut down if the belt is not available, then the stock-out cost can be hundreds of thousands of dollars per day. However, most managers will not use such numbers in the computerized maintenance management system (CMMS) as the basis for inventory calculations, since the calculation is likely to conclude that the reorder quantity level should be absurdly high.
- Maintenance and life cycle history records may need to be maintained on each part moving through inventory, which can be time-consuming and detail oriented, and can require system integration between a separate records management system and the maintenance management information system. In the aircraft industry, parts are not allowed to be used if the complete history of the part since its manufacture cannot be produced, for fear of counterfeit or other safety-related problems that may have occurred during periods that are unaccounted for.

The best way to determine spares requirements is by statistical failure analysis on the equipment itself rather than by looking at inventory consumption patterns. A number of specialist software applications exist to apply probability to historical reliability and failure data, and sometimes to location specific inventory storerooms, to ascertain how many spares should be purchased and where they should be placed to best respond to potential requirements. In addition, predictive maintenance and artificial intelligence are embedding the statistics and reliability analysis principles in algorithms in order to provide earlier visibility to equipment repair requirements.

Vendor Managed Inventory

As Slater aptly says, there are only two ways to reduce inventory—one is to take items out of inventory, and the other is to put less into inventory. Vendor managed inventory is one way to take items out of inventory, by requiring a supplier to be responsible for them.[92]

For discrete product delivery into energy supply chains, traditional models of consignment and vendor managed inventory generally work as long as there are no hazardous goods, chemicals that change characteristics with pressure and temperature, or other peculiarities; consignment (paying the vendor when the item is used rather than when it is purchased) and vendor managed inventory (having the vendor determine the stocking level and holding them accountable for the fill rate) work the same as in any other industry.

One example of VMI is New Pig's Flawless Logistics and Optimized Warehousing system to reduce lead times and inventory holding costs. The program includes warehouse management, transportation network optimization, and customer relationship management (CRM) software to improve customer service and speed the resolution of any complaints. By implementing the new system, New Pig reduced response time to initial queries and complaints to less than 24 hours, from as much as three days, and increased on-time delivery rates from 95% to 99%.[93]

Another example is Baker Hughes' VMI system for oil & gas production chemicals at the well-site, which eliminates the need for operators to set feed rates, check tank levels, and replenish chemicals. Chemicals managed under the program include corrosion inhibitors, sweetening agents (to remove H_2S), biocides, defoamers, clarifiers, and scale inhibitors. Baker's wireless tank monitoring network (SENTRYNET™) determines the right feed rate for a given chemical based on the composition and volume of the production stream. It also keeps track of the amount of chemicals in a tank and sends an alert to Baker Hughes when it needs to be refilled or when the rate of chemical addition differs from the preprogrammed level (indicating a leak or a plug). Baker then dispatches a truck to refill the tank or fix the equipment. This often replaces a manual monitoring system on the part of the operator, which requires a technician to follow a physical inspection route every few days to check tank levels and replenish them if necessary. Operators have complete visibility to the process, as they have access to the same data stream Baker receives, including tank levels, fill dates, and repair logs. The system also allows Baker to bundle the requirements of several operators in an area and optimize efficiency by serving the entire area instead of just one customer. Baker uses the system to send automatic alerts to its own chemical suppliers, according to preset reorder points, alerting them to incoming orders when the tank at the well-site is depleted, instead of waiting until the chemicals ship from Baker's own facility.

Nalco monitors customer inventory levels remotely, delivers new chemical drums to the customer's location, and manages spent drums. The system reduces inventory for the operator by offloading responsibility to Nalco and reduces the likelihood of spills or container damage due to Nalco's greater experience with its own equipment.[94]

JIT

As mentioned above, the second way to remove inventory is to put fewer items in. One of the ways to do this is to replenish only what has been consumed, which is the essence of just-in-time inventory management.

Just-in-time stems from lean thinking at Toyota in the 1950s. Taiichi Ohno, VP Manufacturing, and Shigeo Shingo, head of industrial engineering and factory improvement training, developed the Toyota Production System, emphasizing the minimization of all waste and "doing it right the first time." Important elements of the TPS are: the Kanban production system, uniform plant loading, reduced setup times, and workforce involvement.

Lean is often described as a philosophy, or even a religion. One Lean consultant described Lean as a set of principles that includes putting the customer first, having an end-to-end total value stream perspective, focusing on consistent flow, and eliminating waste.[95] The breadth of scope of Lean makes it both very important (the reason it has had such a profound impact on business) and potentially enigmatic (the sheer number of people involved in implementing Lean leads to a wide series of interpretations that collectively seem to defy boundaries).

Lean can be defined in many ways (for a full background on Lean the reader is directed to a text more fully dedicated to the subject), and the purpose here is just to explain how JIT fits into energy supply chain management. The following tools have yielded exceptional results in waste elimination programs: the "seven types of waste," the "5S" approach to workplace organization, total quality management (TQM), Total Productive Maintenance (TPM), cellular manufacturing, and diagnostic tools such as The 5 Whys, the Plan-Do-Check-Act (PDCA) cycle, value stream mapping, work sampling, root cause analysis, and throughput analysis.

Lean implementations have realized tremendous benefits. One study of 80 plants in Europe found they had reduced inventory by about 50% and throughput time by 50–70%. They cut setup times by similar proportions without major investment in plant and equipment, and increased productivity by 20–50%. Moreover, the payback time for the investment in JIT averaged less than nine months.[96] Chevron extended its internal Lean Six Sigma techniques to its suppliers.

For example, it streamlined the truck routing for KS Industries (a piping installer), which affected more than 200 employees, and standardized the vehicle fleet for Braun Electric Co., which installed Chevron's electrical systems for oilfield pumps. The initiative reduced average drilling time, reduced workover rig cycle time, and reduced the lead time for engineering drawings.[97]

Since the benefits almost always require the joint participation of suppliers and customers, it is common to gain-share, or share the benefits with partners. Some companies split the benefits 50/50, and others have designed more asymmetric approaches. One asymmetric approach allocates the benefits to the buyer in early periods, then fixes a minimum expected amount above which the savings will be shared with the supplier.[98]

Inventory is one of the "seven deadly wastes" identified in Lean thinking (the others are overproduction, waiting time, transportation of any kind, processing downtime, motion, and waste from product defects, which includes scrap, rework, and recalls. Just-in-time (JIT) is one of six core methods of Lean (the other five are Kaizen, 5S, TPM, and Cellular Manufacturing).

Just-in-time (JIT) production systems rely on the use of physical inventory control cues (or kanban) to signal the need to move or produce new raw materials or components from the previous process. It often requires suppliers to deliver components using JIT. The buyer signals its suppliers, using computers or delivery of empty containers, to supply more of a particular component when they are needed. This results in significant reduction in waste associated with unnecessary inventory, work-in-process (WIP), and overproduction. Removing buffer inventory also exposes problems so they may be addressed. As more and more slack is removed from the system, all process inefficiencies become apparent, and are removed through continuous improvement. If applied repetitively, this results in no WIP inventory, short lead times, short order cycle times, and small lots (theoretically, a one-piece flow). The difference between traditional "push" supply chains and "just-in-time" supply chains is illustrated in Figure 35.

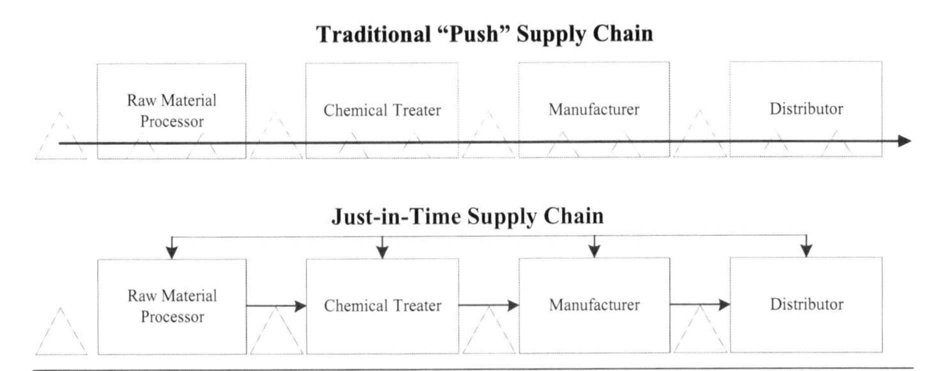

Figure 35. Traditional "Push" vs. Just-in-Time Paradigms
(*Source:* Boston Strategies International)

Transportation and Warehousing Optimization

Transportation routing and scheduling can offer a relatively quick way to reduce operating costs, especially if the traffic or trade management functions are currently handled manually, or if there is variation in service performance either internally or externally. Variations of logistics optimization may include:

- Automation—transportation management systems (TMS) reduce the amount of manual rate checking and booking tasks.
- Mode selection—if expedited air shipments are shifted to ground transport, or change from one mode to another, this can optimize cost and service.
- Cross-docking—a subset of shipments can sometimes be transloaded at an intermediate point without being stored, as part of a rapid-delivery hub-and-spoke network.
- Distribution Center (DC) bypass—moving freight directly to its end destination without storing it at a distribution or even a mixing center can reduce cost and transit time.
- Equipment pooling—sharing fleet assets, usually through a third-party equipment pooling company, can reduce per-mile cost. Pooling is often used for railcars and intermodal chassis.

There are numerous success stories within each of these categories, for example:

- The Middle East division of a major oilfield services company reduced customs clearance time, increased transit time reliability and visibility to cargo status, provided customers multiple cost vs. delivery options, and reduced shipping costs by switching its distribution center hub location and transport mode for imports from Europe and the United States.
- A major international oil company proved that mode selection can be applied creatively. It developed a high-volume crude oil rail shuttle to transport crude oil from an important shale play to refineries 1,500 miles away, thereby increasing the yield from the region, which was constrained by a lack of outbound transportation capacity, and simultaneously increasing the efficiency and operating margins of the refinery by increasing its utilization rate.
- A North American steel company shifted tubular volume from truck to rail, saving nearly $1m in annual freight costs for shipments to Canadian oilfields.
- A pipe distributor saved $300k per year by transferring its tubular freight volume from a truck carrier to an industrial freight broker. The truck carrier charged round trip rates since it ran empty on backhauls, but the freight broker had enough volume to fill backhauls and only charged one-way.[99]

Sourcing Trade-offs

Category Management

The first step in bid slate development is implementing a robust category management process and organization. Category management helps to manage supply risk by establishing formal relationship management plans for categories of goods and services that are of strategic importance to operations.

Category management is a process that defines the desired end state of procurement activities for each family of externally purchased goods or services. It integrates the processes of supply risk management, supply market analysis, and market intelligence to choose how to maximize value from suppliers and supply market conditions. Its scope includes the selection of tendering and Request for Information or Proposal or Quote (RFI, RFP, and RFQ collectively called RFx) processes, bid slate development, bid comparison methods (e.g., total cost of acquisition evaluation methods), risk profile of contracts (e.g., Lump Sum Turnkey, or LSTK vs. cost-plus), the manufacturing strategy (ETO and MTO [engineered-to-order and make-to-order]), and which value-added services should be included in the contract. Through these definitions, management can prepare a strategic negotiating position that helps to mutually achieve the goals.

Some steps in establishing a category management program include:

- Analyzing external expenditure
- Establishing procurement policies (including approval levels, emergency purchases, and sole/single source procurement)
- Ensuring a well-understood and communicated procurement code of ethics
- Dedicating resources to market analysis
- Providing cross-functional career paths that develop people for category management roles and offer them an opportunity to apply the knowledge after they move on to another place in in the organization

Appendix 2 contains a list of commonly sourced commodities and categories that may help in establishing category management families, or groups of externally purchased goods and services that may be addressed within a common set of goals, supply chain strategies, and sourcing tactics.

Category Strategies

Each category should have a documented sourcing strategy, starting with an insourcing versus outsourcing decision, and, for externally purchased services, a determination on whether or not to use a Lead Provider who will manage sub-providers.

Logistics is an example of cases where a Lead Provider is frequently used due to the vast number of transportation carriers which are typically widely dispersed by mode of transport and region or country. When companies hire third party logistics companies (3PLs), they often gain capabilities, flexibility, and responsiveness, and they should not base the outsourcing decision on cost savings alone.

3PLs can save money on operating costs, but the savings is a decreasing part of the equation. 3PLs can often purchase packaging, supplies, fuel, and equipment at lower unit prices due to their economies of scale. They sometimes, but not always, pass these lower costs on to their customers in the form of lower prices, which can reduce operating costs. Historically, they have hired union employees into non-union environments, resulting in an immediate cost savings, although sometimes with some encumbrances as employment obligations lapse or as the union employees quit or retire. Unionization rates have declined from 16% to 11% in the US since 1990 though, dampening this motivation to outsource to 3PLs; and unionization is even less of an issue in developing economies such as those of Asia.[100]

While operating cost savings is often not the driving force, outsourcing logistics can be financially advantageous insofar as it transfers assets to the 3PL. Offloading tangible assets such as warehouse facilities and vehicles, and intangible assets such as information systems, increases return on assets (ROA). Higher ROAs make companies more attractive to lenders and shareholders, making it easier to obtain loans and float shares. Furthermore, a one-time improvement through outsourcing can result in an uptick in stock price.

The main benefit of hiring a 3PL is the capability to serve customers better. Third party logistics companies can provide the following strategic advantages over doing the job in-house:

- 3PLs often have a more global collection and distribution network than many shippers. Their global scale can allow shippers to extend their sourcing and distribution beyond the scope of their existing networks. The sourcing advantage can provide greater cost benefits than shippers may have initially hoped to capture by outsourcing logistics. The distribution benefits may allow them to position for sales in emerging geographies that would have been impossible to reach.
- 3PLs invest heavily in IT platforms that provide measurably higher service levels and service reliability compared to in-house logistics operations. Best-of-breed transportation management systems (TMS) and warehouse management systems (WMS) provide the ability to handle complex pick/pack operations more readily than in-house solutions can.
- 3PLs invest in emerging technologies like RFID sooner than their customers do because they can amortize the investment across multiple customers. This results in earlier adoption of important IT capabilities than would have occurred in-house.

- They also offer value-added services such as custom packaging operations. Examples include clamshell packaging, security packaging, kitting, customized packaging by customer, or personalized packaging by consumer (for example, through individualized insertions).

The logistics outsourcing and selection process should be made in alignment with the overall business strategy. Used correctly, logistics outsourcing can be a key element of effective supply chain management. Therefore, the decision should be made at the senior executive level, and decision-makers should integrate the decision with the company's supply chain management strategy.

When logistics outsourcing agreements fail, it is often due to poor communication and unclear or misaligned goals. Outsourcing arrangements should be based on broad solutions that affect the bottom line in order to avoid unproductive behaviors such as (the possible failure points listed below are an adapted subset based on a presentation by Steve Geary of the University of Tennessee):[101]

- Overemphasis on cost at the expense of service
- Misrepresentation of a partnering program to achieve cost reduction unilaterally
- Expecting or requiring the 3PL to do things the way they were done in-house
- Pay per unit of activity, which motivates activity rather than cost savings
- Expecting the 3PL to succeed without any input or collaboration from the buyer

Petroleum Development Oman (PDO) successfully outsourced its entire logistics function (4PL) to DHL through a 4PL contract with Bahwan Exel/DHL, a consortium of Bahwan Cybertek and DHL Supply Chain.[102] The initial years of the contract were focused on rig moves, but the group eventually took over complete oversight of transport management processes and systems at PDO.[103] The association led to 10–20% improvements in trip planning and journey management, and simplified operations through a centralized organization structure. In addition to cost savings, the partnership brought technological benefits, for example web-enabled booking requests, cargo tracking and tracing, more accurate paperwork, and more meaningful management reporting.[104]

Determining the Optimal Number of Suppliers

When procuring engineered products, having fewer suppliers is generally more effective than having more suppliers. Single or dual sourcing can raise concerns about the competitiveness of the marketplace and of the sourcing event; however, BSI analysis of industry concentration[VII] and prices (see Figure 36 below, which

[VII] As determined by the HHI index. The Supply Market Concentration Index is the sum of the squares of the market shares of the suppliers in each Category (based on the Herfindahl-Hirschman Index methodology).

analyzes data from 2005–2011) shows that there is no correlation between supplier concentration and price inflation, and empirical research shows that price DOES NOT decline with more than four bidders,[105,106] as long as two or more participate in the bid. Single sourcing is a viable option for services that are time-critical, if supply market capacity and in-house capability are limited.

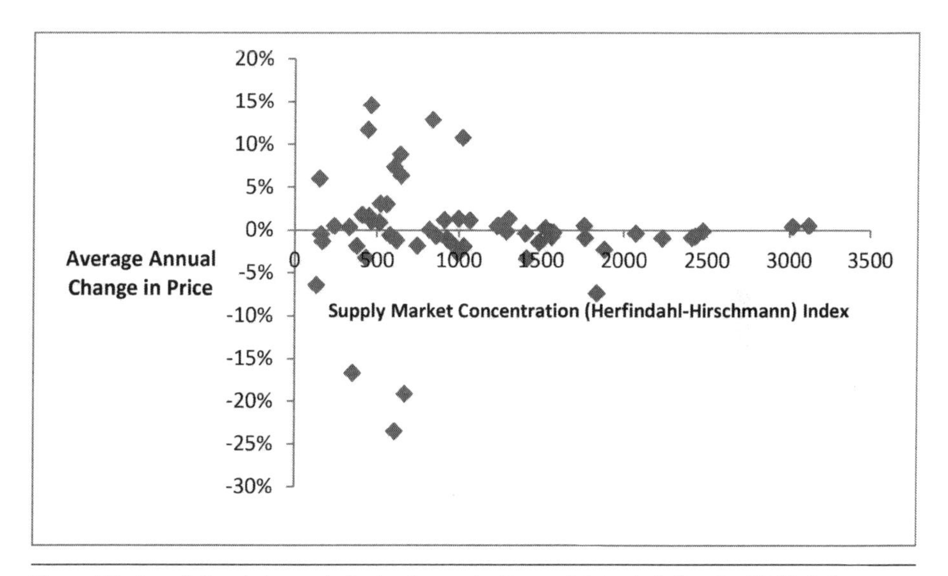

Figure 36. Correlation between Industry Concentration and Price Inflation for 58 Oil & Gas Supply Markets
(*Source:* Boston Strategies International)

Unless there is price signaling or a breakdown in the bidding process, suppliers will bid competitively even if there are only a few bidders. So long as there are at least four suppliers in the market, the market is typically a competitive one. This interpretation is consistent with legal precedent for determining the level of concentration that constitutes monopolistic market conditions where a dominant supplier may exert an influence over prices. Although the law is complex, varied, and subject to interpretation, various interpretations of unacceptable concentration have ranged between about 25% and 35% market share for an individual supplier,[107] which would, simplistically speaking, equate to three or four suppliers in a market.

In fact, suppliers may even underbid (below their cost) in order to win projects. Excessive supplier competition has driven some unsophisticated suppliers, who don't know their actual cost to serve, out of business, as they overreact to extremely competitive situations such as time-based reverse auctions. That being said, the savings from competitive events are extraordinary. On average, companies have saved 6–8% on their largest categories of expenditure.[108] Best-in-class companies save 10%.[109] For individual companies and specific categories, there are many cases of savings in the 30–75% range.

Even if having few suppliers results in higher margins paid to those suppliers, sophisticated buyers will get more than the differential out of the relationship in the form of partnering benefits such as collaborative planning, inventory consolidation, and joint process improvements. Anti-competitive behavior can raise prices by 3–4% in concentrated markets, but efficiency gains that come from lower transactions costs, economies of scale, and better customer service typically outweigh that, resulting in a net 1% price decrease.[110] E-procurement alone, which is implemented more effectively in close collaboration with partner suppliers, can save both buyer and supplier in labor-intensive paperwork and reconciliation of discordant data elements. For example, Qatar Petroleum, BAPCO, and KNPC use an external provider for spend analysis, strategic sourcing, and electronic RFx.[111]

Single and dual sourcing are both viable options, depending on the situation. For a large expenditure on a relatively simple product or service, given no time or capacity constraints, normal market concentration, and many competitive options, operators would select a multi-source strategy (three or more suppliers). However, if the market is capacity constrained and there is little supplier risk, dual sourcing is not unreasonable. Furthermore, if a strategic target supplier is not weak in critical areas (financial vulnerability, R&D capability, etc.), if switching costs are not too high, and if the supplier's price quote is consistent with benchmarks, then single sourcing is a reasonable option as well. Figure 37 contains a flowchart to help know whether to single source or dual source.

Highly concentrated markets—those above 1800 on the HHI index—place a special responsibility on buyers to develop the market and to ensure price-competitiveness. Overly concentrated supply markets raise the possibility of certain types of illegality, which could include vertical or horizontal price fixing, price signaling, predatory pricing, price discrimination, promotional discrimination, and geographic discrimination (carving up markets).[112]

Tools for buyers to avoid overpaying in highly concentrated markets with few suppliers include:

- Dual sourcing to maintain a competitive option and avoid high switching costs. Dual sourcing is a classic method of maintaining a competitive option and avoid high switching costs. In particular, developing smaller or less qualified suppliers can provide options for future development on technically or regionally specialized projects or on a divisional basis.
- Should-cost modeling. "Should-cost" is a process of determining what a product should cost based upon its component material costs, manufacturing costs, production overheads, and reasonable profit margins. A detailed example of "should-cost" analysis is included in the Upstream chapter.
- New supplier development. Buyers should continuously develop new and alternative suppliers to avoid high potential switching

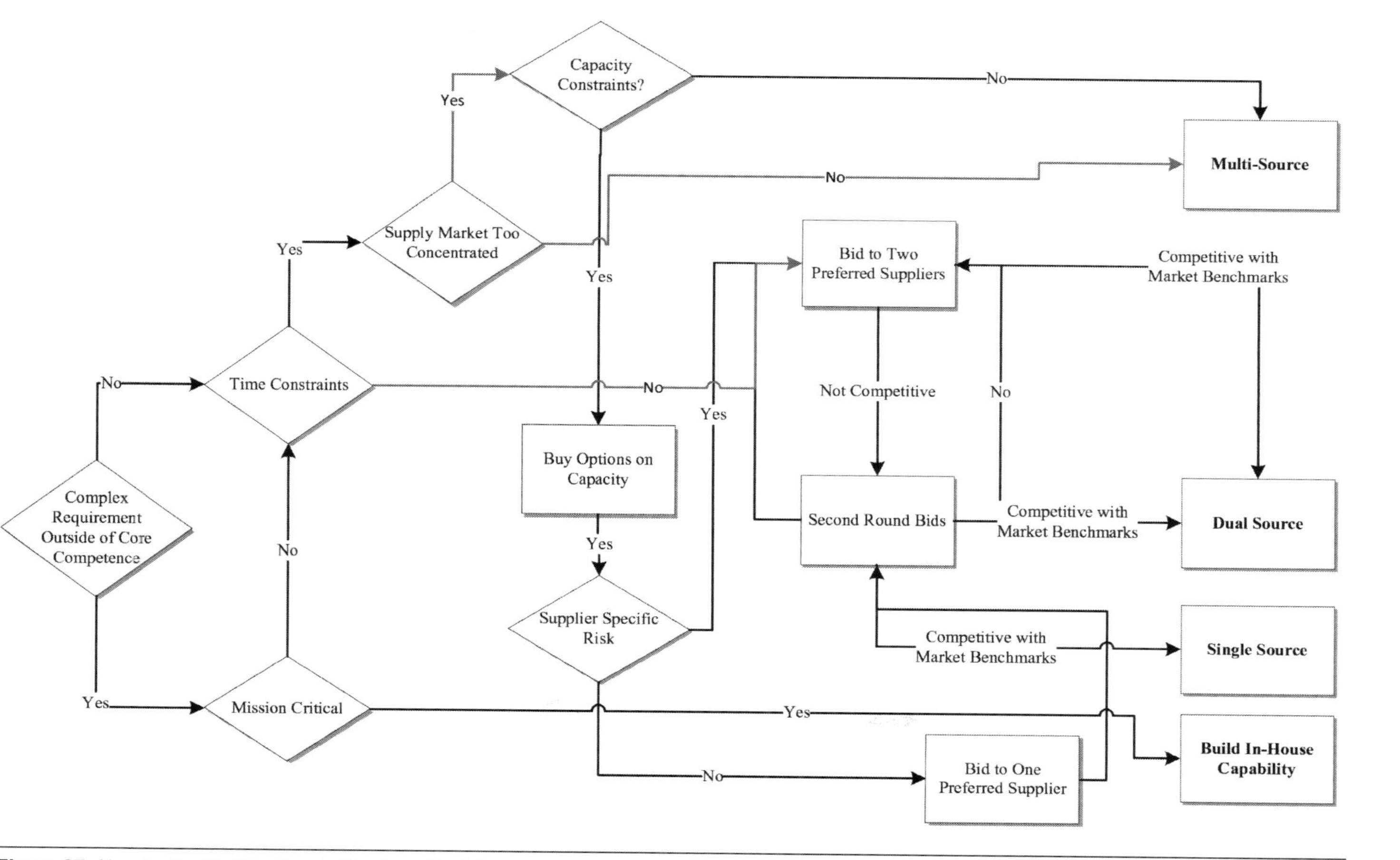

Figure 37. How to Decide Whether to Single or Dual Source
(*Source:* Boston Strategies International)

costs. Supplier development is a long-term exercise, and often yields unexpected innovation and cost reduction opportunities from new potential suppliers. Supplier development activities should first include consideration of existing suppliers of related products or services. Global sources in strategic or target markets can also serve as good benchmarks and potential complements to or substitutes for entrenched suppliers.

- Legal recourse. If evidence or suspicion of price-fixing surfaces, buyers can pursue legal recourse for suspected price-fixing or related crimes. Large expenditure categories have historically seen their share of wrongful activity. For example, in 2007, Siemens, Schneider Electric, Areva, Alstom, Mitsubishi Electric, Hitachi, ABB, and four others were convicted by the European Commission (EC) of fixing prices for electrical switchgear for 16 years. The EC imposed the second largest fine in its history to date on the firms—a total of $977m—although ABB escaped the fines because it reported the offense to the EC.[113] Also, in 2010, the Japanese Fair Trade Commission fined ten companies, including Sumitomo, Furukawa, and Fujikura a total of $180m for setting prices of fiber optic cables and components that they had sold to Japanese telecommunications firm Nippon Telegraph and Telephone.[114] In a preemptive move, the US Department of Justice imposed conditions on the merger of Baker Hughes' and BJ Services' operations in North America. Before completely integrating, Baker was compelled to sell two stimulation vessels, as the merger would have combined two of only four companies providing offshore well stimulation services in the Gulf of Mexico.[115]

If an empirical basis is preferred to decide how many suppliers should be approved or engaged, a decision analysis tool can be used. For extremely large expenditure decisions, BSI has helped clients determine the optimal number of suppliers based on an analysis of the percent probability of occurrence and the monetary severity (and the probability distribution of each of those) of five undesirable outcomes at varying levels of number of suppliers. The five outcomes are: 1) cost of non-standardization, 2) price gouging, 3) output unreliability, 4) missed delivery dates, and 5) project management error. BSI's optimization engine computes the optimal number of suppliers, given those parameters.

Prequalifying Suppliers

Supplier qualification is an especially critical job in oil, gas, and power generation. Projects are time-driven, and the non-performance of even one supplier may delay a whole major project, for example a refinery coming onstream—at very high opportunity cost. Also, safety and environmental risks are high, so anything less than 100% supplier quality could be life-threatening. One mechanical failure could cause a rig blowout (e.g., Deepwater Horizon), and one safety violation could cause an explosion (e.g., lighting a match near flammable vapors). Therefore, qualification is typically a multistage and rigorous process involving the following steps, as shown in Figure 38:

- Pre-screening
- Prequalification as an approved vendor
- Qualification for specific projects and applications
- Technical evaluation
- Approved vendor non-binding contract
- Selection for a project or application
- New supplier/project registration and onboarding

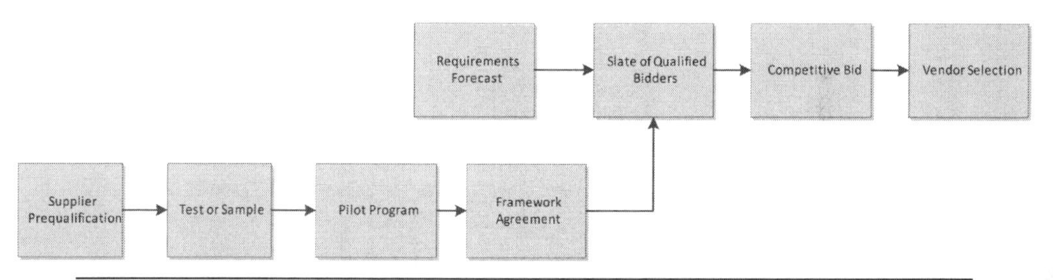

Figure 38. Simplified Representation of a Supplier Qualification Process
(*Source:* Boston Strategies International)

Supplier key success factors

Pre-screening. Pre-screening typically involves a review of the company's technical expertise, experience, and financial strength. At this stage, the financial assessment often focuses on total revenue, profitability, and solvency, sometimes with the use of financial ratios like the quick ratio. A supply market intelligence program can also help pre-screen vendors in a systematic way, for example targeted at certain geographies and product or solution offerings. Since some large investment decisions are based on the availability of suppliers to maintain equipment for 20 years or more, procurement staff should ask suppliers questions that help assess their performance on the key success factors that determine long-term viability. For example, key success factors that define the success of major oilfield service and construction firms might include: 1) demonstrated technological innovation; 2) presence and history in the major oil & gas markets; and 3) international network (especially in the fast-growing areas such as the Middle East and Asia).

Lead times can also serve as a barometer of how much available capacity suppliers have, and so in this respect can serve as a pre-screening criterion if available capacity is a consideration for pre-selection. Figure 39 shows the range of lead times from different suppliers for the same product at the same time, which is indicative of the differences in their capacity utilization rates.

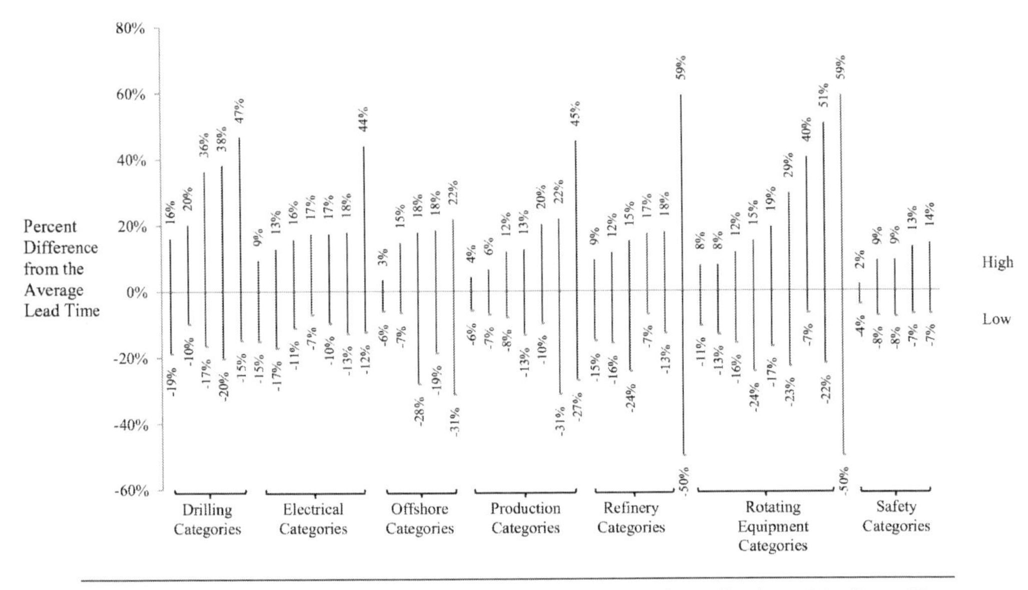

Figure 39. Range of Lead Times from Different Suppliers for the Same Product at the Same Time (*Source:* Boston Strategies International)

Prequalification. Granting 'approved vendor' status is a multistage process in itself. A sales meeting is often followed by a request for a more detailed package, including information that may be requested at various points in time, such as:

- Basic company profile
- Corporate size/financials
- Organization
- Proposed facilities
- Order management procedures
- Order management systems
- Shipping and receiving procedures
- Administrative controls
- Performance along key performance indicators (KPIs)
- Technology
- Reporting/visibility tools
- Customer reference checks
- Pricing

Technical evaluation. Product samples, lab testing, an onsite quality visit, and a pilot project may be involved in evaluating a supplier's technical capabilities and competence. Sometimes the visit is conducted by an independent certifying body or a consulting firm with an approved auditing methodology. Additional factors considered at this stage may include technical capability and fit, and operating cost savings potential value addition.

Approved vendor non-binding contract. If the vendor is approved on technical grounds, a commercial framework agreement is often signed; this locks in a large number of terms and conditions so that future bids can proceed with less (or no) negotiation on these aspects. Basic commercial terms and conditions covered in this contract would include, for example:

- Alteration of terms
- Assignment
- Audit rights
- Changes
- Confidentiality
- Delivery: schedule; delays
- Force majeure
- Governing law
- Independent contractor usage and subcontracting
- Inspection
- Insurance and indemnification
- Intellectual property: design, data, creative work, patents and inventions
- Liens
- Payment, including discounts and refunds
- Price structure
- Safety
- Scope
- Service and/or spare parts if applicable
- Severability
- Substitution
- Taxes
- Term
- Termination (for cause and for convenience)
- Warranty

Selection for a project or application. For specific requirements, the procurement department and the user (Operations, Engineering, etc.) usually ask a subset of the approved vendors to bid on the required services.

New supplier/project registration and onboarding. Once a supplier is technically approved, it becomes eligible to participate in tenders, and if it wins a bid, it may be necessary to further qualify the exact products to make sure it delivers the desired results. For a product or equipment suppliers, the steps necessary for this evaluation may include:

- Determine which items need samples and get samples
- Set up testing program and timetable
- Test product and approve new sources
- Get MSDS sheets
- Register and update vendor information in appropriate ordering system
- Ensure item numbers are updated and mutually exclusive
- Create blanket orders for new suppliers
- Establish controls against maverick spending
- Communicate any changes of suppliers to users
- Communicate sourcing decisions to plants when all else is finalized
- Establish delivery frequencies and time windows, by plant
- Check/update stocking locations, space, and inventory levels per delivery frequencies, by plant

Global sourcing

Despite nationalistic trends and the social isolation forced by COVID-19 having dampened global trade in 2020, the increasingly global nature of energy operations and procurement requires energy companies to expand their manufacturing and sales networks worldwide, and energy consumption is shifting eastward. India is emerging as a substantial oil & gas play due to large-scale drilling projects at state-run oil producers Oil and Natural Gas Company (ONGC), Indian Oil Company (IOC), and Gas Authority of India, Ltd. (GAIL). Suppliers have aggressively migrated to growth areas, notably China, India, and Malaysia.

The regional basis of oil demand has shifted in recent years (2008–2012), causing patterns of trade and transportation to move from OECD toward non-OECD countries and from West to East. Schlumberger now has a presence in more than 120 countries, many with field staff providing installation, troubleshooting, preventive maintenance, and repair.[116] Baker Hughes recently opened up a new multimodal facility in Luanda, Angola,[117] and Weatherford opened a facility in Cairo with the goal of becoming a leading oil & gas service provider in North Africa.[118]

As they globalize, the cost of shipping internationally is forcing suppliers to reevaluate their supply chains, sometimes replacing their raw material sources and reconfiguring their intermediate processing activities and locations. While this sometimes presents opportunities for local suppliers, it can also threaten incumbents. In Saudi Arabia, local manufacturers like Saudi Cement Company and Saudi Yamama Cement have traditionally dominated the cement supply market. However, global leaders such as Lafarge and Cemex have set up plants there.

While cheap labor was frequently the initial justification for global sourcing, the quality and volume coming from the so-called "low-cost countries" is in many cases just as good or better than what is achieved at home, and the costs have risen, narrowing the sourcing cost advantage.

For a long time, China's labor rates were about 4% of those found in Belgium, Germany, and the US, which are the most expensive nations, and lower than those of Mexico and India. Even in 2008, a study found that the average cost of Chinese labor was as low as 10% of the cost of US labor.[119] Even after accounting for the fact that about 20% of the product cost was dedicated to transport,[120] Western manufacturers recently enjoyed as much as a 70% cost advantage by sourcing from China. Second, lenient regulatory controls made low-cost countries attractive sourcing platforms by lowering their costs relative to their Western counterparts. Third, China put unprecedented investment into its infrastructure, thereby addressing some of the conditions that had previously made it unattractive from a logistical point of view. Finally, China joined the World Trade Organization, established accepted legal frameworks that moved toward protecting intellectual property, and trained workers.

Meanwhile, the price difference for sourcing from China has risen toward Western levels, and due to tightening environmental legislation in China, and the COVID-19 pandemic, many manufacturers have shifted at least some of their production from third countries or back to their home countries.

- Labor costs have increased, nearly doubling in urban areas like Shanghai in the 2005–2010 timeframe alone, which eliminated about half of the cost advantage of sourcing from China by 2011. Today the country's labor costs are roughly on par with those of Mexico.[121]
- Overhead costs have increased as the Chinese government imposed environmental regulations.[122]
- Exchange rates have historically been a source of concern for companies sourcing from China, and of debate among politicians and economists. The yuan-to-US dollar rate is set by the government.
- In addition, Chinese authorities have repeatedly imposed quotas and raised export taxes on certain raw materials.

Most of the current market abnormalities and concerns listed above no longer relate to cheap labor costs pure and simple. Therefore, most buyers view China as a complex part of their global supply chains rather than as a place to go for low-cost, low-quality products.

As a result of the global spread of technology, quality improvements, the widespread adoption of international quality, safety and environmental standards, and the erosion of wage differentials, "global sourcing" is not the same game as it used to be. Most buyers don't discriminate between countries in search of cheap labor costs anymore.

Even in the next tier of emerging economies such as Vietnam, Thailand, Hungary, and Turkey, quality and cost are expected, and business will not transact unless fully capable supply conditions exist as they would in the home country.

Managing the Tendering Process

Tendering is a key procurement process that involves five basic sub-processes:

- Work scoping, especially for services procurement
- Bid slate development. Early choices in bid slate development influence the cost of the project, not only by locking in decisions, but by signaling to suppliers the competitive dynamics of the ensuing process of bidding, negotiation, and contracting. Key decision points involve the determination of the number of suppliers and the shortlisting of certain ones.
- RFx process (RFI, RFP, RFQ), including e-procurement
- Bid analysis, including total life-cycle cost analysis
- Negotiation of strategic and non-strategic categories
- Contracting of high- and low-value purchases

In contrast to the apparent simplicity of the tendering process as it is practiced in other industries, complex projects require a variety of tendering options that can dramatically affect the resulting solution and price. For example:

- How many suppliers to interact with at various points in the tendering process (when and how many to shortlist)? Inviting too few suppliers to the bidding may result in a narrower solution set and a higher price.
- How many rounds of proposals to request (zero, one, or two), and how much design and development work to request or expect from suppliers prior to each milestone? Conducting too few rounds of bidding may result in an incomplete specification and a higher price.
- When to discuss and negotiate price (early, later, final)? Discussing price too early may compromise innovation and result in a suboptimal total cost or too high a price.
- When to sign a non-disclosure agreement? Failing to have a supplier sign non-disclosure agreement at the right time could result in giving away information to competitors, as that supplier bids on work for other companies.

E-business can simplify and speed both contracting and payment cycle times, and additionally may help to articulate and clarify versions of the same product. For example, it can specify a downhole tool that can operate at 200 meters depth as opposed to one that can operate at 300 meters depth—a difference that might not be apparent except through an RFP. It can also easily make distinctions between solution bundles (e.g., the difference between a piece of equipment delivered and installed vs. the same equipment delivered, installed, and calibrated). Many facets of purchasing and supply chain management can and should be electronic, for example:

- Vendor pre-registration
- Electronic tendering
- Electronic part and service cataloging
- Contract writing
- Contract administration and compliance monitoring

No matter how much the firm has made the tendering process electronic, process simplification can save time and reduce cost. For example, capital and operating purchasing can be done together, if coordinated, and multiple contract types can sometimes be reduced and simplified.

When to use auctions

There are six types of auctions: English, Dutch, Vickrey, Sealed Bid, Combinatorial, and Buyer-Driven. An English auction is the most frequent and usually the most familiar—the item goes to the highest bidder. In Dutch auctions, the price rises in fixed increments and the first to bid gets the available lot(s) at that price. In Vickrey auctions, the penultimate (next-to-last) bid wins, to avoid a "one-penny-more" pattern. Sealed bid auctions are the classic request for proposal method, with the caveat that there is no negotiation stage. Combinatorial auctions ask all bidders for prices on an array of comparative bundles, and then select the bundle(s) that minimize cost for the buyer. Buyer-driven auctions are essentially English auctions, except that the bidders don't have visibility to each other's bids.

The most common auction for business procurement is a reverse auction, in which the suppliers bid successively lower prices until the clock runs out. Reverse auctions have saved 15–30% for many companies across a range of items where all the specifications are clear and familiar to all the bidders. For example, the US military saved 30% compared to its pre-bid estimates for telecommunication cable assemblies in a reverse auction.[123] A major international oil company saved 16% on the purchase price for one category. Indian energy firm Oil and Natural Gas Corporation, LTD saved 12% auctioning its requirements for 100 km of medium voltage power cable.[124]

Reverse auctions yield above-average results in industries like telecommunications. However, even in the telecommunications industry fewer than 15% of companies use reverse auctions for procurement of materials and services.[125] Fewer than about 5% of all companies currently use auctions, and energy companies and power utilities use them less than most other industries due to the prevalence of high-spec, safety-critical, high-opportunity-cost applications that often require more attention to performance before cost. Despite many efforts to advance the state of the art, auctions do not work well for complex equipment and solutions.

Negotiations and multi-round bidding can greatly enhance the quality and customization of the solution and adapt it to the application, and this adaptation is often worth much more than can be saved on upfront acquisition cost in an auction.

How to tender products under development but not yet commercialized

The project schedule, number of qualified bidders, the degree of partnership required between owner and contractor, and the extent to which the project's final costs can be estimated determine when and how a buyer negotiates price.

Traditionally, major capital projects follow a well-established model. The owner releases information about the project, and potential bidders express their interest. The owner then qualifies the bidders and issues a request for proposal to this "short list." These suppliers then develop a proposal, including a firm price and commercial terms. The owner evaluates these bids, negotiates the final agreement with the leading bidder, and awards the contract.

For a project that requires a partnership between the owner and the contractor, the short list of bidders is typically narrowed down to two (instead of three or more) due to the need to disclose more information about the project, as well as the smaller pool of qualified bidders. Optimally, the project can be defined well enough that the bidders can estimate the total cost to execute it. When this is the case, the owner can follow a "full price" selection model. Under this model, the owner solicits a full proposal from both short-listed bidders and selects the one with the best value for money after negotiations.

Some projects must be negotiated before the technology is fully commercialized, but there is more than one supplier with a solution in the design phase. For example, three OEMs have subsea compressor models in the test phase, to compete with only one commercialized model from a fourth OEM. Projects currently in the planning phase that include a subsea compressor therefore conform to a "partial price" selection model, as the full cost of the project cannot be determined until the compressor design is finalized. Under this model, the owner selects a supplier based on bids for the parts of the project that can be estimated, as well as non-price criteria such as the test results of the designs under development. Owners must be careful using this model, as suppliers can underbid on the priced elements of the proposal, knowing that they can make up any difference on the elements without a price tag.

In some cases, there is a compelling reason to select one preferred supplier and forgo a competitive tender entirely. For example, there might be only one supplier with the necessary technology. To continue with the example of subsea compressors, until 2012 only MAN had a commercial model. When this is the case, the owner must proceed carefully, placing the burden on the supplier to demonstrate that it can execute the project effectively and postponing the actual contract award until a target price has been agreed.

Assuring Local Content Where Needed

Operators doing business in foreign countries often need to observe local content laws, as a result of an increasing trend toward " resource nationalism"—countries with mineral wealth are making a concerted effort to channel the wealth generated by the exports into sustainable domestic economic development. Resource nationalism is propelling a trend toward more local content, which requires developing suppliers that otherwise may not have been prequalified.

Countries such as Angola, Brazil, China, Venezuela, and others have specific laws requiring local content (see Figure 40). In many cases, the laws have been enacted since 2001. Various complaints in the World Trade Organization (WTO) have claimed that at least some of the local content regulations violate the countries' commitments to the WTO.[126] One ongoing debate is centered around the idea that China's local content regulations have illegally allowed it to spawn a domestic wind power industry.[127] Although complaints are still outstanding, several countries like Brazil and China have relaxed some of their rules, including calling them "voluntary" and reducing mandatory percentages of domestic content as a result of the WTO complaints and of declining oil & gas market conditions.

Local supply may be more expensive than conventional supply, especially when considering the fixed cost to onboard any new supplier, and the possible need for upfront training and investment if the available local suppliers do not meet existing quality standards. The challenges can be diverse and significant, for example:[128]

- Limited technical expertise and supply base due to a lack of professionally qualified labor. This can particularly apply to managerial talent for precision manufacturing and engineering for capital projects.
- Local supply market habits, customs, and norms, which may be oriented toward a historically different level of quality expectation.
- Insufficient IT infrastructure, possibly including poor phone and Internet access, inadequate IT infrastructure, possibly including restrictions on content or VOIP, weak Internet security, and outdated or incompatible systems or software.
- Trade restrictions
- Intellectual property loss due reverse engineering or unenforceable contracts.
- Late payment
- Antiquated logistics infrastructure. Developing countries have worse logistics infrastructure than developed ones, according to a survey-based index (the Logistics Performance Index) established and tracked by the World Bank. Aggregate scores for infrastructure, customs clearance, international shipments, general logistics competence, tracking & tracing, and timeliness of shipments are far worse in developing countries than in mature economies. For example, Nigeria has an LPI of 2.6 and Russia has an LPI of 2.6, compared to 3.9 and 4.1 in the United States and Germany, respectively.[129]

Country	Year First Enacted	Scope of Transactions Affected	Obligations Under the Laws	Assistance and Facilitation Available
Angola	2003	• Oil industry • Information technology Logistics activities • Education sector • Industrial hygiene	• Hire Angolan staff • Source Angolan products	• Business Support Center (CAE) provides vocational training to help IOCs meet employment requirements
Azerbaijan	2002	• Production sharing agreements (PSAs)	• Contractual agreements come from PSAs rather than legislation • Onshore operations are under joint venture • Offshore operations work under PSAs	• 2002 Enterprise Center supports local enterprises win contracts with IOCs
Brazil	2003	• Oil production equipment	• Starting in 1997, the ANP used bidder's commitment to local content as a condition for awarding rights to exploration and production. • 18% tax on imported oil production equipment (2003) • Monitoring and reporting local content (ANP Ordinance 180, 2003) • Gradually increasing local content requirements (could go as high as 95% by 2017) • Relaxations on local percentage of manufactured goods in 2018 and 2019[1 2]	• Large contracts may be broken into separate pieces to satisfy local content requirements
China	2002 marked the end of official local content laws	• Joint-venture with Chinese enterprise • Exception: No local content requirements for wind turbines (since 2010) • Per WTO in 2002, Chinese bureaucracy is attempting to purge local content legislation for energy regulation	• Secrecy (Chinese local content requirements are kept in "Neibu" directives) makes precise obligations unclear, inspiring wrath from WTO and EU • The existence of Neibu is not controversial; however, what it contains is. Besides Chinese government, no one is sure what the regulations entail, but they most likely have local content. China acknowledges the existence of Neibu, and so does the IMF. • Local content rules scrutinized by WTO in 2019, and China has reclassified some as voluntary.	• Most energy companies in China use local supplies and labor anyway because of lower cost
Equatorial New Guinea	2004 and 2014	• Production sharing agreements (PSAs) • Local supplies and labor	• Government right to 20% share in contracts with foreign operators • 2004 forum on local content recommended transfer of technologies to state companies and a new National Institute of Technology • 2006 hydrocarbon law puts small emphasis on developing local content requirements • Production sharing agreement terms emphasize using local supplies and labor • 13% minimum royalty on petroleum projects • Mandatory local content plan covering local workforce development, technology transfer, training and social infrastructure development[3]	• Marathon has started a craft training program to help meet local content requirements.
Nigeria	2003/ 2010	• All oil & gas activities	• Specified percentages of local content by activity and by year	• Several government arms to monitor and channel foreign company activity (e.g., Nigerian Content Development & Monitoring Board)
Venezuela	2001	• Exploration and production	• 2001 Organic Law of Hydrocarbons guarantees exploration and production rights to the state • Exploration and production must be carried out by state companies or joint ventures in which the state has a stake of more than 50%	• None (foreign companies must develop local content to do business)

Figure 40. Evolution and Status of Local Content Regulations in Six Countries
(*Source:* Boston Strategies International)

[1] "The Brazilian National Development Bank (BNDES) reduces local content requirements of machinery and equipment from 60 to 30 percent." US International Trade Administration. February 6, 2019. https://www.trade.gov/market-intelligence/brazils-bndes

[2] "Brazil's ANP approves revised local content rules, appeases shipbuilders." S&P Global Platts, April 12, 2018. https://www.spglobal.com/platts/en/market-insights/latest-news/oil/041218-brazils-anp-approves-revised-local-content-rules-appeases-shipbuilders

[3] Equatorial Guinea: How To Avoid Sanctions For Breach Of Local Content. Website post by Centurion Lawyers and Business Advisors. Accessed January 15, 2021. https://centurionlg.com/2019/10/31/equatorial-guinea-how-to-avoid-sanctions-for-breach-of-local-content/#:~:text=The%20local%20content%20regulations%20oblige,plan%20they%20themselves%20have%20prepared.

Combining the "hard" infrastructure risks (like logistics infrastructure) with the "soft" risks (like loss of intellectual property) highlights the fact that many countries that are rich in energy resources and/or are involved in major capital investment in the energy industry involve relatively higher supply chain risks than other countries without those resources. Figure 41 summarizes supply chain risks for fifteen countries.

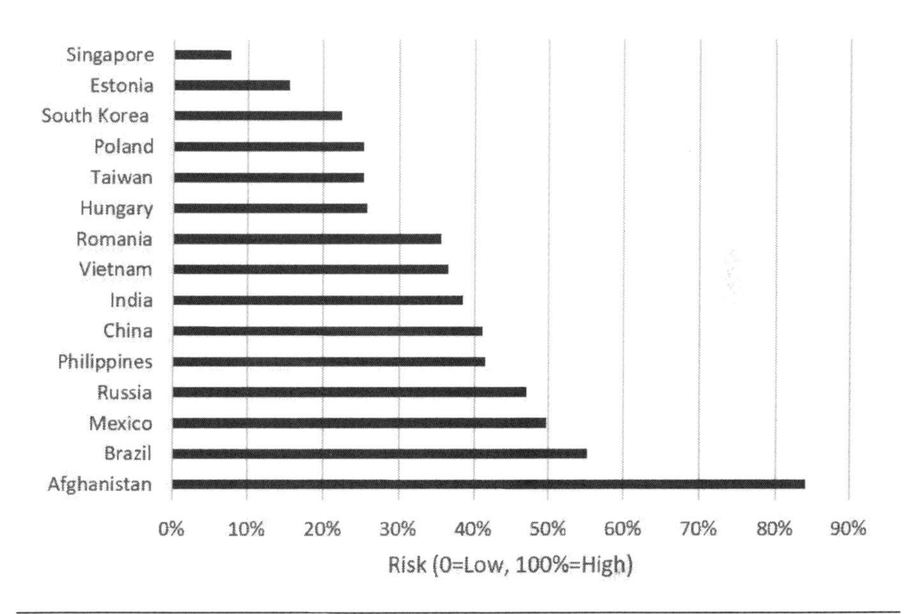

Figure 41. Supply Chain Risks for 15 Countries
(*Source:* Boston Strategies International)

One drilling equipment manufacturer's experience complying with local content restrictions in West Africa is instructive. It needs to employ nine local workers for every expatriate at manufacturing or distribution operations in Nigeria and Angola, with little or no grace period to train the local employees. The official minimum requirement is 70% of the labor force as of 2010,[130] but suppliers must maintain this ratio at all times and therefore typically stay well above the minimum level in case local workers quit, or suppliers need to bring in more expatriate workers on short notice. Combined with a high turnover rate among the local workforce, the net result of this requirement is that foreign suppliers often have several unproductive workers, either actively in training or simply on the payroll to satisfy local content restrictions, for every productive one.

Other companies have had major successes in the same countries, however, demonstrating that these challenges are manageable. FMC bills its Pazflor project as a local content success story.[131] Starting in 2003, Angola required local content for projects in the oil, information technology, education and training, logistics, and hygiene industries. The government's goal is to make all of these sectors "Angolan," in the sense that they are integral to the domestic economy because of local content required in these industries. These sectors are what the Angolan government has targeted as essential to economic development. Therefore, foreign companies doing business in Angola are required to hire Angolan staff and source Angolan products. The government offers a Business Support Center (CAE) to provide vocational training to help IOCs meet employment requirements.

Pazflor is located in deepwater Angola block 17, 150km (93 miles) off the coast of Angola under 600–1,200m (1,968ft–3,907ft) of water. The development extends over 600km^2 (232 square miles) and is operated by Total E&P Angola, a wholly owned subsidiary of Total, with a 40% interest. StatoilHydro (23%), ExxonMobil (20%), and BP (17%) also have ownership stakes. The Pazflor contract was FMC's largest ever, awarded in January 2008, requiring expansion of local production and support capabilities, including installation services and life-of-field support, and extensive local content. FMC delivered 49 subsea trees (25 production, 22 water-injection, and two gas-injection trees) and wellhead systems, along with associated production manifold systems, a production control and umbilical distribution system, gas export and flow line connection systems, and ROV tooling.[132]

Local content is also affecting supply chains in Brazil. The ANP, Brazil's national petroleum agency, put incentives in place for local content in about 2000. It converted the incentives into requirements in 2003 and 2004. The percentage of local content required increased from 15.5% in 1999–2002, to 40% in 2003–2004, then fell back to 20% in 2005–2008. In 2005, ANP imposed the use of certificates of compliance, which defined local content as the total value less anything not produced in Brazil. Rented items are valued by considering their book value, useful life, and the length of time they are active. Suppliers are required to use a certified auditor. According to the law, the supplier must send his credentials to the concessionaire, which must provide ANP with all suppliers' certificates of local content before it will approve payment. If the targeted thresholds are not reached, ANP assesses penalties.[133]

Clearly, companies wishing to do business in these countries are under pressure to identify and develop high-potential local suppliers.

Total Cost of Ownership (TCO) Trade-offs

Total Cost of Ownership

Many configurations of the supply relationship involve upfront costs and ongoing operating and/or maintenance costs. With complex equipment, the only way to know which are the lower-cost options in the long run is to evaluate them on a total cost of ownership (TCO) basis. Although the concept is simple, few companies have a reliable framework for calculating TCO, and many decisions are made based on gut feel, historical precedent, or political expediency.

Total Cost of Ownership (TCO), sometimes referred to as life-cycle cost, is the total cost of acquiring, commissioning, operating, maintaining, and disposing of a product or system. For the technologically complex and frequently expensive equipment sourced by an oil & gas or power producer, this equation includes substantial costs before and after the initial purchase, which dwarf the acquisition cost alone. Before a piece of equipment is bought, it must be specified and suppliers selected, a process that often requires a study in itself for heavy equipment such as injection pumps. It must then be transported to the point of use, which can involve multiple modes of transport, handling insurance, and customs duties. Installation and inventory costs also need to be factored in; these may include hidden factors like property taxes, warranty expenses, and storage costs. Throughout its operating life, maintenance and energy costs eventually outweigh the initial purchase price for most electrical equipment. Finally, decommissioning costs can be substantial, particularly for offshore platforms with subsea structures that must be dismantled and removed.

A comprehensive measurement of the cost of ownership should include ten elements, which are depicted graphically in Figure 42 (in condensed form):

- Forecasting requirements, which includes engineering resources and studies that would not have otherwise been required
- Acquisition, including the cost of the equipment itself and any ancillary equipment required to make it operational. If the equipment is being repurposed, the acquisition cost would consist of the cost of converting or adapting equipment from a previous use.
- Commissioning, including testing and calibration
- User training and engineering support
- Operation, including power, man-hours of any operating personnel, and allocated facility costs.
- Maintenance, including maintenance staff and facility cost and testing or measurement equipment
- Spares and repair parts inventories
- Asset tracking and record-keeping
- Infrastructure adaptation to suit the configuration, including software and networking costs related to the unit
- End-of-life removal and disposal, including environmentally safe disposal costs

Combined Purchase and Operating/Maintenance Agreements

Most suppliers offer, and sometimes require, that the Original Equipment Manufacturer (OEM) maintain the equipment for a period of time. Sometimes warranty coverage is linked to the OEM maintenance contract. This introduces an economic decision to the operator: OEM maintenance is usually more expensive than in-house or third-party maintenance, but that cost needs to be weighed

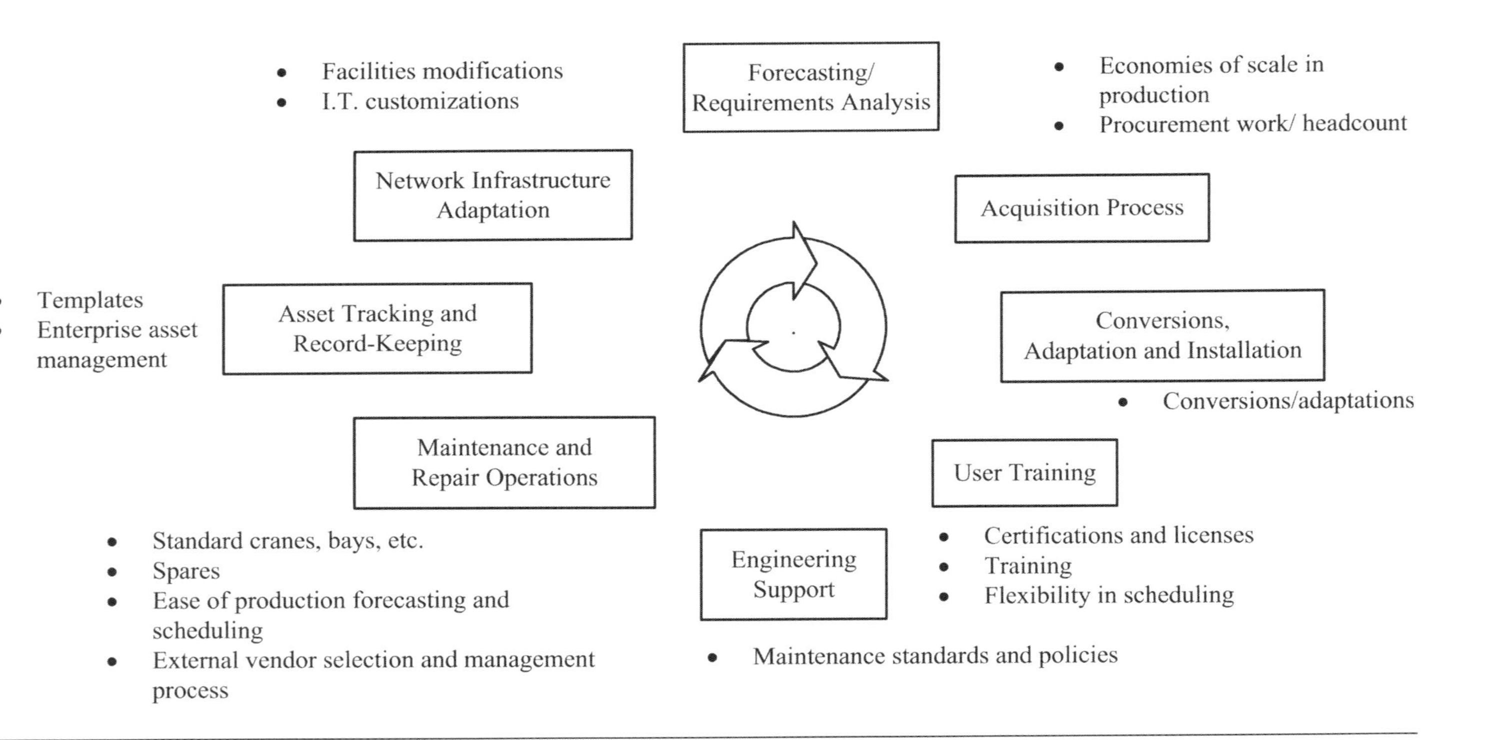

Figure 42. Life cycle Cost Components
(*Source:* Boston Strategies International)

against the loss of warranty coverage and balanced by higher reliability or uptime if the equipment is maintained by the OEM rather than a third-party shop.

The market of third party providers has evolved to encompass a broad range of services such as parts delivery, component reconditioning, spare parts management, rotor repair, testing, remote monitoring and diagnostics, upgrades and conversions, field service, training, inventory management, warranty management, commissioning and startup, and even full operating and maintenance agreements.

The driving factors behind most buyers' decision to use a third party are cost and lead time. A maintenance agreement from a third party is often 25–50% less expensive than one from the OEM, and third parties can often deliver faster. However, buyers must consider several factors before opting to use a third-party repair firm.

- Depending on the manufacturer and the equipment in question, allowing an unlicensed technician to repair the equipment may void any manufacturer warranties. Third parties may also not be qualified to decouple the machinery from the system in which it operates and recommission it.
- Quality and reliability of the components are also a concern, as sometimes third parties reverse engineer the parts and reproduce them without the original drawings and equipment. For example, a third party attempting to re-create a gas turbine blade must scan the inside of the blade (using expensive computerized axial tomography equipment) and go through a trial-and-error process to re-create the serpentine cooling pattern during the casting process.
- Third parties can take up to ten years to gain enough experience repairing a new model to service the entire unit effectively, especially for complex equipment such as gas turbines. As a result, banks are less likely to finance new equipment purchases unless the units are under a long-term service agreement (LTSA) with the OEM until service specialists have proven that they can fix the machines effectively.

If the concerns are addressed, however, third-party repair suppliers, who are also sometimes component suppliers, can provide a useful alternative to an OEM service contract.

When looking for a third-party supplier, buyers should look two tiers down in the supply chain, at the suppliers to the OEMs. As demand growth has outstripped production capacity, OEMs have outsourced production of an increasing number of components. This has caused them to delegate a significant amount of control over product quality and delivery schedules to component suppliers. These component manufacturers may, in some cases, provide an effective alternative for service agreements relating to their area of expertise.

Health, Safety, & Environmental (HSE) Considerations

Risks to life and to the environment needs to be continuously managed during operations and maintenance management, not only within the company but within supply chain partners. This chapter provides frameworks and tools for managing supply chain risk.

Upstream HSE Management

ISO 31000 requires that risk management be integrated into all organizational procedures for particular types of risk or circumstances. According to these guidelines, companies need a proper risk management policy, authorized representatives on the project who are responsible for effectively implementing and analyzing the risk involved at various stages, and procedures, guidelines, and standard practices for managing a crisis.

Unfortunately, most of these conditions were absent preceding the Macondo well explosion below the Deepwater Horizon drilling rig in 2010. This disaster serves as our anchor here for discussing upstream oil & gas HSE risk management policy, risk analysis methods, practices, procedures, and systems.

The US Mineral Management Service (US MMS) was responsible for establishing directions and a deterrent approach to risk management for organizations working on offshore facilities. It gave waivers for legally required environmental impact tests, and investigators later found that it did not have sufficient staff or auditors to monitor the risk of the project, which could have also led to crisis on any other facility by other operators such as Shell, ExxonMobil, and others.

At the same time, BP skipped a time-consuming cement log test that would have determined if there were any weaknesses in the cement. Cameron manufactured the blowout preventer (BOP) that did not activate, while Halliburton laid the cement plugs in the well that yielded under pressure. The spill might have been suppressed earlier if the authorities had comprehended the level of risk and depth of the disaster by assessing all the risk parameters throughout the drilling lifecycle. To evaluate the probability of the Macondo spill and its magnitude, during the planning phase, the MMS utilized the "oil spill risk model" developed by the US Geological Survey. This model relied on data from previous incidents, using historical data on the basis of a low probability of a spill. However, it did not detect and isolate specific faults at an early stage, before they caused a problem, so the model assigned only a 5% probability to large oil spills. The model estimated the most likely size of a large spill at 4.6k barrels, vs. the actual figure of 4.9m barrels.[134] During the three-month period of the spill, the magnitude of the disaster changed significantly with the estimated flow rate increasing from the initial 1k bpd to a final official figure of 62k bpd. So, during the Macondo oil spill, when assessments by regulatory bodies varied greatly over a period of time, the risk evaluation process turned questionable.

BP, despite being the principal contractor on the project, only decided how the well would be drilled and how it would be cemented. However, instead of assessing external and internal uncertainty on objectives while formulating the strategies to manage risk and the scope, it instructed Transocean to remove drilling-muds during cementing to save time, despite Transocean's objections, ultimately damaging the formation and requiring Transocean to start drilling over again.

The companies did not establish any formal risk treatment plan to deal with a crisis of this proportion. ISO 31000 standards have advocated for risk mitigation guidelines to remove the risk source and methodologies to alter the consequences, but in the case of the Deepwater Horizon oil spill, the mitigation strategy was adopted *after* the incident, resulting in a fragmented approach that depended on hasty decisions. In the end, the skimming and booming methods used on the spill recovered only about 10% of the total oil that was spilled.[135] In fact, the use of dispersants at the spill site impeded additional recovery. Not only did BP not have proven chemical technology to disperse oil at such a large scale or at water depths of thousands of feet, but the actual measures to be taken in case of an oil spill were not clearly outlined.

The events suggest that the Macondo crisis could have been avoided if all five parties had embraced well-defined frameworks, processes, and practices to control risk. The documented guidelines for continual checking, monitoring, critically analyzing, or specifying the status throughout the drilling life cycle could have been effective.

Since the Macondo crisis in 2010, well operators and lessees are being held to higher standards. Well operators now need to submit detailed emergency action plans on a well-by-well basis, and stock the required safety equipment on the rig for immediate use in case it is needed. For example, in the US, the BOEMRE (Bureau of Ocean Energy Management, Regulation and Enforcement) has introduced the following rules for improving safety and accident prevention, blowout containment and spill response:

- The Drilling Safety Rule imposes tough new standards for well design, casing and cementing, and well control procedures and equipment, including blowout preventers. Operators are now required to obtain independent third-party inspection and certification of the proposed drilling process. An engineer must also certify that blowout preventers meet new standards for testing and maintenance and are capable of severing the drill pipe under anticipated well pressures.
- The Workplace Safety Rule requires operators to identify risks, establish barriers to those risks, and seek to reduce human and organizational errors. Operators now are required to develop a comprehensive safety and environmental management program that identifies the potential hazards and risk-reduction strategies for all phases of activity, from well

design and construction, to operation and maintenance, and finally to the decommissioning of platforms.

- The NTL-N06 directive requires that operators develop oil spill response plans including a well-specific blowout and worst-case discharge scenario, and also provide the assumptions and calculations behind these scenarios. BOEMRE engineers and geologists will then independently verify these worst-case discharge calculations.

Since the Deepwater Horizon disaster, BP has also self-imposed standards for its equipment, especially blowout preventers (BOPs). The new standards include enhanced response measures for blowouts and oil spills and require that an independent third party test the cement used for deepwater well sealing They call for a BOP with at least one set of extra "blind" rams that will prevent a section of drill pipe from blocking the wellbore open. It requires that a third party perform the testing and maintenance of subsea BOP, whenever it is brought to the surface. BP also included intense measures in its Oil Spill Response Plan (OSRP), based on lessons learned from the Macondo disaster. The new drilling rules and extra measures documented by BP are related to well design, spill containment, and modification in blowout preventers.

Due to the evident possibilities for disastrous accidents in upstream oil & gas, quality standards are critical throughout the entire oil & gas value chain. Equipment suppliers need to follow rigorous quality regimes as much as the operators. As one example, compressor manufacturer Ebara implemented FMEA (failure mode effect and criticality analysis) in new product design and in individual production stages, and developed a work standard to prevent nonconformities, including the "5 Whys" of Lean methodology. It also continues the Lean journey toward lower inventory and lower cost through its Mindora and M Zero Challenge campaigns, which foster collaboration between the design, manufacturing, and sales departments.[136]

Health, Safety, Environment (HSE) Management Systems track metrics to ensure compliance with relevant HSE legislation and regulations. The system also utilizes continuous improvement processes that incorporate Lean Six-Sigma methodology. Cameron's Compression System Division claims that its HSE Management System helped lower OSHA recordable injuries by 43%, lost-time injuries by 58%, and lost-time days away from work by 90%.[137]

Downstream HSE Management

Refiners need to include their contractors and supply chain partners in HSE planning in order to assure a safe and stable supply chain. There are more fatal oil refinery incidents than in the next three highest industries combined. Refinery incidents are to blame for 58 deaths in the United States between 2005 and 2015.[138] However, since employers are not required to include contractors as employees when reporting deaths, the actual totals may be even higher.[139] Such loopholes are leading to a lack of accountability in downstream processes.

In some cases, the accidents may be blamed on outdated regulatory frameworks. Russia's regulations, which govern construction of refineries, date back to the first half of last century and remained practically unchanged since the mid-1960s until Gazprom Neft submitted a draft of Safety Regulations for Oil, Petrochemical and Gas Processing Facilities, in May 2011, to the Ministry of Energy. In 2010, the US EPA proposed amending the Clean Air Act to reduce greenhouse gas emissions from refineries and power plants.[140] The same year, API issued two new safety standards to help increase accountability.

- Recommended Practice 754, *Process Safety Performance Indicators for the Refining and Petrochemical Industries*, provides a set of process safety indicators for identifying events that may predict safety issues, and assessing the magnitude of those risks.
- Recommended Practice 755, *Fatigue Risk Management Systems for Personnel in the Refining and Petrochemical Industries*, provides guidelines to help reduce fatigue risk.

Proper operations management and supply chain management are essential instruments in ensuring safe operations. Tesoro Corp.'s Anacortes oil refinery suffered an explosion in 2010 due to improper maintenance of a heat exchanger that subsequently burst, killing five people. The piece of equipment that blew apart was almost 40 years old. At the time of the accident, the welds that popped had not been inspected using a sophisticated method to detect cracks since 1998, according to an investigation by state regulators. Also, Huntsman has demonstrated leadership in implementing an enterprise-wide operational risk application. The characteristics of the tool are instructive. First, it is integrative, meaning that it assembles risk information from a variety of frameworks and systems in the company. Second, it is web-based, which allows the project to involve employee participation to enter data into the system worldwide, invoking the "worker involvement" principles of Total Productive Maintenance and Lean. Third, it is flexible, allowing for configuration and custom reporting. Fourth, it is to some degree predictive: Huntsman expects the tool to capture knowledge about current and previous risks and apply that to future risks in order to provide a predictive risk outlook.[141]

Power Industry HSE Management

For the most part, health and safety standards in power generation are based on cross-industry standards and country-specific legislation, making it necessary to understand global HSE standards as well as local rules and regulations in the jurisdiction of power generation. For example:

- In the US, a number of government regulations encourage utilities to be prepared for crisis management in the event of disaster. These include the North American Electric Council (NERC)'s cybersecurity standards, which require power plants to continuously access and enhance their security systems; the Homeland Security Act of 2002, which identifies critical infrastructure such as utilities as high-priority; and the Sarbanes-Oxley compliance, which deems utilities to be Level 1 critical systems, which are expected to be restored within 48 hours.

- In India, TATA power's Haldia power plant, implemented ISO 14001:2004 (environmental management system), and OHSAS 18001:2007 (health and safety management system).
- In Australia, the Shaw River power project implemented the Environment, Health, and Safety Management System (EHSMS) for the power station and corresponding gas pipeline. The EHSMS is based on international standards such as ISO 14001:2004 and Australia/NewZealand standard AS/NZS 4801:2001 (occupational health and safety management).

The World Bank has established a worldwide standard, at least in T&D, with its Environmental, Health, and Safety Guidelines. The World Bank Group's Environmental, Health, and Safety Guidelines for Electric Power Transmission and Distribution contain performance levels and measures that it considers achievable at new facilities by using existing technology and at reasonable cost. For example, it provides a table for minimum working distance for employees for different voltage levels (taken from OSHA 1910, subpart R, subpart S). It also states that the guidelines are dependent on the host country and if the measures of both differ, then the more stringent one is to be adopted (unless in view of specific project circumstances).

While the international standards are intended to support ERM approaches, other frameworks may be more useful for identifying and managing risks at the project level.

The Council for Security Cooperation in the Asia Pacific released a memorandum titled *Safety and Security of Offshore Oil and Gas Installations* in January 2011, which identifies new safety and security risks resulting from an increase in offshore drilling activity. The memo urges regional governments to implement safety standards, such as those put forward by the 2005 ASEAN Agreement on Disaster Management and Emergency Response, and to dialogue with operators to agree on their respective responsibilities and develop joint emergency response plans.[142]

For more risk management tools, the reader is advised to consult the Bibliography. For immediately adaptable frameworks dealing with risk, managers may want to adapt the root cause framework, the probability vs. severity cube, or Failure Mode Effect Analysis.

Root Cause Analysis

The most universal problem-solving framework that could be applied to supply chain risk framework may be a root cause framework (see Figure 43).

Another way of looking at supply chain risk involves classifying the likelihood vs. its severity, and then ranking the overall risks based on the combined score.[143]

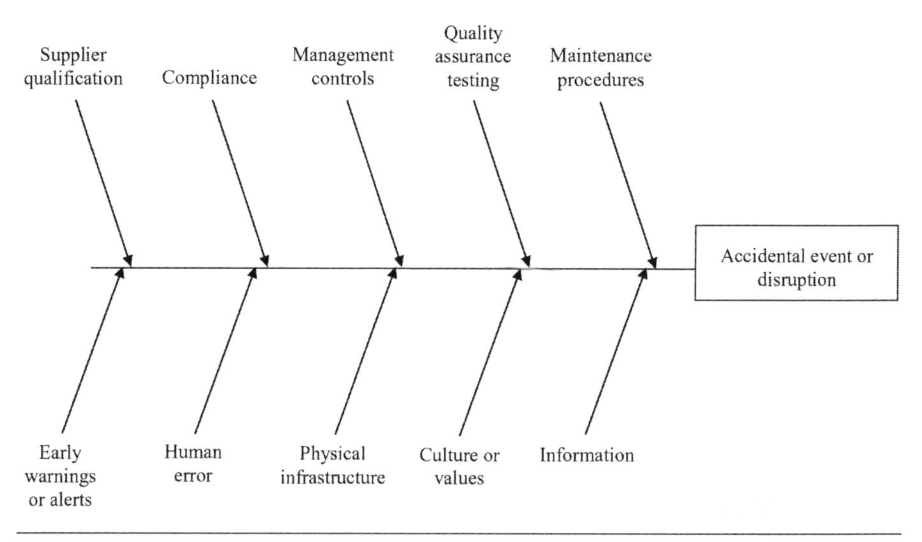

Figure 43. Root Cause Analysis Applied to Supply Chain Risk
(*Source:* Boston Strategies International.)

Failure Mode Effect Analysis

Failure Mode and Effects Analysis (FMEA) is an analytical method to identify and prevent process failure. Planners most effectively use FMEA during product development and process planning. The exact steps may vary depending on who describes the process, but a generalized and comprehensive description of FMEA would include these steps:[144]

- Gather a cross-functional team of people who touch the process at the beginning, the middle, and the end
- Define what output the system is intended to produce
- Identify where it can go wrong (failure mode)
- Identify the effects, or consequences, of failure (qualitatively)
- Develop a scoring sheet, framework or matrix for capturing the group's assessment of severity, frequency, and detection on a common scale (for example a spreadsheet with scoring indices from one to 10) as it relates to the relevant time frame, such as the lifetime of the equipment or process
- Quantify the severity of the potential failure
- Quantify the frequency of the potential failure
- Quantify the chance of the failure being detected early enough to be acted upon (using an inverse scale, so a high score is a low chance of detection)
- Calculate the risk priority number = severity x frequency x chance of detection

- Identify ways to reduce the frequency of failure, to increase the chance of early detection, or to reduce the severity of the failure (the drivers of the risk priority number)
- Establish a tracking log, with responsibilities and dates

Organizations pursuing risk management programs may also want software applications that embed the accountabilities. Here is an example which touches on two areas of risk management that have not been specifically addressed by the above standards and tools: separation (or segregation) of duties, and environmental risk.

Separation of duties can contribute to reducing the possibility for errors that might lead to safety incidents. Any implementation of a risk management framework will involve a careful review of internal policies and procedures, so having a risk management application can help improve the design of the system as well as the ability to trace what should have happened in event of an accident. Valero and Areva both implemented an application for this. Valero Energy implemented a process for identifying and remediating segregation of duties (SoD) conflicts to comply with the Sarbanes-Oxley Act by improving access compliance enterprise-wide. It selected access-control software to standardize the process for compliance assurance, as opposed to the fragmented compliance initiatives that had previously been in place. Areva implemented a tool that allows the flexibility to design customized risk matrices for the company's multiple divisions, and leverage preventive controls built into the business processes to stop future violations.

Environmental risk management monitors and addresses potential environmental breaches. Faced with the need to prove environmental compliance following community complaints, KNPC chose an environmental, health, & safety (EH&S) and crisis management solution program to respond to changing environmental regulations.[145]

Selected International Risk Management Standards

Agencies such as the International Standards Organization (ISO), the International Maritime Organization (IMO), and the UK's Office of Government Commerce (OGC) have developed risk management frameworks. For example, after the BP Deepwater Horizon/Macondo oil spill in 2010, the International Standards Organization (ISO) developed an action plan (ISO/TC 67) for materials, equipment, and offshore structures for the petroleum, petrochemical, and natural gas industries. The action plan provided an inventory of 71 existing standards and related documents available from ISO or other organizations, particularly the American Petroleum Institute (API). It also proposed 31 standards or related documents for development or update by ISO, API, or other organizations. The Directorate General of the European Union also proposed new rules to incorporate technological improvements. These propose removing minimum technical requirements and replacing them with a goal-setting approach that will assess risk on a case-by-case basis and determine the best risk mitigation approach based on the best technology currently available.

Some of the important risk management standards are discussed below.

ISO 31000 (International Standards Organization)

ISO 31000, like the AS/NZS 4360 standard upon which it was built, contains generic guidelines that any organization in any context and scope can use for effective risk management. It considers risk management as an integral part of all organizational processes and an effective part of decision-making. The principle of ISO 31000 is that risk management must be structured, timely, adapted to each organization's needs, and transparent to all in the organization. Insofar as management should define and endorse a risk management policy that is in line with the company's objectives and legal and regulatory compliance, designing a framework for risk management must involve an understanding of the internal and external context of the organization, including its internal and external communication and reporting mechanisms.

ISO 31000 also describes attributes of high-level performance in managing risks and lists indicators for organizations to compare their performance against these criteria.

Key indicators for enhanced risk management include:

- Continuous improvement in risk management—explicit performance goals and review process
- Defined and accepted accountability for risks and risk treatment embedded in job descriptions
- Explicit consideration of risks and risk management in decision making process—evident from records of such meetings
- Communication about risks with internal and external stakeholders

The Government of Alberta,[146] Qatar Petroleum,[147] and mining firm TWP Projects[148] are just a few of the energy stakeholders that have adopted an ISO 31000 enterprise risk management system.

ISO 31000 provides guidance for formulating a policy on risk, offering these elements:[149,150]

- Risk management and internal control objectives (governance)
- Statement of the attitude of the organization toward risk (risk strategy)
- Description of the risk-aware culture or control environment
- Level and nature of risk that is acceptable (risk appetite)
- Risk management organization and arrangements (risk architecture)
- Details of procedures for risk recognition and ranking (risk assessment)
- List of documentation for analyzing and reporting risk (risk protocols)
- Risk mitigation requirements and control mechanisms (risk response)
- Allocation of risk management roles and responsibilities

- Risk management training topics and priorities
- Criteria for monitoring and benchmarking of risks
- Allocation of appropriate resources to risk management
- Risk activities and risk priorities for the coming year

ISO 31010 also provides a large number of tools to quantify the consequences, probability, and level of risk, as outlined in Figure 51. ISO 31100 provides a useful table of risk management tools, which informed Figure 44 below.

Tool	Identification	Assessment	Response
Risk Questionnaires	×		
Risk Checklists/Prompt Lists	×		
Risk Identification Workshop	×	×	
Nominal Group Technique	×	×	
Risk Breakdown Structure	×	×	
Delphi Technique	×	×	
Process Mapping	×	×	
Cause-and-Effect Diagrams	×	×	
Risk Mapping/Risk Profiling	×	×	
Risk Indicators	×		
Brainstorming/"thought shower" events	×		
Interviews and focus groups	×		
"What if?" workshops	×		
Scenario analysis/scenario planning/horizon scanning	×	×	×
Hazard and Operability Study (HAZOPs)	×	×	
PEST (Political, Economic, Sociological, Technological) Analysis	×	×	
SWOT (Strengths, Weaknesses, Opportunities and Threats) Analysis	×	×	
Stakeholder Engagement/Matrices	×		
Risk Register/Database	×	×	×
Project Profile Model (PPM)	×		
Risk Taxonomy	×		
Gap Analysis: Pareto Analysis	×	×	
Probability and Consequence Grid/Diagrams (PIDs)/Boston Grid	×	×	
CRAMM	×	×	×
Probability Trees		×	
Expected Value Method		×	
Risk Modelling/Risk Simulation (Monte Carlo/Latin Hypercube)		×	
Flow charts, process maps and documentation		×	
Fault and event tree modelling; Failure Mode Effects Analysis (FMEA)		×	
Stress Testing	×	×	
Critical Path Analysis (CPA) or Critical Path Method (CPM)		×	
Sensitivity Analysis		×	
Cash Flow Analysis		×	
Portfolio Analysis		×	
Cost-Benefit Analysis		×	×
Utility theory		×	
Visualization techniques, heat maps, RAG status reports, Waterfall charts, Profile graphs, 3D Graphs, Radar chart, Scatter diagram		×	×

Figure 44. Summary of Risk Management Tools
(*Source:* Draft BS 31100 Code of practice for risk management, 2008, p. 24. Note: These tools are still applicable since ISO 31000 was updated in 2018.)

COSO Enterprise Risk Management—Integrated Framework

The Committee of Sponsoring Organizations (COSO) of the Treadway Commission developed a framework for Enterprise Risk Management (ERM) in organizations that came to be known as the COSO–Integrated Framework. COSO defines ERM in terms of strategy followed by an organization for managing risk within the firm's "risk appetite" as it strives to achieve its objectives. The COSO ERM framework identifies risk management as an iterative process as opposed to a sequential process, where each step has a bearing on the others. It aims to augment the internal control structure within an organization for effective ERM.

The COSO ERM framework is built upon 20 key principles in five key components:[151]

1. Governance & Culture
    - Exercises board risk oversight
    - Establishes operating structures
    - Defines desired culture
    - Demonstrates commitment to core values
    - Attracts, develops, and retains capable individuals
2. Strategy & Objective-Setting
    - Analyzes business context
    - Defines risk appetite
    - Evaluates alternative strategies
    - Formulates business objectives
3. Performance
    - Identifies risk
    - Assesses severity of risk
    - Implements risk responses
    - Develops portfolio view
4. Review & Revision
    - Assesses substantial change
    - Reviews risk and performance
    - Pursues improvement in Enterprise Risk Management
5. Information, Communication, & Reporting
    - Leverages information and technology
    - Communicates risk information
    - Reports on risk, culture, and performance

The framework describes the following as components of ERM:[152]

1. Control Environment—The internal environment encompasses the tone of an organization, and sets the basis for how risk is viewed and addressed by an entity's people, including risk management

philosophy and risk appetite, integrity, and ethical values, as well as the environment in which they operate. This step includes the establishment of a process to set objectives and of a situation in which the chosen objectives support and align with the entity's mission and are consistent with its risk appetite.

2. Risk Assessment—Internal and external events affecting achievement of an entity's objectives must be identified, distinguishing between risks and opportunities. Opportunities are channeled back to management's strategy or objective-setting processes. Risks are analyzed, considering likelihood and impact, as a basis for determining how they should be managed. Risks are assessed on an inherent and a residual basis.
3. Control Activities—Policies and procedures are established and implemented to help ensure the risk responses are effectively carried out by avoiding, accepting, reducing, or sharing risk—developing a set of actions to align risks with the entity's risk tolerances and risk appetite.
4. Information and Communication—Relevant information is identified, captured, and communicated in a form and time frame that enable people to carry out their responsibilities. Effective communication also occurs in a broader sense, flowing down, across, and up the entity.
5. Monitoring—The entirety of enterprise risk management is monitored, and modifications are made as necessary. Monitoring is accomplished through ongoing management activities, separate evaluations, or both.

Formal Safety Assessment (International Maritime Organization)

Formal Safety Assessment (FSA) proposes a systematic process for assessing the risks associated with shipping activity and for quantifying the costs and benefits of IMO's standards for reducing these risks. It was initially charted out as a response to the Piper Alpha disaster of 1988, when an offshore platform exploded in the North Sea, taking 167 lives, and serves as the basis for IMO's current risk management plans. Guidelines for FSA for use in the IMO rule-making process were approved in 2002 and updated by MSC-MEPC.2/Circ.12/Rev.2 in 2018. FSA consists of five main steps:

- Identification of hazards (list of all relevant accident scenarios with causes and damages)
- Analysis of risks (evaluation of risk factors)
- Risk control options (regulatory measures to control and reduce the identified risks)
- Cost benefit assessment (determining cost effectiveness of each risk control option)

- Recommendations for decision-making (information about the hazards, their associated risks, and the cost effectiveness of alternative risk control options is provided)

Other Risk Management Standards

There are dozens of other risk management standards, most often at the national level. This book is not intended to provide sufficient coverage to list them all.

One that is rich in detailed tools such as risk heat maps is the UK government's Management of Risk (MOR) framework, which focuses on risk management in public sector organizations. It was first published in 2002 by the UK Office of Government Commerce, which is a part of the Efficiency and Reform Group of the Cabinet Office in response to the Turnbull Report (a UK study that addressed standards for internal control and risk transparency), and was republished in 2017.[153]

Another is DNV's Integrated Software Dependent Systems (ISDS) standard. Data security has become a much larger concern since the advent of smart grids. Information system security standards have become more prevalent as smart grid implementation revealed that data breaches could shut down operations and compromise safety. For example, DNV developed the ISDS standard beginning in 2008, and Dolphin Drilling received the first ISDS class notation.[154] It was updated in 2017.[155]

The Supply Chain's Role in Reducing Environmental Footprint

Increasing pressure from government regulatory bodies, society, and consumers has led to an increasing interest in measuring environmental footprints and reducing carbon emissions. This initial pressure has been compounded by a natural business desire for efficient processes, as well as an awareness that green processes are often more efficient; these trends have been intensified by rising prices of depleting resources, and supply shortages of some raw materials.

A broad movement throughout recent history has compelled energy industry management to monitor and comply with environmental legislation. Here are some major organizations and protocols to be familiar with:

- 1970—the US Environmental Protection Agency (EPA) and Occupational Safety and Health Administration (OSHA) were formed.
- 1976—the Resource Conservation and Recovery Act (RCRA) established laws surrounding the disposal of hazardous waste.
- 1979—the EPA enacted the Toxic Substances Control Act (TSCA).
- 1980—the Comprehensive Environmental Response, Compensation, and Liability Act (CERCLA).

- 1986—the Superfund Amendments and Reauthorization Act (SARA) was amended.
- 1986—the Right-to-Know laws established hazardous communication standards such as the now well-known Material Safety Data (MSDS) Sheets.
- 1992—the UN's Basel Convention (entered into force in 1992), to prevent transfer of hazardous waste from developed to less developed countries (LDCs).
- 1997—the Kyoto protocol was adopted.
- 2003—the Waste Electrical and Electronic Equipment (WEEE) Directive dealt with the problem of persistent electronic waste.
- 2004—the Stockholm Convention on Persistent Organic Pollutants (effective 2004) restricted the production and use of persistent organic pollutants (POPs).
- 2004—the Rotterdam Convention promoted an open exchange of information and calls on exporters of hazardous chemicals to use proper labeling.
- 2006—the International Restriction of Hazardous Substances (RoHS) Directive was instituted.
- 2008—the deadline for pre-registration of substances under Registration Evaluation and Authorization of Chemicals (REACH). See Appendix 3 for more information on REACH.
- 2010—the UN's Year of Biodiversity and other initiatives encouraged a widespread focus on climate change.
- 2015—UN Sustainable Development Goals
- 2016—the Paris Agreement on climate change and mitigation measures
- 2019—the Kigali Amendment to the Montreal Protocol on Substances that Deplete the Ozone Layer

Ecological practices such as green product design (especially in the case of energy-consuming equipment such as motors, generators, and pumps), green materials management (transport modes, packaging, etc.), and reverse logistics or recycling can make a significant impact on efficiency and cost. Companies in other industries have in some cases implemented highly structured models for incentivizing suppliers to be "green," and it is likely that some energy producers will emulate this practice in the future.[156]

Many are measuring their carbon footprint. The Carbon Disclosure Project (CDP) is an independent not-for-profit organization with a mission of accelerating solutions to climate change and water management by disseminating information to businesses, policymakers and investors. Over 3k organizations in 60 countries around the world measure and disclose their greenhouse gas emissions, water management, and climate change strategies through CDP, so they can set reduction targets and make performance improvements. This data is made available for use by institutional investors, corporations, policymakers and their advisors, public sector organizations, government bodies, academics, and the public. CDP's Supply Chain initiative works with global corporations to understand the impacts of climate change across the value chain, harnessing their collective purchasing power to encourage suppliers to measure and disclose climate change information. CDP's Public Procurement initiative is designed to enable national and local governments to ascertain the impact of climate change in their supply chains.

In the area of electricity consumption, demand response has gained traction in recent years, as has the transition from old, inefficient lighting technologies to new, far more efficient ones like compact fluorescent bulbs and efficient motor technologies based on IE2 and IE3 efficiency standards. These trends affect the choice of suppliers and technologies, and ultimately the cost and efficiency of production of oil, gas, and power.

In the area of air emissions, new air emission regulations in the US and EU for engines will affect prices and availability of equipment, necessitating an equipment transition and new maintenance strategies for end users.[157] The US Environmental Protection Agency (EPA)'s air emission standards for spark ignition (SI) engines rated less than 500 hp came into effect in early 2013. The EPA aims to reduce annual toxic emissions by 1k tons, particle pollution by 2,800 tons, carbon monoxide emissions by 14k tons, and volatile organic compound emissions by 27k tons annually. Engine operators will need to retrofit their engines or replace them with new model engines, which will cost over a third more than the existing ones. Similarly, the European Union implemented Euro 6 in 2013, which decreased emissions of particulate matter by 50% and NOx by 75–80% over the previous (Euro 5) standards.

Oil & gas companies have been responding with plans to reduce CO_2 emissions. Saudi Aramco released a carbon management roadmap as early as 2006, outlining five avenues for reducing carbon emissions (CO_2 capture: fixed sources, CO_2 capture: mobile sources, CO_2 EOR, CO_2 sequestration, and developing industrial applications that use CO_2 as a feedstock), both from its operations and from transportation applications that use fossil fuels such as passenger cars. The firm is a major sponsor of the Weyburn-Midale CO_2 Storage and Monitoring Initiative, a research project run by the IEA to study the feasibility and best techniques for combining EOR with CCS.[158]

In the area of chemicals and wastewater, environmental pressures have led to a host of new green biocides, decontamination and neutralization techniques, and chemical, electrical and ultraviolet treatments to remove pollutants and undesirable bacteria from produced water. Green efforts intensified after Nalco and BP came under public pressure for extensive use of the dispersant Corexit 9500, which Louisiana residents claimed is four times more toxic than the leaked oil and other viable dispersant options.[159] A number of innovations demonstrate a strong commitment on the part of the industry to improve the environmental impact of chemicals and wastewater. NexLube, an independent US start-up, produces lubricants entirely from recycled waste oil at its Florida Group II base oil plant. Drillers are recycling wastewater. For example, each well in the Marcellus shale consumes 100–300k gallons of water for drilling (excluding fracturing). One drilling company recycled 80% of its water in 2009, 90% in 2010, and 100% in 2011.[160] Also, thousands of companies are complying with REACH's far-reaching and time-consuming registration process. Appendix 3 summarizes the REACH legislation and what companies need to do to comply with it.

ISO 14001 provides a framework for a holistic approach to environmental policy for organizations. Like ISO 31000, it defines generic requirements for an environmental management system as a reference standard for communicating between companies, regulators, and other stakeholders. Draka, the Dutch cable supplier, implemented environmental management systems for all operations in compliance with ISO 14001.[161]

In order to make a significant impact, project managers need to work with suppliers and customers two or more links away in the supply chain. This brings up trust issues that are more fully addressed in *The Guide to Supply Chain Management*.[162] Customers and suppliers are traditionally loath to share information that may reveal confidential or sensitive business information, especially related to costs and anything that may carry liability, such as carbon footprints or environmental damage caused by their accidents. However, compelling, vocal, and well-informed leaders can leverage a groundswell of common interest to engage trading partners in collaborative initiatives to better secure safety and the environment.

Hydrogen/Fuel Cells

Chapter Highlights

1. Fuel cell technology is being commercialized in vehicles and some industrial applications. Some power plants are moving toward replacing natural gas with hydrogen, but currently this consists of gas turbines running on hydrogen, not of fuel cells per se.
2. One overriding challenge is establishing a clean hydrogen-powered fuel cell value chain from beginning to end ("green hydrogen"). If fuel cells are fueled with hydrogen produced by steam-methane reforming processes fueled by natural gas, the low-emissions advantages of the fuel cell are negated by the CO_2 emitted in the production of the hydrogen. Producing hydrogen with wind or solar power, or through carbon capture and sequestration (CCS), and then using that hydrogen in fuel cells, would provide carbon-neutral energy.
3. Storage mode, purity, and safety considerations are currently throttling development and commercialization, as the eventual high-volume solution must be efficient, effective, and safe.
4. Most early stage commercialization activities are currently handled by labs or are performed in-house, not outsourced, to protect intellectual property.
5. Due to the rapid pace of change and technological development, partnerships and business models are frequently built on bundled manufacturing and distribution; bundled product development, operations, and maintenance; or both.
6. Formal contracts tend to be short-term to keep options open while the technology and competitive landscape matures.
7. Economies of scale are significant in the membrane and gas diffusion layer (GDL) of the fuel cell stack. Supply arrangements should account for this.
8. Precious metals-based catalysts in the Membrane Electrode Assembly (MEA) can be recycled. At least one fuel cell manufacturer has already launched a recycling program.

9. Supply chain innovations need to focus on: 1) reducing the potential for leaks and flammable conditions; 2) realizing economies of scale in the GDL and the fuel stack; 3) sourcing platinum to ensure adequate supply; 4) protecting the FCCU from cyberattack; 5) developing optimal chemistries for various power ranges and environments; and 6) creating a market for replacement membrane elements.

Introduction

Fuel cell technology, with hydrogen as its fuel, is emerging primarily in transport and industrial applications. Only 0.02% of the United States' power was provided by fuel cell–powered power plants as of 2016, and that amount was delivered by 56 >1 MW units totaling a cumulative 137 MW.[163] Crossover applications are common, however, indicating that fuel cell adoption at utilities may evolve through initial applications in transport and industrial applications, or from industrial applications to utility use. Hydrogen-fueled and fuel cell–powered utility-scale power plants seem likely to benefit from the investment in commercialization currently being made in automotive and industrial environments.

Most hydrogen power projects today are fuel cell development programs for transportation vehicle applications, especially buses and heavy trucks. For example, Scania partnered with ASKO to run a pilot project to test commercial feasibility of hydrogen fuel cell–powered trucks. The companies' plan is to deploy four hydrogen-fueled trucks with electric drivetrains that can be refueled at ASKO's hydrogen gas station.[164] Also, MiWay, the municipal public transport agency serving Mississauga, Ontario, Canada, is running a hydrogen fuel cell electric bus pilot project on its journey toward a zero-emission bus fleet.[165]

Industrial companies that have access to the hydrogen as a byproduct of their current manufacturing process are experimenting with using their byproduct hydrogen for commercial power production. South Korea conglomerate SK E&C (an EPC firm), SK Advanced (a propylene producer), and Bloom Energy, a US-based fuel cells technology company, are building an Advanced Propane De-Hydrogenation plant in Ulsan, South Korea, to produce hydrogen as a byproduct of the propylene production process. The result will be used as fuel for solid oxide fuel cells.[166]

Also, Rolls-Royce and Mercedes Benz have partnered through their subsidiaries to pilot the use of vehicle fuel cells for stationary power generation. Rolls-Royce's business unit Power Systems and Mercedes-Benz's innovation lab Lab1886 are developing an off-grid power station based on fuel cell modules built by Mercedes. Their target customers are data centers for whom uptime is mission critical, where the main goal is to replace diesel generators with hydrogen fuel cells.[167]

Currently, utility applications of fuel cells are limited. POSCO Energy of South Korea commissioned a 59 MW hydrogen power plant in 2014,[168] and Daesan

Green Energy, a joint venture between Doosan, Hanwha Energy, and Korea East-West Power, began work on a 50 MW fuel cell utility in 2019.[169] Outside of the utility industry, hydrogen-fueled industrial plants are currently being tested, with fuel cells and thermal processes being used to convert natural gas to hydrogen gas. Some vehicle manufacturers are experimenting with larger-scale fuel cells that can be used for commercial and utility-scale power. For example, Hyundai Motor Company is piloting fuel cell power generation together with Korea East-West Power (EWP) and Deokyang in South Korea. The pilot project will deliver a 1 MW hydrogen fuel cell power plant. Hyundai is building the fuel cell system; Deokyang will supply the hydrogen; and EWP will sell the electricity.[170] Independent of fuel cells, some power plants are moving away from oil & gas toward hydrogen fuel. One power plant in Germany, the Leipzig Süd district heating power plant, recently set a goal of 100% hydrogen fuel, and Siemens Energy is supplying two gas turbine packages that will eventually operate on 100% hydrogen.[171]

Regulatory and fiscal incentives are central to early adoption of fuel cell technology. In the United States, 85% of fuel cell power generation capacity lay in three states that offer such incentives. California had 36% of the nation's fuel cell power-generating capacity, Connecticut had 27%, and Delaware had 22%.[172] Between the EU's Green New Deal and the Biden administration's pro–clean energy policies in the United States, the global regulatory environment may be about to favor clean power, including fuel cells.

Given the dearth of real-world examples in place, the supply chain for utility-scale hydrogen fuel cells is largely hypothetical at the time of this writing. Most likely a utility would co-locate next to a chemical or petrochemical producer that produces a byproduct that can be separated through an air separation plant to yield hydrogen. In non-co-located scenarios, the hydrogen would be piped to the utility. The utility would operate a large fuel cell, which may be provided by a fuel cell manufacturer, which in turn would build it from a proton exchange membrane, a catalysis unit involving platinum and carbon, and a gas diffusion layer that allows gases to pass through pores. Transmission lines would carry the electricity to customers. The hypothetical supply chain might resemble the flow in the schematic diagram in Figure 45 below.

Fuel cells consist primarily of an electrolyzer and, in the most common applications today, a fuel cell proton-exchange membrane (see below regarding the various types of fuel cells). However, these components need to operate within an ecosystem that produces and delivers the required hydrogen fuel to the electrolyzer, and a transportation system that distributes the energy from the PEM to the point of consumption. Making and operating fuel cells is not necessarily complicated, but making them operate in a clean way in a green ecosystem can be elusive, as the production and delivery of the hydrogen fuel to the fuel cell, and the transportation and distribution of the electricity from the fuel cell to the consuming apparatus, may not be clean or green.

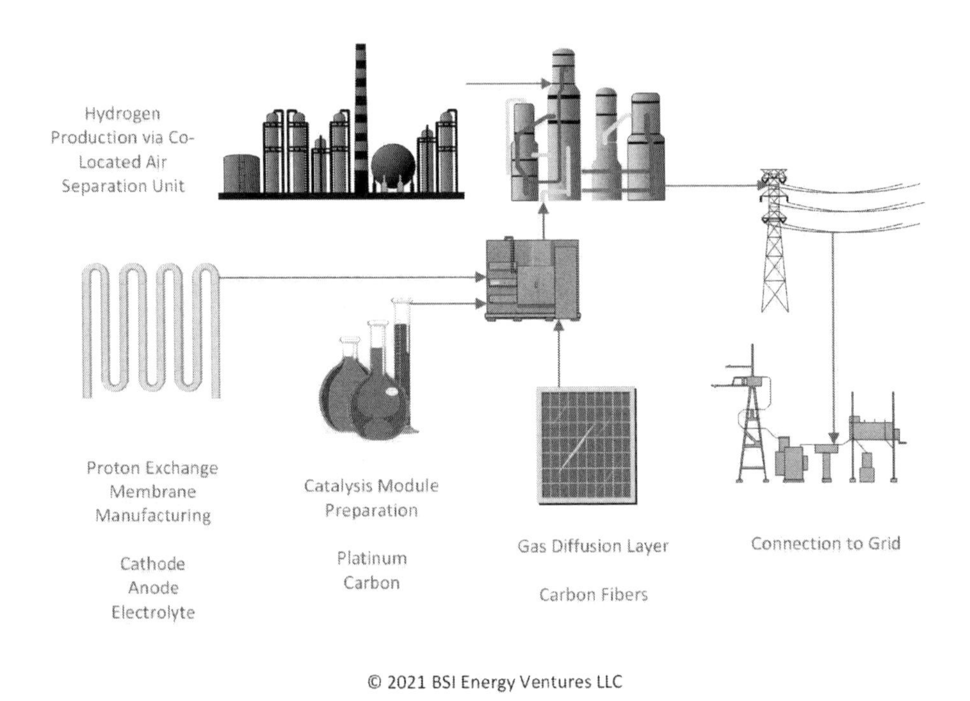

Figure 45. Utility-Scale Hydrogen Fuel Cell Power Generation Value Chain (Hypothetical)
Source: BSI Energy Ventures

The value chain for hydrogen fuel cells in utility scale applications basically consists of three steps:

- Production of *hydrogen* as the fuel, frequently with the help of electrolyzers
- Storage and release of the energy from the electricity in a *battery* (charge and discharge)
- *Transmission and distribution* of the electricity to the point of consumption

For high-volume requirements, the standard components required for a utility fuel cell value chain are hydrogen, an electrolyzer (especially for distributed applications such as refueling fuel-cell powered vehicles), the fuel cell itself (a proton-exchange membrane, or PEM), an energy storage system, and a transmission and distribution network. Concerning the conversion of the hydrogen into electricity, there is a variety of fuel cell types, each of which is suitable for different applications based on the technical specifications and parameters.

- *Proton-exchange membranes (PEMs)* consist of membrane electrode assemblies (MEAs). MEAs are comprised of electrodes, electrolyte, a catalyst, and gas diffusion layers.[173] PEM fuel cells also have subtypes;

for example, one type of PEM is a Direct Methanol Fuel Cell (DMFC), in which methanol is used as the fuel.[174]

- *Alkaline Fuel Cells* contain a solution of sodium hydroxide or potassium hydroxide as the electrolyte. Electrodes are made of carbon and a metal such as nickel. However, zinc or aluminum could be used as an anode if the by-product oxides were efficiently removed and the metal fed continuously as a strip or as a powder.[175]
- *Phosphoric acid fuel cells (PAFC)* are a type of fuel cell that uses liquid phosphoric acid as an electrolyte, saturated in a silicon carbide matrix (SiC). The electrodes are made of carbon paper coated with a finely dispersed platinum catalyst.[176]
- *Molten-carbonate fuel cells (MCFCs)* use an electrolyte consisting of a molten carbonate salt mixture (a mixture of two or more of lithium carbonate, potassium carbonate, and sodium carbonate) suspended in a porous, chemically inert ceramic matrix called beta-alumina solid electrolyte (BASE). The mobile entity that moves from the cathode to the anode through the molten electrolyte is the carbonate ion.[177]
- *Solid Oxide Fuel Cells (SOFC)* comprise four layers, three of which are ceramic. Oxygen breaks into ions at the cathode, where they can diffuse through the electrolyte to oxidize the fuel. In this reaction, two electrons and water are released as a byproduct. Electrons rush through an external circuit, repeating the cycle to generate electricity.[178]

Production of *hydrogen* as the fuel by conventional means produces CO_2, which negates the point of using a fuel cell in the first place. Steam-methane reforming can produce hydrogen, but it produces substantial amounts of CO_2, which defeats the purpose of the fuel cell. Ninety-five percent of the hydrogen produced in the United States is made by steam-methane reforming, which uses methane (CH_4) from natural gas to produce hydrogen with thermal processes,[179] and the CO_2 emissions from steam-methane reforming are significant. Carbon capture and utilization or storage (CCUS) can also potentially provide source of clean ("green") hydrogen,[180] but this is not always economical. *Electrolyzers* can produce hydrogen, but they require a source of electricity, and if that source is powered by fossil fuels, through, for example, gas turbines, the clean nature of the fuel cell is negated or at least offset. If they are supplied by pipelines of hydrogen produced with power from renewable power generation such as wind or solar, then the electricity is essentially carbon-free.

Trade-offs and Management Techniques in Capital Project Management

Technology Choice Risk Management: Choosing the Right Fuel Cell Chemistry

Hydrogen fuel cell technology is still not mature as compared to other renewable technologies, and many challenges across the entire value chain are yet to be understood well as it is scaled up. Fast technological evolution and economies of scale are driving LCOE down at a rapid rate for renewable technologies such as solar and wind for stationary (power generation) or electric vehicles for mobility (transportation) applications. In such a case, technology choice risk management becomes important for investors and operators before investing in a large-scale fuel cell project requiring huge investments. Executing pilot projects in joint venture with other entities is a prudent way to minimize risk while testing the technology.

For those making power engineering decisions, the choice is not simply whether to adopt fuel cells or an alternative power source; it's also what type of fuel cell will best suit the purpose: polymer electrolyte membrane (PEM), alkaline (AFC), phosphoric acid (PAFC), molten carbonate (MCFC), or solid oxide (SOFC).[181]

- Polymer electrolyte membrane (PEM) type fuel cells are among the most commonly used today, but their stack size is less than 100 kW. So they tend to be best suited to smaller applications such as backup power, portable power, distributed generation, transportation, and specialty vehicles.
- Alkaline (AFC) type fuel cells are of similar size and can operate in low temperatures and start up quickly, so they tend to be used in applications such as military, space, backup power, and transportation.
- Phosphoric acid (PAFC)–type fuel cells range up to about 400 kW in modules of 100 kW and tend to be applicable for distributed generation and CHP.
- Molten carbonate (MCFC)–type fuel cells are used in applications up to about 3 MW in modules of 300 kW, which puts them in the range of potential utility scale applications, as well as industrial and military uses. However, at their current stage of development, due to their high temperatures and corrosive electrolytes, they last only about five years, which cannot compete with the typical 20- to 40-year lifespans of conventional thermal power plant technologies.
- Solid oxide (SOFC)–type fuel cells currently range up to about 2 MW, which also makes them contenders for industrial and utility applications. Like MCFC-type fuel cells, they operate at very high temperatures, so they require extensive heat shielding to protect humans from contact

and are currently very expensive due to the exotic materials needed to build them. Even with those issues addressed, the further challenge is to make them durable for commercially acceptable periods of time that come somewhere close to conventional technologies.

Any and all of these technological variants need to meet safety thresholds, purity requirements, and storage limitations in order to become commercially viable. Hydrogen is highly flammable when in contact with air, and hydrogen leaks and even flames from hydrogen fires can be invisible, all of which make safety a looming concern for the commercial development of fuel cells. Moreover, each application has varying requirements for the purity of the hydrogen feedstock, which needs to be precisely measured and managed. Finally, compact storage requires high pressure, to five or ten thousand psi in some cases, and this comes with additional logistical and safety considerations.[182]

Capital Project Risks and Mitigation: Managing an Early-Stage Demonstration Project In-House

At this relatively early stage in the evolution of hydrogen power, most energy conglomerates are delivering projects through in-house sister concerns rather than EPCs. For example, a recent hydrogen powered fuel power plant operated by Hanwha Energy, at Seosan, Korea, has been built by Hanwha Engineering & Construction.[183] Fuel cell projects operated/co-operated by SK Energy have been contracted to SK E&C.[184]

Those going the EPC way are starting off with small-scale projects. For example, FirstElement Fuel selected Black & Veatch as the EPC contractor to engineer, permit, and construct 19 hydrogen refueling stations across California. The project represents the first stage of a comprehensive program for constructing a statewide True Zero hydrogen network, the first such network in the US.[185]

This, coupled with the fact that some EPCs have shifted their primary focus or left the power space altogether, make EPCs less relevant to hydrogen projects at present. For example, in 2020, AECOM and Fluor exited power sector EPC work altogether, and others such as Toshiba and Black & Veatch (B&V) departed from coal to focus exclusively on low carbon technologies.[186] The emergence of a competitive set of EPCs for hydrogen power plants has yet to become clear.

Supply Unavailability Risks and Mitigation: Ensuring an Adequate Source of Platinum

The current state of the art in fuel cell production relies on the use of a proton exchange membrane (PEM) as a polymer electrolyte that is made of perfluoro-sulfonic acid polymers (such as the Nafion membrane). PEM fuel cells require a very active catalyst material on the electrodes, typically made from platinum, which is by far the most suitable element used in this application. Almost all current PEM fuel cells use platinum particles on porous carbon supports to catalyze both hydrogen oxidation and oxygen reduction. However, the metal is highly expensive and is available in limited supply, which sometimes hinders the large-scale manufacturing of hydrogen fuel cells.[187] In fact, the global market for this precious metal experienced supply shortage due to a combination of a drop in mine supply and an increase in demand. A platinum deficit of 1.2 million ounces was recorded in 2020, mainly due to COVID-19 related mine shutdowns. However, this was the second consecutive year of shortage.[188]

Outsourcing Risks and Mitigation: Forming an R&D Partnership for Lab Experimentation

The fuel cell technology and its prototypes for various applications are still at an infancy stage, leaving project owners with a strong incentive to outsource large aspects of development projects to those with specialized knowledge and equipment.

For example, Hydrogène de France (also known as HDF Energy) engaged Ballard Power systems to develop and integrate a multi-megawatt (MW) scale fuel cell system into HDF Energy's power plants which use a combination of solar, wind, and MW scale energy storage systems.[189]

Also, the New Energy and Industrial Technology Development Organization (NEDO) contracted Toshiba Energy Storage Systems & Solutions on commissioned business for a multi-utility pure hydrogen fuel cell module for large modes of transport as a part of technology R&D programs for widespread implementation of fuel cells. The project will develop a compact, lightweight, and high-power 200kW-class standard module of pure hydrogen fuel cells for use by vessels, railroad vehicles, construction machinery, and other large modes of transport.[190]

Supplier Partnering Risks and Mitigation: Licensing Technology from a Vendor

Due to the early stage of evolution of the fuel cell industry, supplier partnering is the only option for most ventures, and equity partnerships and joint ventures, which most purchasing departments disconsider, are commonly used for product development purposes.

To further improve manufacturing of fuel cell stacks, Bosch is working closely with a Swedish manufacturer of fuel-cell stacks, Powercell Sweden AB. Under the terms of the agreement, the two partners will collaborate to ready the polymer-electrolyte membrane (PEM) fuel cell for production. Bosch will then produce this technology under license for the global automotive market. The resulting stack will complement Bosch's portfolio of fuel-cell components, and it is to be launched in 2022 at the latest.[191]

Ballard Power Systems, a Canadian developer of fuel cells, entered into a joint venture with Weichai Power to form Weichai-Ballard Hy-Energy Technologies Co. (Weichai—51%, Balard—9%) to build fuel cell stacks and modules using Ballard's liquid cooled stack (LCS) technology for commercial transportation applications in China, through a multi-year technology transfer program. Based at its global headquarters in Shandong, the JV will leverage from Weichai Power's proficiency in design engineering and powertrain integration.[192]

Procurement Bundling Trade-offs: Negotiating a Package Deal Including Customized Development through O&M and Servicing

Because of their early stage of evolution, contracts for fuel cell products and services very often bundle together manufacturing with distribution, and product development with operations and maintenance. In an example of bundling manufacturing with distribution, one company contracted Plug Power, a US-based hydrogen fuel cell manufacturer, for the supply of hydrogen fuel cell units, storage and dispensing infrastructure across its distribution network for two years.[193] In an example of bundling product development with operations and maintenance, a European shipbuilding company hired Germany-based Proton Motor Fuel Cell GmbH to provide hydrogen fuel cell solutions for marine applications from development to implementation, including power backup and storage.[194]

Contract Term Risk and Mitigation: Working with Short-Term Agreements Until Requirements and Technologies Stabilize

The short-term nature of the agreements cited in this chapter, including research and development and early-stage commercialization initiatives, is indicative of the infant stage of the current industry worldwide. Many product development initiatives are focused on a time horizon of one to a few years, and most commercialization agreements are only a couple of years in length. As the industry stabilizes around some core product/market technologies and applications with commercial runways, the length of collaborative agreements will increase substantially. Eventually, assuming continued growth, the contractual landscape would resemble that of solar and wind power, with 20–30 year PPAs and corresponding operations and maintenance agreements.

Trade-offs and Management Techniques in Operations & Maintenance

Cybersecurity Risk Management: Protecting the FCCU

Because fuel cells are getting a strong start in vehicles and especially transit buses, cybersecurity will be key to their successful evolution. Smart vehicles can be expected to be a target of cyberattacks, and electronically equipped transit buses and trains can expect to be a particularly rich target for terrorists. The fuel-cell control unit (FCCU), the central control unit for operation of the fuel-cell system, or "electrical power plant" of an electric vehicle equipped with a fuel cell, needs to be especially secure from malicious intervention.

The Automotive Software Performance Improvement and Capability Determination (ASPICE) reference protocol and the AUTomotive Open System ARchitecture (AUTOSAR) standard are useful references and standards that, while not directly pointed at cybersecurity, should provide improved security if followed carefully. In time, cybersecurity standards specific to fuel cells, and specific to energy production from fuel cells, may emerge from organizations such as the Society of Automotive Engineers (SAE), the National Institution for Standards and Technology (NIST), North American Electric Reliability Corporation (NERC), and International Organization for Standardization (ISO) and their European counterpart, the European Committee for Standardization (CEN).

Bosch, which makes FCCUs for hydrogen fuel tanks, uses modular software architecture aligned with the ASPICE and the AUTOSAR standards; Bosch's system consists of hardware and software components to monitor the tanks to enhance security and monitoring.[195]

Internet of Things (IoT) and Artificial Intelligence (AI) Technology Choices: Piloting AI Technology in Autonomous Vehicle Applications

Smart technology including AI is playing an increasing role in optimizing operations and maintenance parameters. This includes thermal conditions including the likelihood of fire, security including the possibility of intrusion, monitoring of power generation, and real-time monitoring and optimization of operational parameters. Doosan Fuel Cell and KT (formerly Korea Telecom) are jointly developing an artificial intelligence–driven, unmanned self-driving platform for fuel-cell-powered vehicles. KT brings the AI technology and Doosan brings the fuel cells and fuel cell operating system.[196]

Peak Capacity Strategies: Achieving Economies of Scale to Gain Effective Capacity

Economies of scale are highly significant in fuel cell manufacturing operations, and this will encourage a high degree of industry concentration (and a high degree of fallout from those who cannot scale competitively and need to sell out to the market leaders), centralized purchasing, and the formation of a few large production plants instead of many smaller or national facilities (which might conflict with local content desires of host countries).

The fuel cell stack, which accounts for 71% of the total system cost at 1,000 systems, drops to 48% at 500,000 systems per year, respectively. What's driving the economies of scale? The membrane costs decrease with higher volumes, declining from 28% of system cost at 1,000 systems per year to 10% of system cost at 500,000 systems per year. The gas diffusion layer (GDL) costs also decrease with higher volumes, declining from 20% of system cost at 1,000 systems per year to 5% of system cost at 500,000 systems per year. Meanwhile, catalyst and bipolar plate costs increase from 20% and 13% to 45% and 27% of total stack cost, respectively.[197]

As fuel cell production volume increases, manufacturers will face challenges scaling up the complex manufacturing process, though. Manufacturing planning systems can smooth production peaks and valleys, and manufacturing execution technology such as robotics and RFID scanners can help manage production in real time. However, manufacturing membrane electrode subassemblies and fuel cell stacks, and then assembling complete modules at the higher volumes, has not yet been demonstrated.

In preparation for expected growth in the demand for fuel cell engines to power automotive, bus, truck, train and marine vessels, Ballard Power Systems has been scaling up capacity (sixfold) to make membrane electrode assemblies, fuel cell stacks, and assemblies of complete modules. Radio frequency (RF) scanners will feed data into its manufacturing execution (production management) system.[198]

Sourcing Trade-offs: Avoiding Intellectual Property Disputes in Fuel Cell Supply Agreements

Due to the embryonic nature of the industry, and to the fact that very little can be considered "common knowledge," purchasing agreements frequently involve acquiring unique intellectual property, or at least trade secrets, so in the absence of clearly drawn boundaries of their companies' strategic footprint and direction, management might be inviting trouble by forming agreements that inadvertently give away valuable intellectual property to suppliers who could then become competitors. Typically then, supply agreements, and especially partner shipping agreements, are supply chain activities of the highest order and must involve executive input and signature.

For example, Hyundai Motor and LS Electric signed a memorandum of understanding to develop hydrogen fuel cell–based power generation systems for automobile applications. Under the cooperation pact, Hyundai brings in the core technology for hydrogen fuel cell systems and will be responsible for supplying them. LS Electric brings in integrated power system solutions. The joint effort will leverage experience of both companies in mass production of hydrogen electric vehicles and power systems, respectively.[199]

TCO Trade-offs: Making (or Spending) Money in Recycling and Refurbishing

Because hydrogen fuel accounts for only 35% of the total cost of ownership for stationary applications and 21% of the TCO for automotive applications, manufacturers will have a strong incentive to reduce the initial sale price of fuel cells in order to sell vehicles and power plants in the first instance.[200] In addition, there may be a strong opportunity for them to make money on the back end through proton exchange membrane (PEM) replacements and hydrogen itself, which is likely to create strong downward pressure on initial fuel cell prices to capture the market. This dynamic is likely to force purchasing managers to be aggressive in negotiating materials and component prices, and supply chain partners to be ruthless on year-on-year (YOY) cost reduction.

In a bid to reduce total life-cycle cost, Ballard has developed recycling and refurbishing processes for components such as catalysts made of precious metals, which occupy a significant share of the production cost. It aims to recycle catalysts, reuse bipolar plates, and refurbish fuel cells. For fuel cell stacks that have reached the end of life, the customer returns the fuel cell stack to Ballard, which replaces the membrane electrode assembly (MEA) and reuses the hardware and plates. Ballard then sends the used MEAs to a metals recovery specialist for recovery of the platinum and other precious metals. This process can generally be done for fuel cell stacks that are ten years old or less and will typically save customers 30% of the cost of purchasing a new fuel cell stack. Typically, more than 95% of the precious metals in the MEA are reclaimed during this process. Electronics, pumps, valves, hoses, and metal for housing and frames can also be recycled in conformance with standard regulations such as the EU's Waste Electrical and Electronic Equipment Directive (WEEE).[201]

Supply Chain Roadmap for Fuel Cells

As "green hydrogen" is essential to reaping environmental benefits from fuel cells, both the hydrogen supply chain and the fuel cell manufacturing and distribution supply chain need to be developed in parallel. High-impact supply chain efforts are likely to focus on: 1) reducing the potential for leaks and flammable conditions; 2) realizing economies of scale in the GDL and the fuel stack; 3) sourcing platinum to ensure adequate supply; 4) protecting the FCCU from cyberattack; 5) developing optimal chemistries for various power ranges and environments; and 6) creating a market for replacement membrane elements.

More broadly, the fuel cell supply chain needs supply chain help in all four areas of supply chain impact: *rationalization, synchronization, customization,* and *innovation.*

Cost reduction actions need to focus on:

- Achieving economies of scale in the membrane and gas diffusion layer (GDL) of the fuel cell stack
- Developing recycling and refurbishing processes, workflows, and physical shipping logistics for inputs such as catalysts

Supply chain–related support for Innovation needs to address:

- Ensuring safety by designing transport and logistics infrastructure that minimizes the potential for leaks and flammable conditions, including by measuring and managing the purity of the hydrogen feedstock so it conforms to each application
- Creating a market for replacement membrane elements

Supply chain-driven customization of fuel cells for various applications needs to accomplish:

- Configuration of product and service delivery per the application; for utility applications more likely phosphoric acid (PAFC), molten carbonate (MCFC), or solid oxide (SOFC)
- Development of storage compression distribution modes that balance the cost benefits of high compression against the risk and severity of potential explosion

Supply chain actions to balance supply and demand (*synchronization*) need to encompass:

- Sourcing platinum to ensure adequate supply
- Protecting the FCCU from cyberattack

Figure 46 summarizes key emerging supply chain value creation opportunities for fuel cells.

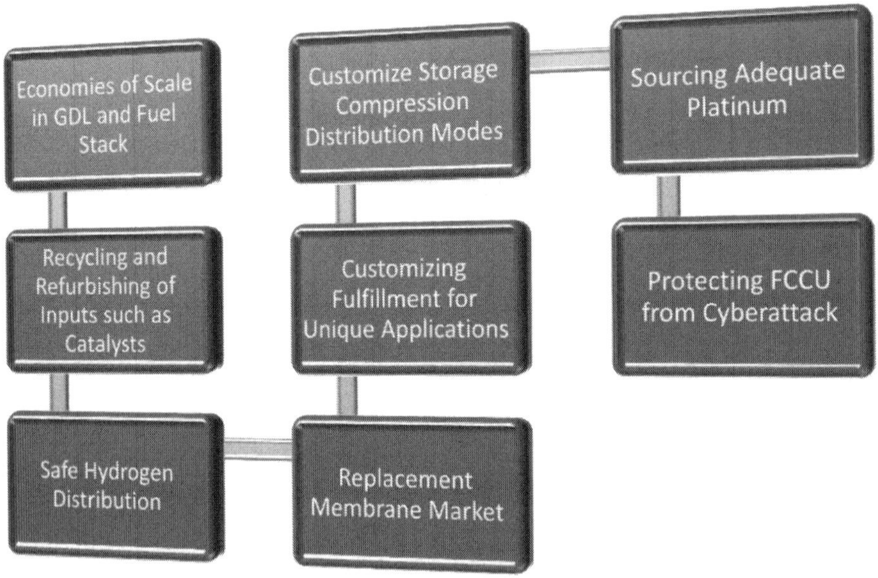

Economies of Scale in GDL and Fuel Stack

Recycling and Refurbishing of Inputs such as Catalysts

Safe Hydrogen Distribution

Replacement Membrane Market

Customizing Fulfillment for Unique Applications

Customize Storage Compression Distribution Modes

Sourcing Adequate Platinum

Protecting FCCU from Cyberattack

Figure 46. Emerging Supply Chain Opportunities for Fuel Cells

Utility-Scale Energy Storage

Chapter Highlights

1. Pumped Storage - Hydropower (PSH) storage is by far the most common type of utility-scale energy storage in the US today (aside from Impoundment-type Hydropower - water stored behind dams).
2. Battery energy storage is being rapidly adopted and is likely to be considered integral to solar projects in the future.
3. Lithium-ion battery storage technology is rapidly increasing its footprint due to cost declines of nearly 90% over the past ten years. For utilities, its power capacity and discharge rate are best suited to transmission & distribution grid support, and reserve and response systems. Its steep cost declines are being assisted by scale achieved through rapid adoption in commercial and residential power systems as well as electric vehicles and consumer electronics.
4. Overheating and fire risks associated with lithium-ion batteries need to be managed through choice of battery chemistry as well as system engineering including thermal control systems, and compartmentalization and separator technologies.
5. Buyers are negotiating long-term agreements for lithium-ion battery supply to offset risks of capacity shortage due to strong demand and manufacturing capacity bottlenecks.
6. Flake graphite, a raw material used in lithium-ion battery anodes, is short supply and is controlled mostly by China, resulting in a search for alternative supply options. Ethical sourcing constraints also impact the availability of numerous raw material minerals, especially cobalt, which currently comes primarily from one country which crosses the line drawn by governments and industry associates regarding slave and child labor.
7. The lithium-ion battery supply chain involves up to eight levels, and outsourcing is extremely common, nearly ensuring that supply chain managers dealing with energy storage will need to set and execute a logical supply chain strategy including which activities to outsource, which to keep or build in-house (if any), how to bundle the

procurement of the various components and services required from raw material through installation, and which companies to partner with.

8. Smart systems including artificial intelligence and machine learning are extensions of the storage features, in order to optimize charge and discharge timing and rates to maximize revenue and minimize cost within constraints including tariff structures, demand response incentives and penalties, and equipment warranties.
9. Cybersecurity protection is essential, for hackers can inflict damage, and potentially shut down the power grid by penetrating energy storage systems.
10. Important supply chain advantages can be gained through: 1) ensuring availability of critical minerals such as cobalt; 2) minimizing or eliminating the risk of overheating and fire; 3) proactively managing battery costs throughout the supply chain; 4) stacking multiple benefits simultaneously to increase the ROI; 5) reducing the cyber security attack surface; 6) deploying AI for battery operating system and load management; 7) integrating energy storage with vehicle charging networks; and 8) developing economical battery recycling approaches.

Introduction

Utility-scale energy storage is both an optimization vehicle for conventional power generation and an enabler of renewable but intermittent power generation technologies such as solar and wind. It has the economic leverage to decrease the effective Levelized Cost of Electricity (LCOE), with measurable economic and social benefits at the national and global levels. International lending agencies such as the International Finance Corporation (IFC) are actively working to make energy storage more accessible and less expensive, especially in developing economies, because of its economic leverage.

Proper characterization of the supply chain for energy storage depends on the technology involved, which in turn depends on the application, the power range, and the discharge rate required. There are generally 12 applications, of which seven commonly recognized utility-scale applications are depicted in Figure 47 below.

- Primary frequency regulation
- Renewable energy capacity firming (renewable integration)
- Energy arbitrage
- T&D upgrade deferral (postponement)
- End-user demand management (including "peak shaving" and demand response programs and demand charge reduction programs for commercial and industrial customers)

- Batteries and Uninterruptible Power Supplies (UPS) for distributed generation
- Virtual Power Plants (VPP) (load leveling across power-generating assets). By 2025, 80% of residential systems will connect to virtual power plants (VPP) networks, according to one source.[202]

Other applications include transmission congestion relief, supply and voltage control, black start capabilities, and reduction of T&D line losses.

Primary Applications	Primary Users
Frequency Regulation	• Power Generators • Commercial & Industrial Users
Renewable Energy Capacity Firming and Integration with Grid	• Intermittent Renewable Energy Generators • Demand Response Service Providers
Energy Arbitrage	• Power Generators
Demand Charge Reduction	• Industrial and Commercial Users • Residential Consumers
Transmission & Distribution (T&D) Upgrade Avoidance or Deferral	• T&D Owner
Reduction of T&D Line Losses	• T&D Owner
Distributed Generation	• Industrial and Commercial Users • Residential Consumers

Figure 47. Energy Storage Applications
(*Source:* BSI Energy Ventures)

The technology for these applications depends on the power range required and the discharge duration. Some of the prevalent technologies in utility scale energy storage are depicted in Figure 48. For bulk power management (high-power, high-discharge) applications, the options are normally pumped hydropower storage, compressed air energy storage, fuel cells, and flow batteries. There are also many less frequently used, and emerging, technologies for bulk power management, including cryogenic energy storage and new variants on gravity-based, thermal, and ocean wave energy storage. For Transmission & Distribution and local reserve and response, battery storage is typical, and the battery chemistry varies depending on the output requirements. Supercapacitors, flywheels, and other technologies are currently infrequently used in utility applications.

While pumped hydro storage has been around since 1907[203] and currently represents 95–96% of all storage in the US,[204] the use of battery storage is surging, lithium-ion battery storage in particular, because by comparison to pumped hydro storage, it requires almost no capital or time to deploy and can be deployed anywhere.[205]

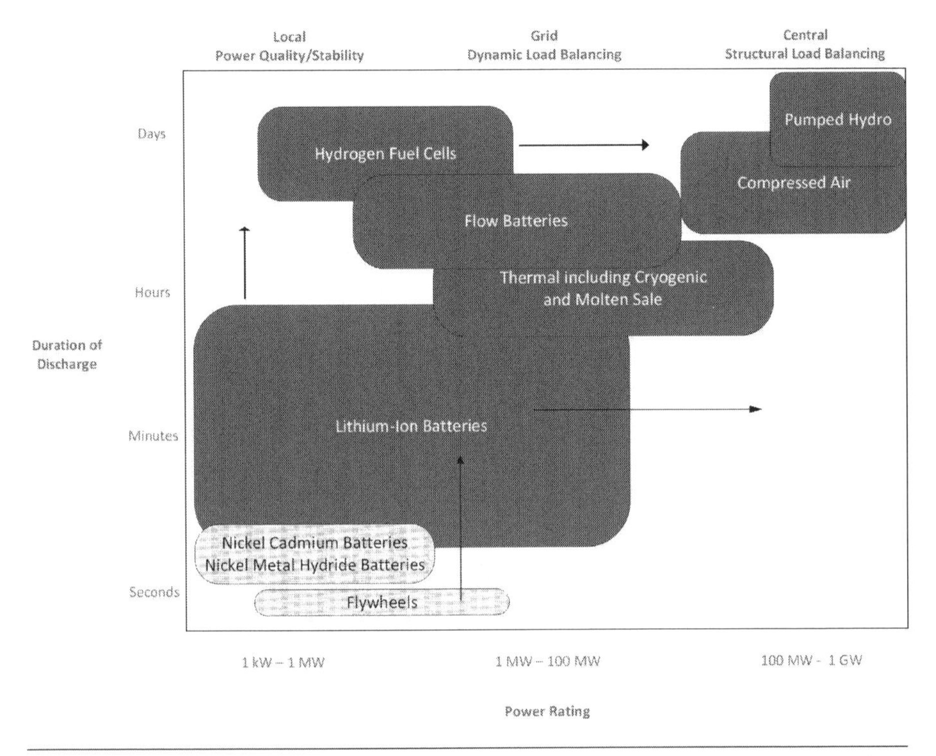

Figure 48. Selected Utility-Scale Energy Storage Technologies by Power Rating and Discharge Time (*Source:* BSI Energy Ventures, after EPRI[206])
Note: Blue bubbles represent relatively widely recognized current utility-scale applications.

The equipment requirements, and hence the supply chains, vary widely according to the technology of the systems.

- Pumped Storage Hydro systems are based on pumping water between two reservoirs at high- and low-demand periods. Their power plants are based on variable speed motors, torque convertors, generators, pumps, and turbines. Procurement and supply chain management is analogous to that of a combined cycle plant in that it centers around rotating mechanical and electromechanical equipment.
- Compressed air energy storage (CAES) systems are based on compressing air to high pressure during low energy demand and then releasing it during high load demand. Although CAES often comes up in discussion of utility-scale energy storage, only a handful of utility-scale

compressed air energy storage systems are operational today, mostly in Germany, the US, and the UK, and in one plant planned in Canada.[207]

- Flow batteries, which have larger capacity than lithium-ion batteries and do not have the same flammability risks, are based on two chemical solutions, each held in a separate tank, producing electricity as they pass over a membrane. There are relatively few flow batteries in service today; the total market for flow batteries is valued at less than $300 million worldwide today, and the growth rate for flow battery sales is modest, at 14% per annum.[208] Costs of flow batteries have not decreased at the same rate as for lithium-ion batteries.[209]
- Lithium-ion battery storage is in widespread use today and is the prevailing favorite for growth into a broader range of power applications due to its impressively decreasing costs and its potential to find applications across the power, automotive, and consumer electronics industries, just to name three big ones, which presages continued cost reduction per kwH based on massive economies of scale already being realized at many gigantic manufacturing plants. The cost of lithium-ion batteries decreased by 87% between 2010 and 2019,[210] which spurred a dramatic and concurrent increase in adoption, with the market projected to nearly double between 2018 and 2025, according to BSI Energy Ventures.[VIII] The maturation of solid state batteries such as those being pioneered by QuantumScape, which appear to be able to recharge three times faster and last four times longer than "wet" cells, could extend the application range and further lower the cost of lithium-ion batteries, making them central to utility and grid applications in the future.[211]
- Other types of storage such as cryogenic energy storage, new variants on gravity-based and thermal storage, and ocean wave energy storage are continuously being developed, but are not in widespread use.

The supply chain management tools and techniques presented in the Coal and Gas-Fired Power Plants chapter of this book can be applied to manage supply chains for PHS/PSH and CAES, which are similarly centered around the procurement, installation, and integration of rotating electro-mechanical equipment like turbines, compressors, motors, and expanders (for CAES) and control systems, as well as static high-pressure equipment. CAES also necessitates an air tank or cavern (e.g., a salt cavern), which may involve earthworking; for this problem the Upstream Oil & Gas chapter should be useful.

SCM tools for cryogenic, gravity, and ocean wave and ocean thermal energy systems are not discussed in this edition as these technologies are in limited commercial use today. Supply chain management approaches for these may be

[VIII] Per output of BSI's global lithium-ion battery and components supply-demand model

addressed in future editions of this book as they scale and become relevant to a broader audience.

Among battery storage systems, this chapter will focus on supply chain management for lithium-ion battery systems rather than flow batteries due to the much larger penetration and rapid uptake of lithium-ion batteries compared to other battery storage technologies.

The main components of a lithium-ion battery storage system are a battery, a monitoring and control system, a power conversion system (inverter) that converts DC power from the battery to AC power for the grid, and a housing. The batteries themselves consist of a module, which contains packs, which contain cells.

The energy storage value chain starts with the manufacturing of anodes, cathodes, and electrolytes for the battery cells. The cells are assembled to form packs, which are further arrayed into stacks. Batteries are connected to the energy source (solar or wind power), and the set-up includes monitors and controls, a power conversion system, and an AC transformer to convert stored energy into alternating current to supply to the end use point or grid. Figure 49 shows this sequence graphically.

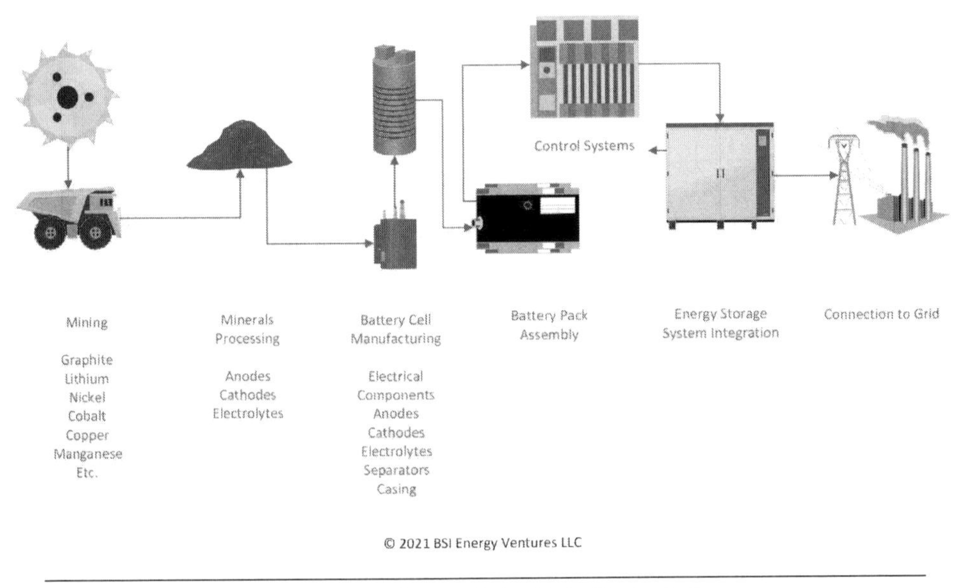

Figure 49. Lithium-Ion Battery Energy Storage Value Chain (Illustrative)
(*Source:* BSI Energy Ventures)

Energy storage systems are a relatively low-cost investment that leverage the efficiency of the existing generation, transmission, and distribution infrastructure. The supply chain cost of pure wholesale storage, not as an accessory to a power project but for frequency regulation and other benefits to existing plants, totals about 12% of the cost of electricity ($0.003/kwh), according to analysis by

BSI Energy Ventures. This splits about evenly between integration costs and the carrying cost of the capital investment in the hardware elements listed above. In exchange for that investment, the systems achieve a 59% return on investment, based on a "value stacked" set of benefits including deferral of upgrades and avoidance of losses due to curtailment.[212] Energy storage may or may not provide an attractive commercial return as a standalone investment, but if there are existing power generation assets that can be made more efficient through energy storage, the investment in energy storage usually offers a clear payback.

Trade-offs and Management Techniques in Capital Project Management

Technology Choice Risk Management: Minimizing Thermal Risks in Batteries

One of the major risk factors in battery-based energy storage systems, and lithium-ion systems in particular, is the potential for overheating and the risk of fire. The same design problem experienced by Samsung Galaxy owners occurs in large-scale lithium-ion battery systems. In fact, there are many recent examples of energy storage systems catching fire.

- A fire broke out during the commissioning phase at an Ineo battery container at Engie's test site in Drogenbos, Belgium, in 2017. The 6MW project was the first time an energy storage system (ESS) was to be used for grid Frequency Containment Reserve (FCR) services (operating reserves to contain fluctuations to maintain balance in power systems) in Belgium. While a 1MW Ineo Scle container was heavily damaged, two other containers suffered only light damage.[213]
- A battery storage system exploded in the 2-watt battery array at Arizona Public Service Co.'s Battery Storage System, leading to temporary shutdown of two other plants.[214]
- Between May 2018 and January 2019, 21 battery storage systems caught fires across South Korea, which accounts for a third of the global installed battery storage system capacity.[215] The government asked 522 out of 1,490 Energy Storage Systems (ESS) facilities across the country to suspend operation and new commissioning.[216] LG Chem's energy solutions business lost $124 million in the first quarter of 2019, following seven straight quarters of profits.[217]

A "tabless" lithium-ion battery design developed by Tesla reduces the distance that electrons have to travel within the battery cell, thereby reducing the ohmic resistance and thus the heat produced and also the efficiency loss. Less constrained by overheating concerns, Tesla will be able to more than double the size of the battery packs from 2,170 cells (21 mm by 70 mm) used in its Model 3 and Powerwall solution to 4,680 cells (46 mm by 80 mm).[218]

Until and unless tabless lithium-ion batteries actually solve overheating conclusively, a number of measures to mitigate the risk of fire and explosion should be part of supply chain management initiatives. One option is to use heat-resistant separators inside the battery cell to prevent the heat from one component from igniting other components. A second option is to improve the battery control system that regulates voltage so as to prevent overheating. And a third option is to specify a battery chemistry that is less flammable. Each option has its pros and cons. Separators are not always infallible, depending on temperature levels and other factors. Control systems are expensive and may not act quickly enough to prevent fire. Alternative battery chemistries are usually not as high performing, which would force compromises that customers may find unattractive. While some of these issues might typically fall within the purview of an Engineering department, the Supply Chain department is integral to these choices due to the fact that these components are nearly always outsourced, and the cost versus performance trade-offs can depend on procurement terms like volume commitments, which impact economies of scale and contract terms such as quality tests and approvals.

Capital Project Risks and Mitigation: Insuring System Performance Reliability

The possibility of business interruption has become a top concern since the COVID-19 experience exposed how susceptible most supply chains are to unplanned external disturbances. To minimize the risk of business interruptions, such as COVID-19 supply chain disruptions, ESS Inc., a manufacturer of large energy storage systems, has begun offering ten-year insurance coverage on its long-duration energy storage products through Munich Re, one of the world's leading insurance companies. The policy provides a warranty for ESS Inc.'s flow batteries, supporting system performance guarantee. The insurance coverage, which also addresses technology risk, is designed to provide long-term assurance of project performance to system owners, investors, and lenders in the energy sector.[219]

Supply Unavailability Risks and Mitigation: Securing the Battery Raw Materials Supply Chain

In 2020, the EU added graphite, a key raw material for anodes in lithium-ion batteries, as a critical raw material, over concerns of a potential supply shortage. The EU considers graphite to be a strategic resource because of the anticipated growth in demand for lithium-ion batteries, especially for electric vehicles but also for other applications including power management in generation, transmission, and distribution. Demand for graphite anode material spiked from 20,000 tonnes in 2011 to nearly 400,000 tonnes in 2020 and is expected to hit 600,000 tonnes by 2023.[220]

Government mandates are adding to the already strong growth in demand based on an attractive value proposition on an unsubsidized basis. For example, India mandated storage capacity for 50% of generation installed in 2019 for a 1.2 GW auction of solar plus storage.[221]

Another factor contributing to materials availability concerns centers on over-reliance on China. More than 90% of world supply of amorphous graphite, and around two-thirds of flake graphite supply in 2019 came from China. Almost 100% of the anode precursor material—flake graphite processing, which produces lithium-ion battery grades (spherical graphite)—happens in China. But even China became a net importer of graphite in 2019 because of tightening restrictions and state controls on quotas. For example, one Chinese flake producer, South Sea Petroleum, reported that its Chinese production lines have been able to operate only for a few months a year because of a graphite shortage, citing limited graphite ore resources and production quotas. Even miners outside China have witnessed high costs and environmental barriers to entry.[222]

Buyers have signed long-term procurement agreements to mitigate the supply risk. For example, Thyssenkrupp Materials Trading signed a ten-year agreement for graphite from Australian graphite explorer EcoGraf to supply its battery manufacturing operations. The agreement covers the sale of half of the planned output of purified spherical graphite and byproduct fines from the plant.[223]

Outsourcing Risks and Mitigation: Licensing Strategies

Outsourcing is common in battery manufacturing. Its frequency is an extension of the extraordinary growth of the Electronic Contract Manufacturing industry, which sprang up in the 1990s due to the exploding sales of electronic devices, which has only accelerated as mobile phones and tablets have come onto the scene. Take the example of FREYR AS, a Norwegian battery manufacturer, which buys its components from 24M Technologies. FREYR is developing over 40 GWh of scalable, modular battery cell production capacity via a phased development approach utilizing its strategic partnership, including in-licensing of next-generation technologies.[224]

The lithium-ion battery materials supply chain consists of many layers, each of which may be partially or wholly outsourced. Battery manufacturers (OEMs) buy from pack manufacturers, who buy from cell manufacturers, who buy cathodes, anodes, and electrolyte materials. Smaller components such as separators and electronic components are often sub-sub-contracted. For example, Talga, an Australian advanced materials company focusing on battery anode and graphene additive products, sources anodes for its graphite mining and battery anode production facility in northern Sweden from ABB.[225]

Companies like Iberdrola and Vattenfall in the power industry, and new entrants such as Tesla, which is competing with them for distributed energy platforms installed on residential rooftops, are at the top of the food chain, while major electronics manufacturers supply them with battery packs and energy storage management systems. They in turn decide how much of the supply chain to insource and how much to outsource.

Supplier Partnering Risks and Mitigation: Crafting Flexible Partnerships

Due to the many layers in the battery supply chain and the importance of speed to market, there is substantial opportunity for strategic partnering, and also substantial risks of making the wrong partnering decisions or ceding too much commercial, financial, or technological advantage to the partners.

For example, Tesla partnered closely with Panasonic for the batteries it uses in its electric vehicles (EVs). Tesla had formed a very close relationship with Panasonic, but the relationship degraded as a result of production delays, so it has aligned instead with Chinese suppliers for the batteries and energy storage systems used in its rooftop tile systems.[226] Tesla retained sufficient independence to alter its purchasing and supply chain strategy, even after having committed an extraordinary amount of capital and intellectual property to the Panasonic relationship.

Procurement Bundling Trade-offs: Architecting a Supply Chain of Specialists

Are there economic benefits to bundling battery system integration with pack assembly, or pack assembly with cell manufacturing, or cell manufacturing with anode or cathode materials supply? While the answer depends on each application, technology, and scale of operations, each step is highly specialized and technical, so *prima facie* it's not obvious that it could be absorbed by an adjacent layer in the supply chain.

Contract Term Risk and Mitigation: Matching Contract Term to the Product Roadmap

Contract term is an especially tricky question in the battery supply chain. Rapid growth of electrification raises the clear potential for material supply shortages in future years, when demand may outstrip supply. Since the establishment of commercial-scale capacity takes years, this sets up a chasing game in which there is always the possibility for either overcapacity or shortage, based on gaps between the timing of demand and the timing of capacity additions. Such gaps have arisen with nickel and other raw material resources that analysts predicted would be in shortage. In some cases, manufacturers added massive capacity until it was in oversupply, initiating a cycle of boom and bust.

Supply managers must forecast their specific material requirements. In one case, a battery manufacturer decided to sign a long-term agreement for lithium. SQM signed an eight-year lithium supply contract with LG Energy Solutions.

Chile's SQM (the second-largest global producer of lithium) plans between 2021 and 2029 to supply ultralight metal lithium, a main ingredient in powering electric vehicles. The agreement is for 55,000 metric tons of lithium carbonate equivalent.

Trade-offs and Management Techniques in Operations & Maintenance

Cybersecurity Risk Management: Reducing the Attack Surface

Since electric grids have been the focus of repeated cyberattacks, cybersecurity safeguards must be in place when procuring energy storage systems. The attack by Industroyer malware (also known as Crashoverride) on Germany's and Ukraine's power grids serves as just one example of the constant threat.

There are two ways in which such attackers can hit storage systems via inverters. One way is by manipulating the grid-connected power of PV and storage to cause unwanted fluctuations in voltage frequency in the distribution grid. This can ultimately lead to the grid going down. Another way is by manipulating the battery management system make a battery catch fire or explode.

Not only could a hacker potentially blow up a system battery, but they could do so rapidly, compromising the energy storage system in seconds. Lack of encryption in communication protocols, along with weak passwords, could expose a major security vulnerability for inverters.

Certifications such as ISA/IEC 62443 help manufacturers and system integrators safeguard against cybersecurity threats.[227]

Internet of Things (IoT) and Artificial Intelligence (AI) Technology Choices: Learning from Demand Fluctuations

Utilities (Front-of-the-Meter, or FTM) systems use AI to forecast co-located solar production and energy market prices to maximize the value of energy sold through the grid into energy markets. Such strategies can be complex due to the multiple parameters and value streams involved, such as, for example, forward capacity, day-ahead and real-time energy, and ancillary services, such as operating reserves and frequency regulation that each have a different price value and curve, risk and wear and tear on the ESS that must be optimized.[228] Also, utilities can use AI to optimize the split of sales to grid customers under those programs, and sales to specific high-usage industrial customers under other rate structures. Each of these tariff structures is unique and complex. Making energy storage decisions that account for all of them simultaneously is nearly impossible if done manually. Furthermore, even if manual computation was feasible, the analysis would need to be recomputed on a continuous basis, which is uneconomical. Therefore, AI serves a very important purpose in energy management systems at utilities.

For commercial and industrial companies (Behind-the-Meter, or BTM) that draw electricity from the grid, AI can be used to forecast customer site load, solar production, and other co-located generation, which can in turn be used to modify charge and discharge rates to avert costly energy spikes. Intelligent energy storage can also do a much better job than static demand-response programs at optimizing the amount of energy drawn at various times of day and at various tariff rates, adjusting the quality of electricity sold back to the grid, and operating equipment within warranty parameters.

Peak Capacity Strategies: Optimizing Reliability with Control Systems

AI can also help to optimize preventive maintenance and performance tracking for energy storage systems. Network operations centers (NOCs) work 24/7 and track the ESS, its components, and related ancillary equipment, such as non-export relays, load meters, and data feeds from solar and other onsite generation. The NOCs also collect information about site load, weather conditions, and market pricing. With such a high volume of data continually coming in, human operators cannot process everything. Automation tools can read and parse the high volume of data and route it to the appropriate location. Machine learning tools can read and understand which messages should be considered high priority and automatically handle the alarm, or route it to the appropriate triage team. A combination of process automation and AI can automatically and remotely resolve issues that can be dealt with via network communications. It can auto-create trouble tickets that dispatch repair crews. And algorithms can analyze trends and anticipate similar failures or performance constraints.

AI also supports asset optimization, which looks at the resting state of charge, cycling, and depth of discharge, and at how these factors degrade the battery over time. By understanding operating conditions, AI can predict wear and tear on the ESS and help maximize the life of the battery system.

Peak Power, Inc., an AI solutions developer, integrated 375 kW / 940 kWh of battery energy storage with AI technology at the site of a commercial customer in New York. The energy storage system features an intelligent software platform that optimizes charge and discharge rates by forecasting moments of peak demand on the grid through the use of Big Data and Machine Learning.[229]

Sourcing Trade-offs: Ensuring Responsible Sourcing of Cobalt and Other Minerals

Cobalt has been an ingredient in lithium-ion battery chemistries for decades. Lithium Nickel Manganese Cobalt Oxide, or NMC for short, is used in best-selling battery storage products, including the LG Chem Resu and the Tesla Powerwall.[230] Another popular lithium-ion battery chemistry is Lithium Nickel Cobalt Aluminum Oxide (NCA).

Because of labor issues including slave and child labor in the Democratic Republic of the Congo, where 60% of the world's cobalt comes from, a number of international protocols have discouraged the sourcing of cobalt from this country. Many multilateral organizations have established process-based recommendations and guidelines to discourage "irresponsible" minerals sourcing. For example, the OECD authored a Due Diligence Guide designed to help companies respect human rights in their sourcing practices. The OECD also hosts a minerals forum centered on responsible and ethical minerals sourcing.[231] Also, the International Council on Mining and Metals (ICMM) established a set of ethical mining principles to which its members, such as Rio Tinto, Anglo American, Glencore, and BHP, must adhere. The ICMM's requirements are process based, not prescriptive solutions, so members can theoretically satisfy the requirements solely by compiling various internal documentation.[232]

Some of these have been codified into law. Although not specifically related to cobalt, a 2010 US law, Section 1502 of the Dodd-Frank Act, has heightened legal scrutiny of minerals sourcing. It requires American companies to check their supply chains for any tin, tungsten, tantalum and gold sources that might be under militia control in the Congo region, and to take steps to avoid supporting such regimes through their sourcing practices. Many people feel that cobalt should be added to the list. Effective starting in 2021, EU legislation imposes a similar limitation on conflict minerals, which is directed at the same four minerals—tin, tantalum, tungsten and gold—because of their frequent connection to forced labor or the financing of armed conflict.[233] The targeted minerals are generally not used in popular lithium-ion battery chemistries by large manufacturers today (they are used in some alternative and less frequently used batteries), but there is a strong enough connection to cobalt and the Congo region that they constitute a very real factor for anyone connected with the lithium-ion battery supply chain.

TCO Trade-offs: Understanding the Levelized Cost of Storage

Most energy storage investments are complex assets with multiple revenue streams and multiple cost streams attached to them, and this can complicate TCO analysis. The energy storage industry calls the commercial evaluation of energy storage systems "value stacking" because these different revenue and profit (revenue-cost) streams can be realized simultaneously. For example, wholesale energy storage systems can be justified on the basis of revenue streams that stem from energy arbitrage, frequency regulation, spinning or non-spinning reserves, and resource flexibility. Each "use case" can deliver very different financial results. The same energy storage system can be used in transmission & distribution networks with no revenue streams and be justified instead on the deferment of investment in T&D upgrades. Both have the same or similar acquisition and operating costs, but very different paybacks.

For this reason, the Rocky Mountain Institute illustrated the benefit of energy storage systems through a set of four use cases instead of coming up with a specific return on investment percentage.[234] The Institute of Electrical and Electronics Engineers did the same when comparing lithium-ion battery payback.[235] Lazard prepares its Levelized Cost of Storage based on six use cases—one for each of Wholesale, Transmission & Distribution, Utility Scale (PV & Storage), Commercial & Industrial (Standalone), Commercial & Industrial (PV & Storage), and Residential (PV & Storage). Furthermore, within each use case the cost of lithium-ion battery storage is compared to the analogous cost of flow batteries and lead-acid batteries of different chemistries. The associated Levelized Cost of Storage varies from $0.03 to $0.10/kwh.[236] The Indian Energy Transitions Commission further estimates that co-located battery energy storage systems are 7–8% more cost-efficient than standalone battery energy storage systems.[237]

Project management professionals working on energy storage projects need sufficient financial background and business acumen to fully evaluate the range of options and the most effective technologies and applications.

Supply Chain Roadmap for Energy Storage

The energy storage supply chain has tremendous opportunities for improvement as it matures. The companies that seize the opportunities will have a distinct competitive advantage.

Most importantly, excellent supply chain management can support innovation which can capture more of the eligible market. Three specific advances will move the industry forward, and supply chain management can play a role in each of them:

- Working with suppliers to minimize or eliminate the risk of overheating and fire
- Integrating energy storage with vehicle charging networks
- Developing economical battery recycling approaches
- Supply chain management can ensure adequate supply and balance demand with supply in two ways:
- Ensuring availability of critical minerals such as cobalt
- Reducing the cybersecurity attack surface

Finally, supply chain management can reduce costs in three ways:

- Proactively managing battery costs throughout the supply chain ("should-cost")
- Realizing multiple benefits simultaneously to increase the ROI of energy storage investments ("value stacking")
- Deploying AI for battery operating system and load management ("Smart Load Management")

Figure 50 summarizes the emerging supply chain value creation opportunities for energy storage.

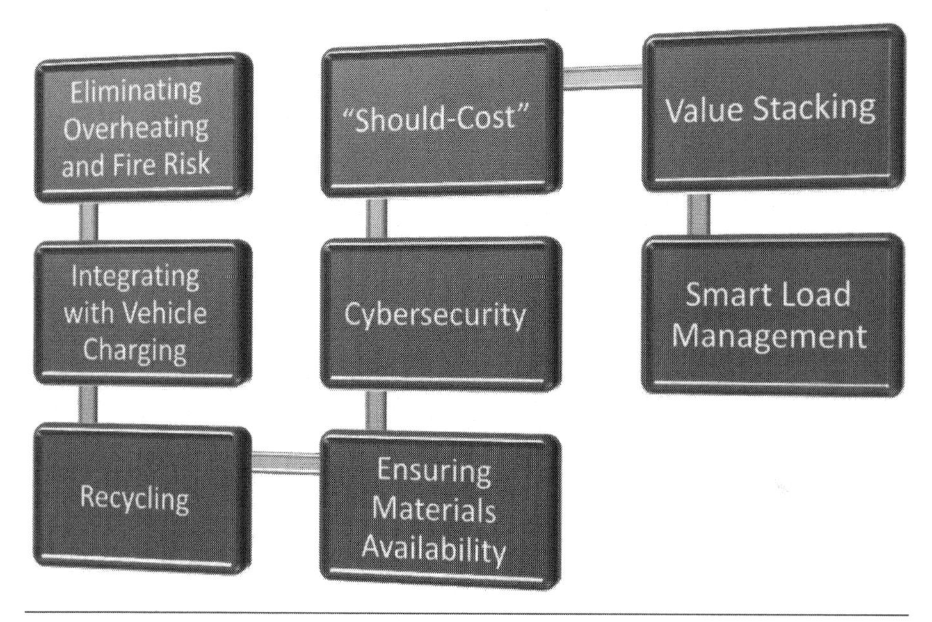

Figure 50. Emerging Supply Chain Opportunities for Energy Storage

Wind

Chapter Highlights

1. The development of IEC standards for offshore wind should help to reduce technology risks, which have largely been mitigated in onshore projects but are still a major factor in offshore wind and especially offshore floating wind.
2. Wind+storage has not been as effectively developed as solar+storage, which is arguably one of the reasons capital investments have sought solar projects more than wind projects, and are expected to continue doing so. However, efforts to integrate long-term energy storage into wind farm operations are under intense study and could change the growth vectors of solar and wind to the extent that they are successful.
3. Meanwhile, wind power owners are using optionality in supply chain contracts, and emerging financial instruments for hedging the risks of power unavailability at peak periods are allowing them to offset some of the risk of mismatch between supply and demand in operations.
4. Local content and knowledge transfer through local supply chain development is a central element in achieving alignment with public interests in wind farm projects. Integration with local manufacturing, skills development, and jobs pave the way for stakeholder buy-in and government consents and approvals.
5. Cybersecurity and AI are accepted large-dollar capital expenditures for wind projects, including long-term multi-year contracts with IT specialists. The high value of these contracts is justified on the basis of the avoidance of shutdowns, in the case of cybersecurity, and decreases in revenue and cost volatility as well as improved operating ratios in the case of AI.
6. End-of-life disposal of blades and other components needs to be considered at the outset of projects and included in the capital cost justifications, to avoid environmental and financial surprises in

later years. The effort required for such considerations should not be underestimated, as it could involve consideration of the cost of chartering specialized vessels for the transport of oversized marine cargo and the intensive study of disposal and recycling technologies such as grinding blades, and removing and disposing of the resulting chemicals from the materials.

7. Impactful supply chain levers include: 1) adopting construction standards to reduce capital cost, 2) developing and scaling Wind+Storage solutions, 3) ensuring cybersecurity, and 4) fully utilizing the power of Artificial Intelligence and machine learning to optimize equipment parameters in real-time.

Introduction

Wind energy composes 5.3% of total electricity generating capacity worldwide in 2020 and is set to more than double to 12.3% by 2030. Today, 95% of that is onshore and 5% is offshore, but as more offshore fixed and offshore floating wind farms are planned to be erected over the next decade, 14% of the total share is expected to be deployed offshore.[238]

Wind is at a disadvantage versus solar when it comes to the timing of energy production. Solar plus short-term energy storage tends to be compatible with the "duck curve" (generation during low-usage periods during the day, and discharging during high-usage periods in the evening), whereas wind is more intermittent and harder to predict, and storage solutions may have to be longer-term, which raises technical and economic complexities. As a result, in recognition of the higher bankability of solar power combined with storage over wind power without storage, regulatory incentives have nearly universally been designed for solar power rather than wind.

As measured by cumulative capital expenditure, wind power is gradually being overtaken by solar power, as shown in Figure 51. Between 2004 and 2019, global investment in wind power fell 20% behind global investment in solar power. The trend is expected to continue, but with a narrowing differential. Between 2020 and 2030, wind power is projected to grow by 7.0% while solar is expected to grow by 8.7%.

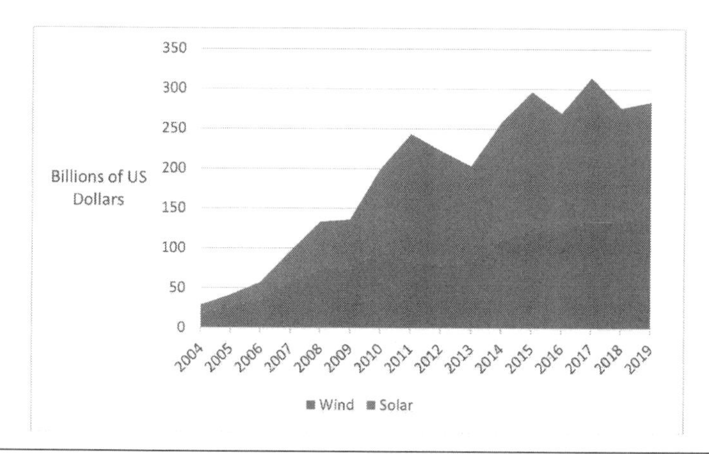

Figure 51. Global Investment in Wind and Solar Power
(*Source:* Author's analysis of data from Statista)

A wind turbine consists of a rotor, blades, yaw motor & drive, brake, gearbox, controller, and generator all contained in the nacelle, which is included in the broad representation of solar PV value chain in Figure 52. The structure stands on top of a tower, the height of which is very significant to the efficiency of the turbine. The tower is made of tubular steel, concrete, or steel lattice, for resilience to the structure at high-velocity rotation. The material of the blades is crucial in determining weight, power, efficiency, and cost of the system. Turbine blades, which must be light in weight yet strong, are often made of materials such as fiberglass-reinforced polyester, carbon fiber, or balsa wood. Cables gather power from multiple turbines in an array to feed inverters and transformers which send power onto the grid.

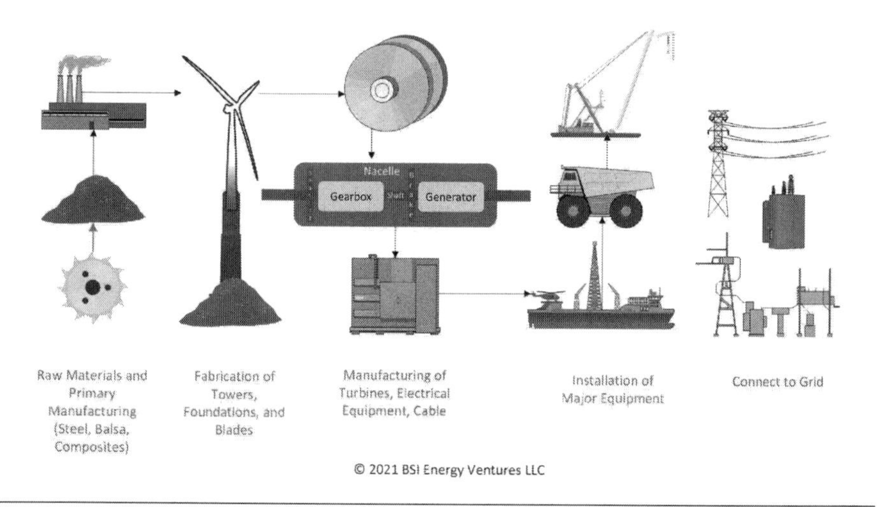

Figure 52. Wind Power Value Chain (Illustrative)
(*Source:* BSI Energy Ventures)

For onshore wind, supply chain costs are a key driver of project cost and profitability. They represent 72% of the final cost of electricity delivered through the system (3 cents/kwh, globally). Onshore wind supply chain costs are concentrated almost entirely (99%) upstream, at time of construction of the tower and turbine. Aside from the financial carrying cost of the tower, turbine, and electrical equipment, the main supply chain costs are onshore assembly and installation including site access, EPC fees and profit margins, and logistics and freight transport of the components—principally the rotor and shaft, generator including gearbox, and tower—to the installation site.

For offshore wind, supply chain costs are the single largest cost in setting up an offshore wind farm. They represent almost all of the levelized cost of electricity produced at the site (11.7 of 12 total cents/kwh). The main costs are: Offshore assembly and installation, including port staging, heavy lift, and the specialized installation vessels that are required for it; transmission line loss; logistics and transport; and the financial carrying cost of the capital investment.

Accordingly, the scope of this chapter includes: onshore assembly and installation, including site access; offshore assembly and installation, including port staging, heavy lift and specialized installation vessels; sea freight of oversized towers, foundations, and turbines; land freight of oversized towers, foundations, and turbines; and line loss (and ways to reduce it).

Trade-offs and Management Techniques in Capital Project Management

Technology Choice Risk Management: Avoiding Wrong Technology Choices in Today's Rapidly Changing Wind Power Landscape

Onshore wind farms have become so ubiquitous that most significant technology risks have been addressed and minimized by now. In the early days, manufacturers and developers experimented with the spacing between turbines, the length of the blades, regenerative braking, and other design features which at the time were viewed as technological variables. As the number of onshore wind farms grew, the permutations of technology reduced, or more aptly, commonized, and owners, developers, and operators could take on these projects with few concerns about the underlying technology.

Offshore wind farms experimented with significant design alternatives made since 1991, when the first offshore wind turbines were installed, including widely differing foundation designs, a wide range of turbine size choices for the same array, and various degrees of bundling of tower, foundation, blades, turbine, and warranty. By now some of these design choice options have narrowed, and standards have been developed around the more robust designs. Although offshore wind turbines had been conceptualized and anticipated for a long time, the first offshore wind turbines were not installed until 1991. A wide range of design variations, some of which proved to be bad ideas, inspired the development of design guidelines for offshore wind turbines. An IEC standard was published in 2009 (IEC, 2009), and a second edition was issued in 2019 (IEC 61400-3-1).[239]

Offshore floating wind is a technology choice that in its current configuration is relatively financially risky due to high commissioning and maintenance costs. If productivity is compromised for any reason, most projects become economically infeasible. This explains why auction pricing for floating wind was twice as high as conventional offshore wind projects in 2019.[240]

For example, the Fukushima FORWARD floating offshore project was dismantled after a few years of work and significant investment. Funded by the Ministry of Economy, Trade, and Industry, Government of Japan, the floating offshore wind farm demonstration project consisted of 2, 5, and 7 MW turbines. However, the overall operational efficiency was not encouraging, as the turbines were spinning only about 4% of the time. They needed to be at least 35% utilized in order to pay back, according to the Ministry, and modern wind turbines are sometimes designed to operate close to a 65% capacity factor, depending on the array.[241] Only one of the three turbines has actually reached 30%: the 2MW turbine even achieved a slightly higher result of 34%, while the 5MW turbine managed only 12%, and the 7MW turbine just 2% between 2016 and 2018.[242] Due to this suboptimal operational performance since its first phase of commissioning since 2013, one of the three turbines was removed in June 2020, while other two will be removed in early 2021, ultimately leading to complete dismantling of the project.[243] In keeping with the evolution of standards, and in an effort to keep the industry focused on robust designs that will withstand the test of time, the IEC published a standard for the design of floating offshore wind turbines, IEC 61400-3-2 in 2019.[244]

Wind+storage is being worked on, for example by one of the leading European wind farm developers, Vattenfall,[245] which if successful may allow wind+storage to compete more effectively with solar+storage, and perhaps even overtake solar+storage in the future.

At least one operator is engaged in a small project to combine offshore wind with hydrogen generation for fuel cells, creating a "green hydrogen" value chain—hydrogen production generated through renewable energy forms—and potentially "value-stacking," or creating multiple revenue streams—electricity from the wind farm, plus hydrogen from the electrolyzer—from common assets. Danish utility Orsted A/S is planning a two-megawatt offshore project that it expects to produce up to 1,000 kilograms of hydrogen daily. The output will be used for vehicular transport in and around Copenhagen as part of the EU's Green New Deal.[246]

Capital Project Risks and Mitigation: Hedging Against Financial Risks

Because of variability in wind speed and fluctuating auction rates at which power can be sold, operators perpetually face production and revenue risk. Much of this risk has been traditionally mitigated by incentives and subsidies. However, the number of projects supported by these benefits will shrink as governments remove subsidies in the coming years. One financial industry source calculates the subsidy rate at 75% today and forecasts that it will decrease to 6% by 2030. Feed-in-premiums and the stability provided by power purchase agreements (PPA) absorb price risk to some extent; however, operators still face the output variability issue.

A novel instrument for wind projects to safeguard against revenue uncertainties is "hedging" to cover the volume risk. Hedging is emerging as an instrument to cover the resource risk of variable wind generation, or volume risk. Hedging allows operators to transfer risk to entities with higher risk appetite looking for a window of profit. Hedging solutions for wind volume risk can be executed in either derivative or (re)insurance formats.[247]

Supply Unavailability Risks and Mitigation: Finding Alternative Sources of Balsa Wood for Blades

The spread of COVID-19 caused a shortage of key components for wind turbine projects. The pandemic led major manufacturers such as Vestas Wind Systems and Siemens Gamesa to shut down their factories and deliveries due to lockdowns. Furthermore, shortages of raw material and components from turbine blades to ball bearings were created due to delays at ports and borders. A severe supply crunch experience for wind projects in 2020 is likely to hamper completion of farms through 2021.[248]

Evidence of this shortage could be seen as China's renewable energy industry faced a supply bottleneck from the COVID-19–led lockdown in Ecuador, a key market for the balsa wood used in the production of wind turbine blades. Ecuador supplies 95% of the world's commercial balsa wood, which is known for its strength and light weight, which potentially impacted up to 10% of the planned capacity additions.[249]

With the increase in risks arising from pandemic, geopolitical factors, climate change, etc., global supply chains are increasingly opening up to high degree of vulnerabilities. In some cases, the best option is to build a strategic inventory of key materials and components. Boston Strategies International forecast the lead times and supply risks for hundreds of strategic materials, components, and services for numerous major power operators. The forecasts had "traffic light" alert systems and recommended actions associated with flagged shortages.

Outsourcing Risks and Mitigation: Dealing with an Increasingly Experienced and Competent Market of O&M Vendors

In what amounts to a sort of Design-Build-Operate arrangement, Arise, a top operator in the Nordic wind-power market, moved from in-house servicing of its wind farms to outsourcing it.[250] Arise awarded Vestas, a wind turbine manufacturer, a 15-year service contract for seven wind farms covering 86 MW in Sweden. This agreement gives Vestas full responsibility for optimizing wind performance and energy output based on a servicing arrangement. The contract is designed to maximize uptime and energy production.

Supplier Partnering Risks and Mitigation: Extending a Turbine Manufacturer Relationship

Positive experiences with a wind turbine manufacturer can sometimes be extended to O&M and technology partnerships. For example, German wind turbine manufacturer ENERCON formed a strategic alliance with Dutch wind turbine manufacturer Lagerwey to widen technological capability and broaden their product lines for turbines in all wind classes.[251]

Partnerships do not always go this well, however. At the Greater Gabbard wind farm, Fluor Daniels absorbed losses of $343 million in 2010 when monopile foundation welding defects caused a project delay. Subsea cable installation company Subocean went bankrupt over the ordeal. The incident signaled a need for better financial control and risk reduction. Greater Gabbard is just one of many examples. One offshore wind operator had to deal with suppliers going bankrupt when it took over responsibility for a project from its general contractor.

Procurement Bundling Trade-offs: Bundling Turbine Acquisition and Service Agreement

As the turbine is the most central component in the wind power system, it is typically preferable to engage a long-term service agreement from the OEM to ensure reliability. In some cases, turbine manufacturers can also be good maintenance partners for the entire wind power system. The Kingebol wind farm (37 MW) in Sweden also includes a 20-year service agreement from Siemens Gamesa. Siemens will install six SG 5.8-170 turbines from its 5.X platform. They will provide 6.2 MW of nominal power and have the largest rotors in the onshore segment (170 meters).[252]

Contract Term Risk and Mitigation: Using Real Options to Achieve Low Cost with Flexibility

Due to the sometimes irregular evolution of wind power technology and regulatory environments, developers and operators are often uncomfortable signing onto long-term agreements. The hesitation sometimes relates to unproven technologies such as, for example, the artificial intelligence that makes blade angle adjustments. More often, however, the hesitation stems from regulatory and stakeholder uncertainties. Even when the technology is stable, stakeholder risks can threaten to slow down or shut down wind farm projects. Projects are at risk of being shuttered entirely if they don't receive the necessary consents, approvals, and permits, and in some cases certain external stakeholder groups are opposed to the projects no matter what the data shows, no matter how safe or how clean or how energy-efficient the project is.

Options contracts allow owners, developers, and operators to protect their financial investment if these stakeholder events unfold unfavorably, or if technology doesn't live up to expectations, or if suppliers fail to perform or deliver, thereby threatening the financial viability of the project. In this author's opinion, options are more commonly used in wind farm development and operation than in other forms of energy except oil & gas.

Options can be achieved by buying or selling financial instruments on a market. For example, an operator could buy options on electricity that would allow it to buy electricity at a future date at an agreed price. For example, the CME GRoup, originally the Chicago Mercantile Exchange, offers PJM Western Hub real-time peak power options such as 5 MW and 50 MW calendar strip and calendar-month options, and a 5 MW average price option on the New York Mercantile Exchange.[253]

Alternatively, and more easily and simply, options can be structured in contracts, so-called "real options." For example, a power agreement can be structured in five-year increments.

If, on the other hand, certainty is required rather than optionality, companies can reserve capacity at suppliers, or combine a capacity reservation with a capacity option. In one case, TPI Composites arranged a five-year contract to supply wind turbine blades from its facility in Izmir, Turkey. As part of the agreement, Siemens reserved two mold capacity positions, one for scheduled supply and the other one for optional production.[254] Siemens has excelled at creating real options in its supply contracts: in another contract with TPI, Siemens traded a minimum volume obligation for regional exclusivity, thereby reducing the risk of supply while giving a supplier upside potential on large orders.[255][256]

Trade-offs and Management Techniques in Operations & Maintenance

Cybersecurity Risk Management: Treating Cybersecurity as a Major Part of O&M Expense

Without proper security precautions, wind farms may suffer due to insecure programmable automation controllers (PACs), a lack of encryption of control messages, and no network segmentation between wind turbines.[257] These inadequacies can potentially lead to an unchecked cyberattack that could halt production of a wind farm altogether, resulting in significant loss of income. According to estimates, the average single-day downtime of a 100MW operating at a conservative 35% capacity could equate to over $50,000 of lost revenue, based on an average power price of $60/MWh.[258]

The growing concern and need to integrate cybersecurity measures led North American Electric Reliability Corporation (NERC) to set forth strict norms for wind farms above 75 MW capacity as to complying with cyber security standards.[259]

Besides these norms, even operators are actively working towards addressing cyber threats. For example, GE Renewable Energy signed a $13 million agreement with Invenergy to secure its entire fleet of wind turbines. The agreement includes upgrading controls and protecting its network with the Opshield intelligent security device. During the ten-year agreement period, GE will provide Invenergy with software maintenance, updates, and patches.[260]

Internet of Things (IoT) and Artificial Intelligence (AI) Technology Choices: Real-Time Automated Bidding Based on Market Price Patterns

Artificial intelligence or machine learning, as it is sometimes called with little to no distinction except that machine learning is, in general, used to mean an application-specific subset of AI, is and will be used in many ways in wind power.

At the operational level, AI can train and optimize the conversion of power to energy storage and the converse discharge from energy storage to the grid, depending on anticipated grid demand and tariff structures. AI can also inform physical characteristics such as anticipatory turbine and blade directionality. In a more holistic application, AI can interpret pricing trends, signals, and patterns, and use them to optimize production and transmission operations and thereby cost. For example, Advanced Microgrid Solutions (AMS) enabled Tilt Renewables' 100 MW Snowtown 1 Wind Farm with AI software capable of executing "precision trading" to boost revenue. This is possible since Australia has a real-time competitive energy market. Since market pricing may result in unacceptable profit margins or even losses depending on the marginal cost of production, AMS's AI software precisely predicts factors such as prevailing prices and supply levels so that bids can be placed at the time when higher rates allow the bidder to maximize returns.[261]

A German consortium called SmartWind is developing an AI-based "Multi Criteria Decision Support System" that will gather inputs on multiple variables and feed them into a multivariate algorithm that determines optimal operating and maintenance parameters and intervals. The math engine coordinates the operation of multiple power plants and minimizes counterproductive electrical and aerodynamic interactions caused by wake effects.[262]

Peak Capacity Strategies: Using Dedicated Production Capacity to Guarantee Delivery Dates

Capacity management involving oversized items that need to be produced using a technical manufacturing process is a unique challenge, whether it is casting or milling or heat treatment, etc. Capacity is constrained by the size of the item, the sequence of prerequisite processes, and the lead times to acquire specialized materials that are usually made to order for order-specific chemistries and pre-treatments.

Despite the capacity constraints, wind turbine blade manufacturers must maintain sufficient production capacity to remain competitive and deliver finished products within customer time windows, or else they risk causing delays in the customer's assembly schedule. Once a project gets funded and is underway, delays usually incur stiff financial penalties that can quickly turn a profit into a loss.

TPI, a composites manufacturer that makes wind turbine blades, dedicates capacity at its facilities to its customers with long-term supply agreements to avoid exceeding capacity and incurring delays. The company dedicates capacity to customers on the basis of contractual commitments. As it explains in its annual report, "We have entered into long-term supply agreements pursuant to which we dedicate capacity at our facilities to our customers in exchange for their commitment to purchase minimum annual volumes of wind blade sets, which consist of three wind blades. This collaborative dedicated supplier model provides us with contracted volumes that generate significant revenue visibility, drive capital efficiency and allow us to produce wind blades at a lower total delivered cost, while ensuring critical dedicated capacity for our customers."[263]

Sourcing Trade-offs: Building Local Value Chains in the Normal Course of Business

Local content remains a hot button in wind power equipment fabrication and installation, given the magnitude of the capital expenditures, the technical know-how involved, and the corresponding potential for knowledge transfer and "smart manufacturing" to pull economies forward into the next generation, and the political and social currency of clean green nationalism.

In Taiwan, for example, MHI Vestas is working towards establishing local supply chains to support its future projects in the country. MHI Vestas signed a purchase agreement with KK Wind Solutions for local assembly and production of technical components to support the turbine manufacturer's future projects in Taiwan, particularly the Changfang and Xidao Offshore Wind Project (CFXD), a 600 MW capacity project located off the western coast of Changhua, Taiwan. The purchase agreement includes power conversion module (PCM) assembly, and manufacturing of low-voltage cabinets and uninterruptible power supply (UPS) systems, all to be carried out locally at the planned production facility near Taichung Harbour by KK Wind Solutions.[264] Part of this is accomplished through a contract with KK Wind Solutions for local power conversion module assembly, and local manufacturing of low-voltage cabinets and uninterruptible power supply systems to support future wind projects in Taiwan. The PCM is a power electronics system that is critical for converting wind energy to electricity. As part of KK Wind Solutions' investment, it will hire and train local workers who will be trained to engineer and manufacture sophisticated offshore wind turbine components. KK Wind Solutions will open a production facility near Taichung Harbour and will be responsible for the assembly of the PCM low-voltage cabinets and UPS systems, which ensure optimal power supply and the maintenance of critical load levels in the turbine.

Besides engagement with KK WindSolutions, MHI Vestas has been involved in signing multiple contracts for local procurement of a wide range of components including blades, blade materials (bonding glue, resin, pultruded carbon plates), towers, switchgear, rotor hubs, hub plates, and nacelle base frames.[265]

TCO Trade-offs: Planning for End of Life and Disposal of Turbine Blades

As the first cost of a 1 GW wind array is typically about 70% of total life cycle cost, and OpEx accounts for between 25% and 35% of a project's overall Levelized Cost of Electricity (LCOE),[266] including development, project management, and installation costs, and the investment can rise to $3 million per MW.[267] There is extraordinary pressure on procurement to find ways to cut upfront investment costs. Unfortunately, this results in a tendency to ignore end-of-life environmental issues and dismantling costs. When trying to secure funding, owners and developers tend to "kick the can down the road" and leave the disposal problems to be dealt with later. Unfortunately, this can lead to unpleasant environmental and economic consequences 10–20 years after project inception.

For example, the highly specialized composite materials used in the fabrication of advanced wind turbine blades often precludes recycling them. As a result, they could end up in landfills without possibility of biodegrading. This begets a responsibility to specify an environmentally responsible way to dispose of blades—and also foundations and other non-biodegradable wind turbine components—at the early stages before their placement in physical locations. While turbine blades have a life up to 20 years, many are replaced with more efficient designs in just a decade, ending up in landfills. The fiberglass construct of the blades and size as long as a football field make them difficult and expensive to transport.[268] However, companies like GE and Veolia are finding ways to address this issue. For example, GE Renewable Energy is working on improved turbine blade recycling and developing alternative materials to increase productivity, resulting in waste reduction.[269] In a pilot project, Veolia tried grinding them to dust to extract chemicals.[270]

Supply Chain Roadmap for Wind Power

The most impactful supply chain techniques for wind power generation, transmission, and distribution tend to reduce cost, enhance the reliability of power, or both.

For cost reduction, certain supply chain techniques can significantly reduce capital and lifetime operation and maintenance costs, as well as operational and budgetary variability. These include:

- Construction standards like those from IEC
- Bundled turbine acquisition and service agreements
- Design-Build-Operate ownership structures
- Contracts based on options
- Machine learning that intelligently adjusts energy storage, turbine and blade directionality, and electricity trading actions based on real-time physical and market conditions

For reliability, certain supply chain practices will gain increasing focus due to their impact on industry economics, in particular their tendency to assure reliability of systems and output when needed, especially in the face of unpredictable consenting, approval, and permitting processes. These include:

- Developing and scaling Wind+Storage solutions
- Ensuring cybersecurity (through standards such as those from NERC for wind farms over 75 MW)

Outside of cost and reliability, environmentally friendly end-of-life disposal of blades is becoming an increasingly apparent problem that will require a supply chain solution.

Figure 53 displays these principal axes of emerging supply chain value creation potential for wind power.

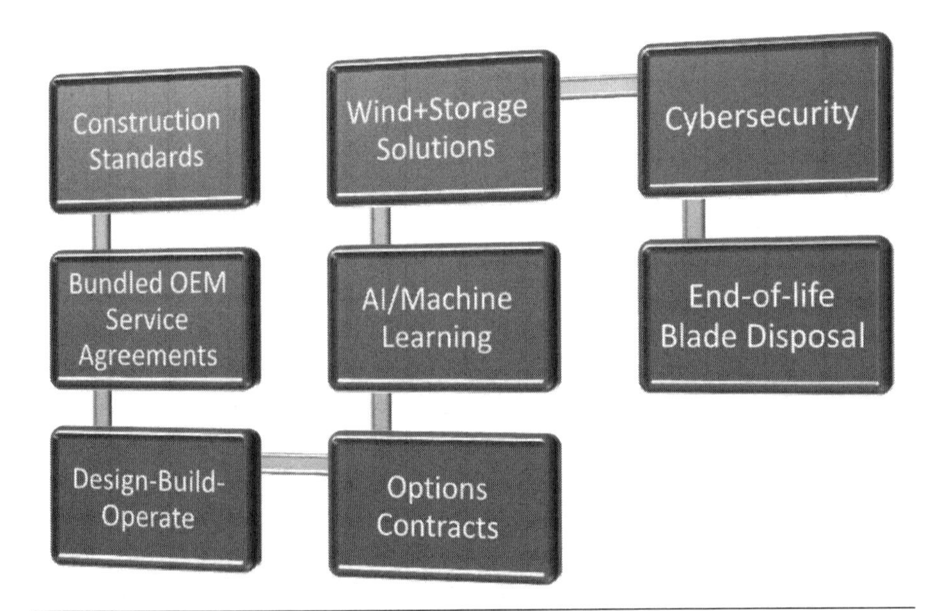

Figure 53. Emerging Supply Chain Opportunities for Wind Power

Solar

Chapter Highlights

1. Solar is a significant power-generating segment, representing as much new electricity generating capacity as coal or gas-fired power, or wind power, taking 2019 in the United States as a reference point.
2. Utility-scale solar is as pervasive (in new megawatts installed) as residential and C&I. Half of that new capacity was for utility-scale projects, and half was for C&I and residential, both of which directly affect utilities as customer, supplier, or competitor.
3. Supply chain costs are an important component of solar power—more than they are for oil & gas.
4. Managing supply chain cost for solar installations involves managing the supply and life-cycle cost of PV modules, inverters and other systems and components, freight and shipping, installation and permitting, operations, and maintenance.
5. Utility-scale solar projects involve numerous important technology choices with widely varying cost and operational consequences. For example, for PV, whether to use crystalline or thin-film PV, and if thin-film, which thin-film materials; whether to use energy storage in conjunction with the solar PV, and if so, where in the network to use it and which energy storage technology to use in which location; whether to set up a central solar "farm" or establish a network of distributed generation sites, and if distributed, where to position them; and whether to use CSP and if so, which CSP technology to rely upon. Furthermore, all of the technology choices inter-relate, so one cannot apply a simple decision tree to arrive at a choice—the study has to go fairly deep and may warrant hiring outside consultants.
6. Several equipment decision specifications have important implications for operations and maintenance cost, especially the capacity of the inverter and whether or not to use a PV panel cleaning system in dusty areas.

7. Due to the recent increase in number and competitiveness of EPC firms, more utilities are opting to hire EPC firms rather than manage projects in-house. The right answer may be tied to the technology choice. Some EPCs may offer more end-to-end services than others when it comes to the bundling of products (e.g., inverters with panels), freight, installation and maintenance services, balance-of-system costs, and permitting and regulatory services.
8. Systems should be designed with IoT and AI capabilities like tracking algorithms, monitoring, and controls, as well as cybersecurity safeguards to prevent hacking and minimize the potential for grid disruptions, denials of service, or shutdowns. Since these factors extend beyond the scope of traditional supply chain management, staffing the procurement operation for a sizable utility-grade solar project should ensure an adequate technology skill set.
9. Due to the rapid growth in demand and the global nature of the supply market, there is a lot of low-cost, low-quality product and many unqualified service providers on the market, even for utility-scale projects. Therefore, choosing the lowest cost equipment and services, which could be the best choice in a mature market, might be a risky path for solar projects.
10. There are numerous opportunities for supply chain management to create cost-saving and revenue-generating opportunities in solar power generation. Among the potential game-changing ones are: 1) IoT/AI, including smart tracking to optimally align the modules toward the sun, 2) securing the polysilicon supply chain; 3) integrating energy storage; 4) managing and minimizing life-cycle cost; 5) enabling distributed solar; 6) expanding the safe use of floating solar; 7) expanding the use of embedded solar; and 8) Integrating rooftop solar with vehicle charging systems. Also, from an environmental and social point of view, social exploitation and CO_2-emitting production processes should be avoided, and end-of-life disposal of toxic PV waste should be anticipated.

Introduction

Utility-scale solar power is becoming increasingly common as solar photovoltaic (PV) costs rapidly decline. Although solar power was responsible for only 3% of global electricity production (627 GW) in 2019, of which 80 GW was in the United States, solar represents about the same share of new capacity being added as wind and natural gas-fired thermal power—which are split roughly evenly with about one-third each in the United States.[271]

Solar PV systems make use of semiconducting material to convert light from the sun into electricity. The value chain starts from the semiconducting material, silicon, which is transformed under heat into polysilicon ingots. Phosphorus deposition on this base material results in wafers. Application of anti-reflective coating on the surface completes a crystalline photovoltaic cell. These PV cells are sealed together in a protective laminate to make solar modules. The modules are mounted in tens or hundreds of thousands in a large megawatt-scale solar plant, interconnected with cables to produce power when the sun is shining. A typical PV system also includes Inverter and battery system attached to the solar panels, along with meters and controls. The power thus generated is fed to the grid, microgrid or any other load (factory). Figure 54 shows a value chain illustration of this process.

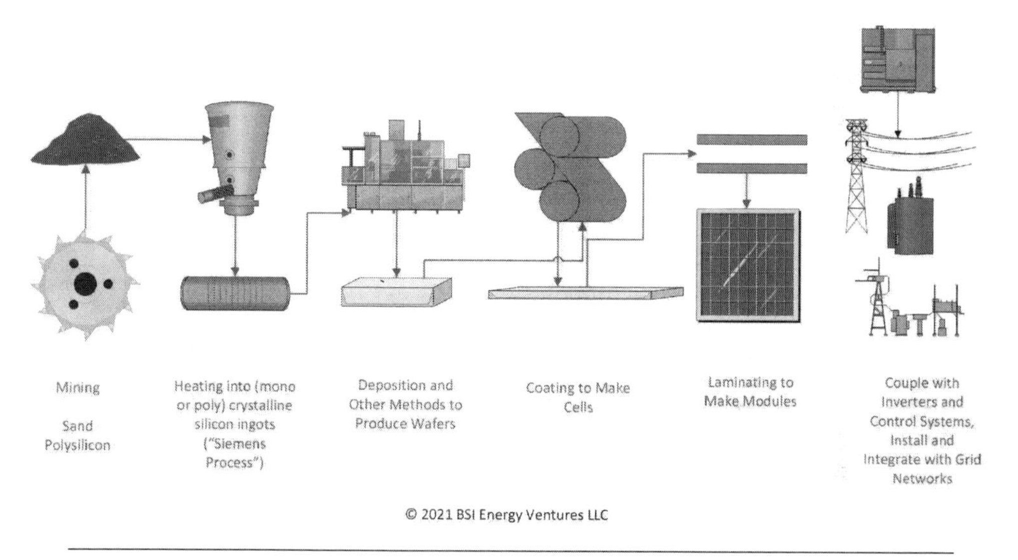

Figure 54. Utility-Scale Solar Photovoltaic Power Plant Value Chain (Illustrative) (*Source:* BSI Energy Ventures)

Inverters may be divided into Central Inverters (for utility-scale or large-sized commercial Solar PV systems) and String Inverters (for small- and medium-scale installation).[272] The control system consists of a router/firewall, a security server, a SCADA server, a historian, a gateway, a fiber switch, a network drop data logger, and a fiber network that connects the panels, tracker, PV inverter, battery storage, substation, and revenue meter.[273] Utility-scale systems also need system design, engineering, installation, and integration.

Every permutation of solar power technology and financing—crystalline or thin film, subsidized or unsubsidized—is less expensive on a Levelized Cost of Electricity (LCOE) basis than any permutation of conventional energy—gas, nuclear, coal, and gas combined cycle—except when coal or gas-fired plants dump their power at marginal cost.[274]

Half of all cumulative US solar installations have been for utility use,[275] and an increasing share of solar utility installations is involving energy storage. One forecaster estimates that 30% of the PV plants will have integrated energy storage by 2025.[276]

Supply chain costs represent 5% of solar power costs—$2.40 per million BTU ($0.007/kWh) of electricity sold for rooftop solar, $4.80 per million BTU for utility-scale solar with battery storage, and $7.90 per million BTU for solar utility-scale installations without energy storage—compared to 8.4% of oil costs ($1.75 per million BTU of oil sold). Supply chain costs for PV + Storage are high, at 45% of the total cost of electricity ($0.016 cents/kwh), including installation labor & equipment, EPC overhead, and developer profit in supply chain costs, and about a third, when excluding structural and electrical balance of system (BOS) costs.

Storage adds significant cost to solar PV, approximately doubling the cost in some cases examined by NREL. Of the incremental storage system cost, more than half are for the structural and electrical balance of system (BOS), installation labor & equipment, EPC overhead, sales tax, and developer profit—not for the batteries or inverters.

Both the cost of solar power and the supply chain intensity of solar generation, transmission, and distribution can be expected to decrease over time and become more competitive with oil & gas. This will be largely driven by growing economies of scale at the end of equipment producers, and expansion in the numbers and competitiveness of service providers. Additionally, solar adoption will further boost as regulatory frameworks on energy storage pricing and permitting procedures become more standardized.

The preponderance (82%) of utility-scale solar supply chain costs accrue upstream at the utility—consisting of freight, EPC fees, and capital project financing costs. A smaller but not insignificant share (18%) of utility scale solar supply chain costs relate to T&D CapEx and OpEx.

Trade-offs and Management Techniques in Capital Project Management

Technology Choice Risk Management: PV, Storage, Distributed, CSP, and Floating

PV Technologies

PV technology comes in many flavors, and each one has unique cost and performance attributes. Crystalline-silicon PV consists of various sub-technologies such as Mono-crystalline silicon (c-Si), Poly-crystalline silicon (pc-Si/ mc-Si), and String Ribbon. Thin film technology consists of various materials such as Amorphous silicon (a-Si), Cadmium Telluride (CdTe), Copper Indium Gallium Selenide (CIG/CIGS), and Organic photovoltaic (OPV/ DSC/ DYSC).

Advances in thin film technology have eaten into the Crystalline silicon (c-Si) market, but both have their places. Each type has different attributes such as efficiency, heat tolerance, cost, surface area required, and ease of installation.[277]

Since the choice of technology has ramifications for not only initial acquisition cost but also operating and maintenance cost and functionality, managers should make a careful study before committing to a particular technology.

Solar + Storage

The benefits of solar + storage for a utility are significant: the benefits provide about a 10–15% cost advantage compared to regular solar PV.[278] Storage adds all the benefits that are described in the Utility Scale Energy Storage chapter.

From a supply chain point of view, exactly how and where to set up storage with solar can make a significant difference in total cost, and should therefore be the main focus of supply chain management attention. Location of the storage assets is as critical a cost driver as the number and configuration of warehouses in a goods distribution operation.

Furthermore, spreading storage throughout the value chain, at the power generating plant, within T&D, and locally, can save massive midstream CapEx and OpEx costs, changing the whole supply chain management game and expanding the required supply chain best-practice toolbox. Midstream (T&D) supply chain management costs, which without storage are 72% of total supply chain management costs, can be substantially eliminated with strategic and widespread use of storage.

In PV + Storage configurations, co-locating the storage system with the PV array can reduce costs for site preparation, land acquisition, permitting, interconnection, installation labor, overhead, and profit.[IX]

Distributed Solar

For distributed solar (mostly rooftop) schemes, the potential for elimination of T&D CapEx and OpEx at a centralized plant reduces total supply chain costs dramatically. Even though it costs more to set up solar on a rooftop than at a utility, due to economies of scale, the potential for savings from eliminating transmission and distribution costs, including line loss, is a value chain game-changer for solar power.

If all rooftop solar installations are still connected to the grid and drawing on power, especially during peak hours, this savings may not be realized. However,

[IX] Per the author's analysis of data from multiple sources.

microgrids and tariff structures could and probably will solve this problem over time as distributed solar penetration increases. Total SA is on top of the opportunity, having recently constructed one of the largest microgrids in Cambodia, consisting of multiple distributed solar power plants.[279]

Distributed solar power plants can capture the economies of scale in PV acquisition and energy storage costs, while also avoiding the infrastructure investment cost of transmission & distribution, thereby benefiting from the best of both central and distributed solar power generation. Energy Efficient Services Ltd (EESL) in India announced plans to build 1,500 MW of decentralized solar power plants by the end of 2020–21.[280]

Concentrated Solar Power

Concentrated Solar Power (CSP) has received a burst of investment over the past ten years. And for good reason: the cost, as measured by LCOE, has fallen from 21 cents per kWh in 2010 to 9.8 cents per kwH in 2018,[281] and major new CSP projects are underway around the world. In addition, CSP is now being tested in conjunction with AI (smart angle adjustment, among other parameters) as a power source for green hydrogen production, which could change the game for both solar and hydrogen.

The technology is not yet without technical and commercial risks, however. Early CSP developments that ran into now-infamous failures can now be used as case studies.

SolarReserve developed CSP plants on behalf of Tonopah Solar Energy in California's Mojave Desert and Tonopah, Nevada, between 2011 and 2015 that became abject failures for reasons of management of new technology—failures that cost billions of dollars of investment. Details appear below. That being said, these US experiences have not deterred CSP projects overseas. The Dubai Electricity and Water Authority (DEWA) is constructing a 700 MW CSP project that is slated to deliver electricity at 7.3 cents per kWh, a record low for CSP.[282] Also, China proceeded with four new CSP projects that totaled 200 MW (half the world's total CSP capacity) in 2019. Notably, China has diversified the underlying technologies. Of the seven CSP projects that China has ever embarked upon, five use molten salt tower, one uses traditional parabolic trough, and one uses Fresnel with molten salt.[283]

The 110 MW Crescent Dunes Solar Energy Project in Tonopah, Nevada, built by SolarReserve on behalf of Tonopah Solar Energy, is an example of a concentrated solar power (CSP) project that went awry because the technology wasn't sufficiently tested. As the first CSP plant with a central receiver tower and advanced molten salt energy storage technology, the plant's operating principle was to soak up enough heat from sun's rays to spin steam turbines and store energy in the form of molten salt. Crescent Dunes began operation in September 2015 but went offline in October 2016 due to a leak in a molten salt tank. It returned to operation in July 2017. While its average monthly production was expected to exceed 40 GW, the plant failed to deliver its intended 50% capacity factor, achieving only about a 20% capacity factor in 2018. As of May 2019, it had never reached that value and exceeded half of it only during nine months.[284]

Compared to the present cost of less than $30 per MWh today at a new Nevada photovoltaic solar farm, SolarReserve's power cost Nevada about $135 per MWh. The steam generators at its solar plant required custom parts and at least 12 people to conduct regular maintenance and keep the plant running. The plant shut down in April 2019 when the project's sole buyer, NV Energy, terminated the Power Purchase Agreement for failure to produce the contracted power production.[285]

The failure of the 110 MW Crescent Dunes Solar Energy Project in Nevada is an example of the significant risk accompanying the choice of technology, and the importance of long-term cost assessments in making the decision.[286]

SolarReserve also developed the $2.2b Ivanpah CSP plant on behalf of Tonopah Solar Energy in California's Mojave Desert in 2011. Seemingly every type of error was made: faulty design, wrong weather forecasts, operational error, and unintended adverse environmental impact. The solar power project in California's Mojave Desert plant was designed to be generating more than a million megawatt-hours of electricity each year, but 15 months after starting up, the plant was producing just 40% of that. The technology itself did not fail. The technology was solar collection via mirrors coupled with water-filled boilers atop three 450-foot towers, with preheating from a gas-fired plant,[287] which was generally not blamed for the subsequent failures. However, the developers' calculations were far off.

The power plant requires far more steam to run smoothly and efficiently than originally determined. Instead of ramping up the plant each day before sunrise by burning one hour's worth of natural gas to generate steam, Ivanpah needed more than four times that much help from fossil fuels to get the plant humming every morning. As a result, the plant's clean energy vision was negated by an immense consumption of natural gas. In 2014, the State of California approved Ivanpah to increase its annual natural gas consumption from 328,000,000 cubic feet (9,300,000 m3) of natural gas, as previously approved, to 525,000,000 cubic feet (14,900,000 m3). In 2014, the plant burned 868×109 British thermal units (254 GWh) of natural gas emitting 46,084 metric tons of carbon dioxide, which is nearly twice the pollution threshold at which power plants and factories in California are required to participate in the state's cap and trade program to reduce carbon emissions.

Another unexpected miscalculation related to forecasts of the amount of sun that would be available to power the PV modules: there was not enough sun. Weather predictions for the area underestimated the amount of cloud cover that has blanketed Ivanpah since the plant went into service in 2013.

A third mistake was a matter of operational error—misaligned mirrors. On May 19, 2016, a small fire was reported when misaligned mirrors reflected sunlight into a level of the Unit 3 tower not designed to collect power, causing a fire that required the tower to shut down for repairs. As another of the three power-generating units was already offline for scheduled maintenance, the plant was left with only one third of its installation functional.

Finally, after it began operation, biologists found the plant's superheated mirrors were killing birds. The Ivanpah plant was delayed several months and had millions of dollars in cost overruns, as the project attempted to conform to wildlife protections for the endangered Desert Tortoise. Once it was built, US government biologists found the plant's superheated mirrors were killing birds. In April 2016, biologists working for the state estimated that 3,500 birds died at Ivanpah in a year, many of them burned alive while flying through a part of the solar installment where air temperatures can reach 1,000 degrees Fahrenheit.[288]

Floating Solar

Floating solar could end up being a widely used form of solar power, under the right conditions. It has been used extensively in commercial reservoirs where there is one owner who is also the off-taker. It is also being used in high-altitude lakes where snow reflects the sun, creating a sort of natural CSP effect.[289] Offshore floating solar, in the open sea, is being pioneered off the Dutch coast.[290]

Considering the potential floating solar seems to have, it has its challenges. Several installations have caught fire due to electrical malfunctions, including a large fire at the Yamakura Dam in Japan and a smaller one at a floating array in the UK owned partly by BP.[291] In order to become a mainstream technology, the technology needs to be demonstrably cleared of these repeated fire hazards.

Embedded Solar

Solar Earth Technologies Ltd. (Canada) has developed a type of photovoltaic cell technology that is designed to be embedded in roads, walkways, and buildings. The cell material is 30% more efficient than flat solar panels. The technology enables clean power for new construction in less developed areas, and serves remote, industrial and off-grid power.[292]

Pico Solar

Very small, even tiny, solar modules can be used to power individual devices like mobile phones. While utilities would not use pico solar modules, their customers might, especially if grid power is expensive or unreliable. Large-scale adoption of pico solar modules could decrease total energy demand from central power plants.

Capital Project Risks and Mitigation: Setting the Scope of EPCI Services

Overall, managing solar supply chain costs means getting control of PV procurement costs, installation costs, and freight costs, in whichever type of solar is involved. Solar PV and the inverters are among the most significant cost components, and their procurement as well as the supply chain to get them to the site should be evaluated carefully during the project evaluation phase. Freight costs and EPC fees are often the next-largest cost elements.

Solar + storage developers must additionally manage installation labor, EPC overhead, and balance of systems costs related to the storage modules. Distributed (rooftop) solar developers must manage ocean and truck freight transport of PV panels to prevent these costs from cutting into their return on investment.

Faced with falling solar tariffs, many developers have tended to use internal resources for project execution to cut costs and improve project returns. However, two factors—increasing number of EPCs and consolidation in the EPC industry—have led to a reversal in this trend. The EPC market itself has been intensely competitive with sharp fall in costs and wafer-thin margins. Hence, there is little cost advantage in self-EPC.[293]

Some of the developers that were previously keen on self-EPC are switching to third-party EPC services. Cypress Creek Renewables, a utility-scale solar power company in the US, has decided to completely shut its internal EPC division over the coming months and shift towards a third-party model. This will allow the company to focus on its core business of developing, financing and operating solar and storage.[294] First Solar, Inc., transitioned away from its internal Engineering, Procurement, and Construction (EPC) execution model in the US, and moved towards leveraging the capabilities of trusted EPC partners.[295]

Supply Unavailability Risks and Mitigation: Securing Polysilicon

Polysilicon and the glass that coats PV panels have been in short supply, and many buyers are using long-term supply agreements to eliminate uncertain availability. Three top suppliers of polysilicon had coincidentally been affected at the same time by accidents, flooding and maintenance issues, leading to shortage of supply and an increase in average silicon wafers costs by an estimated 10%.[296]

In response, Trina Solar signed an upstream manufacturing long-term agreement with Daqo New Energy for the supply of high purity polysilicon. Under the terms of the deal Daqo will supply Trina between 30,000 and 37,600 tons of polysilicon between November 2020 and December 2023. Prices will be negotiated on a monthly basis according to market conditions, with Trina making an advance payment to Daqo. Price setting on a monthly basis will allow Trina to hedge supply price risk.[297]

Also, prices for glass that coats photovoltaic panels rose 71% from July to November 2020, and manufacturers are struggling to produce it fast enough to keep more than a week's worth of sales in inventory, according to Daiwa Capital Markets. The shortage comes as the solar industry turns toward bifacial panels, which increase both power output and glass requirements.[298]

Outsourcing Risks and Mitigation: Focusing on Systems or Modules

Solar PV technology has been evolving so fast that a growing number of manufactures prefer to outsource or even sell their production lines in order to focus on sales and marketing. Specialist PV manufacturing equipment supplier Amtech Systems, an American manufacturer of capital equipment, including thermal processing and wafer handling automation, divested solar equipment production in China to avoid losses.[299] Solar module maker United Renewable Energy reduced in-house solar module capacity by 80% and secured contracts for major solar farm construction projects. In an effort to save itself from falling prices, the company scaled down in-house solar module production and outsourced the bulk of its capacity to reduce the risk of price volatility.[300] Meyer Burger, the Swiss solar PV equipment manufacturer signed a cooperation agreement with Spanish competitor, Mondragon Assembly, to outsource production of SWCT (Smartwire Connection Technology) by the end of 2018. SWCT™ is an exclusive patented Meyer Burger technology for connecting solar cells with each other.[301] Chinese solar manufacturer and lithium-ion battery storage company Comtec chose to outsource manufacturing, citing overcapacity in the market, in favor of project development and new business opportunities. Comtec noted that its rivals have been aggressively ramping up manufacturing output ahead of what they anticipate will be an imminent global solar gold rush.[302]

Supplier Partnering Risks and Mitigation: Using Auctions and Competitive Bids

For solar farm developers, rapidly falling prices of EPC services have clarified the issue of how to engage with EPC firms. The answer is that in this downwardly cycling market the most cost-efficient way to engage with suppliers is through a transactional, price-driven tender, and auctions.

EPC contractors have been discovered to have quoted the lowest-ever prices of around $0.076 a megawatt-hour for solar power projects of NTPC in Rajasthan (260 MW) and Madhya Pradesh (250 MW) in the first-ever domestic reverse auctions. The development comes at a time when global and domestic solar power developers have been aggressively participating in the auctions, pushing down solar power tariffs to the lowest-ever levels of $0.059 a unit in the latest auctions in Rajasthan, India.

Procurement Bundling Trade-offs: Coordinating EPC with Project Funding

Developers rely on EPC firms to aggregate purchasing of the disparate elements of solar PV farm construction, such as PV panels, inverters, cables, structures, and batteries.

However, bundling of funding, development, and EPC services is an emerging combination of services that are being increasingly offered as a package thanks to the competitiveness of the EPC marketplace. In a bid to become more competitive and valuable by bringing a full package solution, Talesun, a China-based cell and module maker, has ventured into delivering more value to operators with funding, development, and an EPC role. The manufacturer intends to play a more comprehensive role in the solar industry by associating with projects at an early funding and developing stage to provide client a turnkey solution. A functional department has been established internally for the development and EPC business of Southeast Asia market including Vietnam, Thailand, and Malaysia with a targeted pipeline of 500MW for solar commercial & industrial project in next three years. In 2020, it delivered first of this kind of a project (although small ~ 570 KW) in Vietnam by playing the roles of engineering, procurement, and construction (EPC) of an entire project, instead of just supplying solar panels, thereby delivering more value to the customer.

For rooftop solar, which may be of interest to utilities providing solar equipment and services to Commercial and Industrial (C&I) and residential customers, bundling and unbundling of certain costs can have a significant cost-reducing impact on the total cost of installation and hence the Levelized Cost of Electricity (LCOE).

- Unbundling of freight transportation. Freight may ultimately be ripe for cost savings, as it was for most of the 1990s following deregulation of the transport industry when shippers sharpened their negotiation toolbox and practices. For rooftop solar, more than 90% of the supply chain costs relate to freight transportation to get the solar panels to the distributed sites. Pallets of solar panels are heavy, distances can be long, and homeowners and C&I customers often don't have discounts on freight.
- Bundling of roof tile and PV panel. Both Tesla and a European competitor, Roofit, are selling integrated offerings that start with roof tiles made of solar PV. Not only is it freight efficient, but many consumers prefer the aesthetics over classic solar panels, which are often ugly modules that lie on top of nicer-looking tiles.
- Bundling of soft costs for economies of scope. For C&I and residential solar installation, installation-related ("soft") costs are another potential source of supply chain cost optimization, efficiency, and savings to be managed. Relevant costs include labor, permitting, inspection, and

interconnection. Tesla offers permitting and installation coordination included with the sale of its solar roof tiles.

- Bundling of vehicle charge station and solar roof. Tesla and Enel X offer both rooftop solar and vehicle charging stations, or at least cross-selling of both.

Contract Term Risk and Mitigation: Coping with Tariff and Regulatory Interference

The solar PV module supply market was recently thrown out of balance by two regulatory actions in the United States. First, the renewable energy Investment Tax Credit (ITC) was seemingly left to decrease on December 31, 2020 (it was extended[303] for another year at the last minute, but the lurching and uncertainty has not had the same effect on the industry as a stable ITC would have). Second, the US government under the Trump administration applied a 30% import tariff under Section 201 of the US Trade Act of 1974 (P.L. 93-618).[304] The US import tariffs had the effect of reducing the US solar PV investment by $19 billion.[305]

As the US solar market bet on continued solar incentives (even if stepped down year-on-year) and now faces a Biden administration that is poised to staunchly support renewables, operators are forming long-term supply agreements in anticipation of a significant rebound of demand.[306] EDF signed a five-year procurement agreement with Canadian solar for the supply of 1,800 MW of high-efficiency poly solar modules. Under the terms of the agreement, Canadian Solar will deliver its bifacial enhanced wafer BiHiKu (CS3W-PB) and enhanced wafer HiKu (CS3W-P) modules to EDF Renewables' multiple solar projects in the US, Canada, and Mexico. This module supply agreement is the largest single module supply agreement signed in Canadian Solar's 18-year history.

Whatever the regulatory climate, Power Purchase Agreements (PPAs), which generally range from ten to 25 years in length, are instrumental in mitigating financial risks involved in constructing utility-scale power plants of any technology. Here are some examples of recent 20–25 year power purchase agreements:

- ENGIE, alongside its investment partner Meridiam consortium and Fonsis, the Senegalese Sovereign Fund, signed a 25-year power purchase agreement with Senelec, the Senegalese off-taker for two solar photovoltaic projects in Senegal, Africa. Developed by the Senegalese Government and the International Finance Corporation (member of the World Bank Group), the projects have a combined installed capacity of 60 MW and are part of the wider Scaling Solar initiative in Senegal.[307]
- SunEdison, Inc., a leading solar technology manufacturer and provider of solar energy services, signed a 20-year PPA with the Southern Nevada Water Authority (SNWA). The agreement enables SNWA to lock in a significant portion of its energy costs at a fixed rate, providing an effective hedge against future increases. As part of the agreement, SunEdison will

develop, construct, own, and operate a 14-megawatt (MW) AC solar photovoltaic (PV) power plant, located in Clark County, Nevada.[308]

- EDF Renewables North America has announced the signing of a 20-year PPA with CleanPowerSF for a 100-megawatt (MW) tranche of the Palen Solar site known as Maverick 6 Solar Project. The project expects to come online by the end of 2021 and deliver enough clean electricity annually to power 49,000 average California homes.[309]
- Solarpack awarded a 25-year PPA in Rajasthan, India, and it will be commissioned during 2022. The project will represent a total investment of approximately € 129 million.[310]

Trade-offs and Management Techniques in Operations & Maintenance

Cybersecurity Risk Management: Assessing Firewall Adequacy

Attacks on control systems for critical infrastructure rose by more than 250% between 2012 and 2016 in the US, as web-linked communication systems have proliferated and nation-states seeking geopolitical and economic supremacy added to the incidence of amateur hacking. Of all the critical infrastructures targeted, power grids have become the ripest target because most or all sectors of economy depend on them: cyberattacks on power grids can be exponentially effective by crippling vast swaths of the industrial and commercial sectors. Furthermore, the absence of power paralyzes many national security systems, making physical terrorist attacks much more effective and more likely.

sPower was the first US provider of solar and wind renewable energy to have been the victim of a cyberattack. The attack left operators at the company, sPower, unable to communicate with a dozen generation sites for five-minute intervals over the course of several hours on March 5, 2020. Each generation site experienced one communication outage.[311] Unidentified hackers attacked the company with a denial-of-service (DoS) attack whose root cause was tracked down to an unpatched firewall—an attacker used a known vulnerability in a Cisco firewall to crash the device and break the connection between sPower's wind and solar power generation installations and the company's main command center. In computing, a denial-of-service attack is a cyberattack in which the perpetrator seeks to make a machine or network resource unavailable to its intended users by temporarily or indefinitely disrupting services of a host connected to the Internet.

The attack did not affect sPower's more critical control systems and did not impact its power generation. In response, Cisco recommended a firmware update, which sPower has been deploying across its system, after testing for compatibility. Given the lack of identified follow-up actions by the attacker, this would appear to be someone testing or scanning for this vulnerability and inadvertently hitting utility infrastructure in the process—but nothing can be said with certainty. The attack, however, highlights the fact that critical infrastructure continues to be fairly exposed and needs much greater security.[312]

Internet of Things (IoT) and Artificial Intelligence (AI) Technology Choices: Using Smart Tracking

AI is transforming solar power efficiency, and project managers need to integrate knowledge about the path of scientific evolution because it dramatically impacts cost and productivity of the system, and consequently drives the nature and form of supply agreements. Manufacturers such as Huawei expect that 90% of the PV plants will involve remote access and machine learning by 2025, using AI at every stage from system design to storage—modules, trackers and inverters—enabled with IoT. Given this situation, operational data is rapidly becoming the most fundamental resource in the solar package.

Smart Tracking is the main axis of AI progress. Adjusting the angle of the panels to catch the most sun has a significant effect on the system's efficiency, so an increasing number of PV plants deploy trackers. Currently, the trackers rely on astronomic algorithms to ensure that the modules are aligned perpendicular to direct sunlight. However, the incline of the land creates shading between different rows of PV panels that could reduce the yields. In cloudy days, the diffused light accounts for a higher ratio than that in normal days. And the tracking angle without optimization can't harvest as much power when the light is scattered. Huawei used AI to develop a tracking algorithm to optimize the angle, which the company claims increases energy generation by at least 1%.

AI can also help shorten maintenance times by producing I-V curves that ascertain current (I) and voltage (V) data that can be used in diagnosing the health of the panel. A reliable and pre-constructed I-V curve diagnosis can shorten site inspection time frame from weeks to half a day, according to Huawei. In addition, a deep and reliable I-V curve diagnosis can identify potential resolutions to solar module health, thereby facilitating quick and comprehensive maintenance interventions. Smart I-V curve diagnosis, aided by AI, has been deployed in 7 GW of PV worldwide so far[313]

The intelligent component of solar PV installations may not be captured unless procurement and engineering staff conduct adequate research, supplier identification, and total cost analysis, and ensure installation allows for configuration of AI dimensions.

Capacity Strategies: Using Solar + Storage to Tame the Duck Curve

Energy storage is the peak capacity strategy for solar PV. It allows energy to be produced at all available times, without curtailment as would otherwise be required, for example in wind farm operation when the farm would produce more energy than the grid needs or can handle.

The now-famous "duck curve," shown in Figure 55, illustrates the peak capacity dilemma of most power producers. Demand for power on most grids tends to rise in the morning when customers wake up and get ready for work, then decline throughout the midday hours, and rise again when they return to their homes especially in the evening hours as they use home appliances. In contrast, in conventional power plants, power is produced on a more or less constant basis. The gap between the demand and the supply leaves overconsumption in the morning and evening, and overproduction at midday. Without any way to store the energy, the power that is produced is wasted.

"Solar Plus Storage," as it is called, is the most rapidly emerging and one of the lowest-cost power options available today, thanks to the elimination of peak capacity loss dilemma that has perennially plagued power operations.

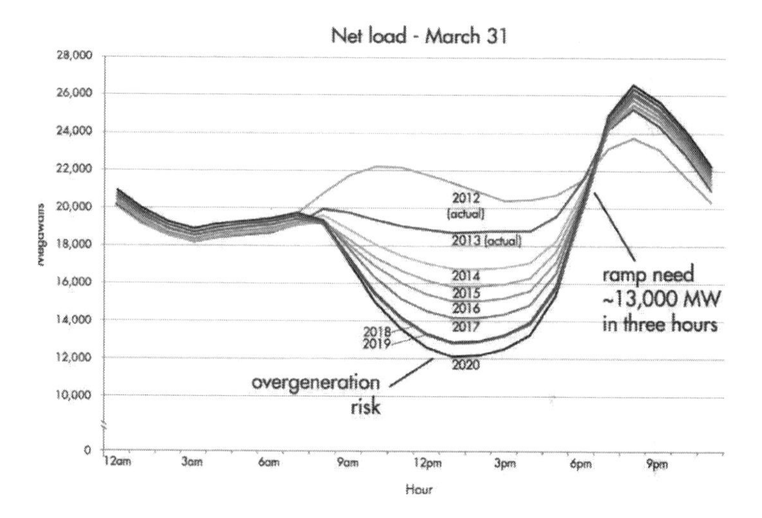

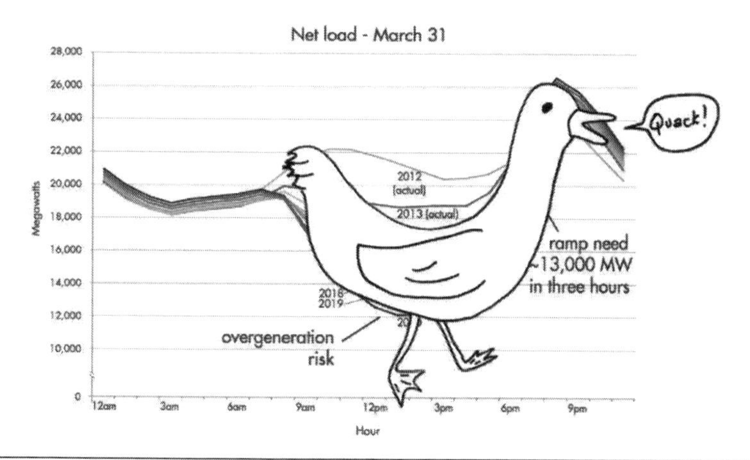

Figure 55. The Duck Curve

(*Source:* Wirfs, Jordan. Learning How To Adapt To More Renewables As "Duck Curve" Deepens. October 25, 2016)[314]

Another technology that is useful for managing capacity is drone-based surveillance. Drones that send signals via digital connectivity to sensors is allowing speedier identification of faulty modules, which can maximize available capacity. Aerospec, which uses drones to conduct aerial inspection of critical infrastructures, determined that a 100 MW (450k PV modules) solar farm it inspected had 3,500 faulty PV modules, or approximately $1 million in replacement cost, a huge cost saving achieved by identifying and replacing these modules within the warranty period. In addition, if left unnoticed, these faulty PV modules would result in close to $350,000 annual revenue lost due to generation lost. Most importantly, drone inspection could be done 6–10 times faster than traditional methods, which give customers ample time to make corrective decisions.[315] Although there is a cost to acquiring and operating the drones, the reduction in replacement costs may be worth the cost.

Sourcing Trade-offs: Identifying Alternative Sources for Competition and Supply Assurance

China's dominance of the solar PV supply chain poses a problematic set of interrelated challenges for sourcing in the solar power industry.

Economically, the supply market is highly concentrated. China controls 80% of global polysilicon capacity and a third of its supply, and 70% of global solar PV production. Five companies in mainland China and Hong Kong control approximately two-thirds of the world's polysilicon market. Hence, it is very difficult for buyers not to become dependent on Chinese suppliers without major investment and time to build alternative supply chains, which raises the possibility that in the meantime Chinese sources could raise prices while their customers would have no credible alternative sources. Sure enough, as demand for solar panels increases rapidly, especially with the rising popularity of bifacial panels that require twice the silicon, while China has hesitated to increase its capacity stepwise to keep up, prices for the glass that coats photovoltaic panels have increased 71 percent. Some forecast a 20–30% shortage of glass for solar PV in the medium term.

Socially, a third of the polysilicon used to make solar panels came from Xinjiang, home of the Uighurs and other Muslim minorities, which some groups claim have been forced into slave labor by the Chinese. Legislators and trade associations have issued statements condemning such actions and encouraging or mandating their members to ensure ethical supply chain management.

Politically, the United States has placed import tariffs on Chinese PV panels, with certain exemptions, with the intent of encouraging the development of a domestic solar PV supply chain. So far this has not happened because of the investment required and companies' inability to achieve a comparable production cost in the United States. Meanwhile, however, importers of solar PV to the United States are at the whim of legislators, who could dramatically impact their business from one day to the next through regulatory trade actions.

Environmentally, the industry in China runs on coal-fired power plants, which defy international attempts to mitigate global warming. That is why China can produce the panels more cheaply than Western countries. It is also the reason why Western countries cannot replicate China's cost advantage.

Given the economic, social, political, and environmental problems involved in sourcing PV panels from China, there is intense pressure on buyers of solar PV to find alternatives,[316] including building a domestic solar PV supply chain in the United States.

In addition to the confluence of issues surrounding sourcing from China, from a sourcing point of view there are also two technical solar system design subsystem choices that can make a significant difference in project economics, and thus should be considered when selecting suppliers and installers.

The first is inverter capacity. Optimal sizing of the inverter can improve array economics. A smaller inverter can reduce power limiting and therefore maximize the output of the panels, but a lower-capacity inverter can reduce the fixed costs and hence accelerate the financial payback of the system. The optimal capacity of the inverter is the capacity that maximizes Return on Investment of the system over its lifespan, which depends in turn on the cost of solar panels and the cost and price of electricity bought and/or sold. In other words, it's not necessarily a good idea to have a one-to-one ratio of array to inverter capacity. Determining the optimal inverter capacity is a system engineering question.[317]

The second choice relates to the panel cleaning system. In dry and dusty or sandy climates, dirt can accumulate on panels rapidly, decreasing power output. Cleaning the panels especially makes a difference in utility-scale power generation in desert and dusty areas, conditions that are found in many environments and are particularly common in the Middle East and Africa. A range of automatic solar panel cleaners is available at various price and quality points. The automatic cleaning system is in itself a modular subsystem that requires design, engineering, installation, and maintenance. Its cost needs to be balanced against the incremental net revenue that can be generated through its beneficial impact on power generating efficiency.

TCO Trade-offs: Avoiding, Replacing, and Disposing of PV Modules

Assuming a 20-year system lifespan, the upfront cost of solar panels and electrical apparatus (including inverter) typically constitutes about 50% of the cost of a 3–25 MW system, with civil work and installation cost making up about 20% and the remaining 30% coming from operation and maintenance cost over the life of the system.[318]

As with cheap electronics of all kinds, cheap solar panels and systems can come at a very high price, and buying cheaper upfront may cost more in the long run. Because of the large number of manufacturers and the strong demand, buyers can get burned by buying cheap materials and equipment. Here are some of the more common blowbacks of spending less upfront:

- Lower output than advertised or illustrated
- Shorter lifespan than expected
- Fire hazard due to poor quality thermal control system
- Ugly appearance
- Limited warranty coverage
- No after-sales service available in your area
- Manufacturer goes out of business
- Equipment failure, especially the inverter
- Higher-than-expected installation costs due to design incompatibilities with standard racks, mounts, and cable systems
- High disposal costs at end-of-life

Once a low-quality system is installed, it may need to be uninstalled in order to install an alternate system, with the result that the total project could cost twice as much as it would have with a high-quality system in the first place.

Another TCO trade-off that will be a major supply chain management concern is solar panel recycling. South Korea and some US states are already mandating the recycling of solar panels. South Korea's Ministry of Environment plans to introduce new rules for PV waste recycling in 2023. Several recycling facilities are already being built, including one by the government, with a combined recycling capacity of 9,700 tons.[319] And in California, starting in 2021, waste generators of decommissioned solar panels (photovoltaic or PV modules) will be allowed to manage them more economically than general waste, under which they would be subject to hazardous materials waste disposal regulations. California is the first state in the United States to identify solar panels as "universal waste" to reduce management burdens and facilitate recycling.[320] In France, Veolia, the wastewater treatment company, is developing a solar PV recycling plant in southern France, with an expected capacity of 4,000 tonnes by 2022.[321] These efforts are aligned with SDG 9 because promoting safe disposal of solar panels contributes towards building sustainable value chains.

Supply Chain Roadmap for Solar Power

There are numerous opportunities for supply chain management to create cost-saving and revenue-generating opportunities in solar power generation, as illustrated in Figure 56 below.

To lower the cost of solar (a *rationalization* supply chain strategy), IoT/AI benefits must be maximized. This includes the use of smart tracking to optimally align the modules towards the sun. Also, given the early stage of maturity of the industry and the wide variety of modules and solar technologies available, it is essential to consider and manage total life-cycle cost to ensure that upfront savings aren't eaten up by poor operating performance and system and component failures. This is a very difficult task at the current time because many of the life-cycle cost factors are not yet known, confirmed, or proven in the field.

To ensure a balance between supply and demand, a multi-stakeholder effort is required to secure the polysilicon supply chain. Also, integrating energy storage will generate critical benefits and must be done with proper attention to equipment parameters and specifications. This requires assumptions and estimates, so it is not easy.

To customize solutions to various applications, solar needs to include both central and distributed power generation solutions. Enabling distributed solar will require alignment of incentives, regulatory frameworks, and installer capacity.

Three innovations open up new horizons for solar, and will require supply chain analysis and support: safe floating solar, safe embedded solar, and integrated rooftop solar and home vehicle charging systems. This last one may be beyond the scope of many utilities, but it cited here because it may become a differentiating competitive factor between solar electricity providers.

Also, environmental and social responsibility require extensive participation of supply chain experts in three areas. Procurement needs to avoid social exploitation (e.g., of the Uighurs) and CO_2-emitting production processes (e.g., coal-fired power). Also, HSE, Engineering, Regulatory, and Logistics need to manage the end of life disposal of toxic PV waste.

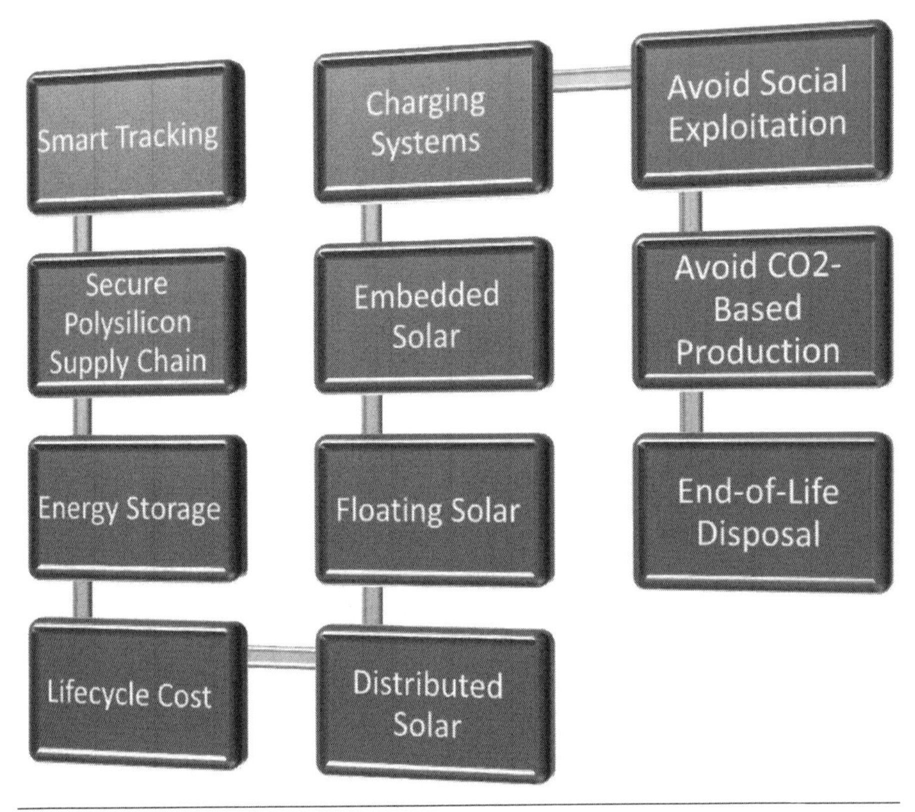

Figure 56. Emerging Supply Chain Opportunities for Solar Power

Biomass

Chapter Highlights

1. Biomass includes many permutations of feedstock and conversion technologies, including hybrid and multi-stage process routes such as co-firing. A wide number of possible process combinations should be evaluated at the outset to minimize the Levelized Cost of Energy.
2. Decision analysis tools, including capacity modeling, network modeling, and scenario modeling, should be used in the planning stages, to evaluate the wide variety of possible feedstocks, processing technologies, and financial scenarios.
3. Biomass plants are significantly less efficient than most other technologies, so quantitative comparisons should be made before concluding that the availability of cheap raw materials makes it the best option.
4. Biomass plants are not necessarily based on renewable energy principles if they consume more feedstock than is regrown.
5. The environmental friendliness of biomass plants may be offset by heavily polluting trucks hauling inbound raw material for feedstock, so inbound raw material logistics should be planned carefully.
6. Partnering with specialist engineering firms is a good way to ensure reliability and precision of technology choices and implementation. However, for large projects, allying with a large international engineering or EPCM firm can also ensure full bundling, compatibility, and integration benefits.
7. Depending on the facility and application, potential supply chain opportunities to explore could include: 1) Hybrid Conversion Technologies, 2) Multi-Stage Production Process Routes and Pre-Treatments, and 3) Replenishment of Natural Resources.

Introduction

Compared to the other renewable technologies covered in this book, biomass power generation includes a much wider range of technologies and fuel raw materials. The principal raw materials are agricultural residues such as corn stalks, crops such as bamboo, trees such as fir, forest debris such as pine needles, and urban waste such as used newsprint, cardboard, and municipal solid waste. The main conversion methods are burning and gasifying the feedstock. Burning can be high-temperature complete combustion that generates steam which powers a steam turbine, or partial combustion that generates an offgas which can be directly used as a fuel. Anaerobic digestion, either through natural decomposition or catalyzed by chemicals, and pyrolysis (partial burning that releases a vegetable oil which can be burned as a fuel) are less common process routes because they are less efficient and produce a less volatile fuel, respectively.[322] Environmentalists appreciate the fact that biomass feedstocks are renewable, but critics point out that burning the raw materials produces both particulate emissions (air pollution) and CO_2, and in the absence of regulatory oversight it may, and sometimes does, deplete the natural resource faster than it is renewed, which under certain conditions makes biomass-based power generation essentially non-renewable.

Biomass-fueled power generation is much less efficient than burning natural gas or refined fuel oils: combustion power cycle efficiencies are 23–25%, compared to more than 60% for natural gas turbines. Co-firing (adding 5–10% bio-waste to a fossil-fuel powered generation system) produces a compromise of efficiency and renewable energy, as do ethanol blends of refined fuels. Combined heat and power, or co-generation, increases the overall efficiency of conventional power plants by 27% (from 60% to 76%).[323] Multi-stage hybrid processes, for example pyrolysis followed by co-firing, can similarly increase process efficiency. Anaerobic digestion systems require extensive land area and can be even less efficient than combustion-based biomass power generation—as low as 20%—and can be boosted by various pre-treatments.[324]

Globally, biomass represents 2.7% of power generation, which reflects the low efficiency rates and the air pollution and depletion problems that can occur in the absence of regulatory oversight. It is almost nonexistent in the Middle East, where fossil fuel production is plentiful and forests are nowhere to be found. It is highest in Europe, where it constitutes 9.7% of regional electricity production when Eastern Europe is included, and 7.5% of EU energy production.[325]

From a construction point of view, the value chain is not dissimilar to other thermal power plants such as those fueled by coal, as can be seen in Figure 57. A fluid bed boiler unit is constructed and piped to a steam turbine powering a generator, which is connected to electrical cables, transformers, and transmission and distribution network. From an operational point of view, it is also similar in that there is a regular (daily) inbound flow of raw materials to burn, a loading operation, and a regular outbound flow of waste materials such as ash. Maintenance on the transmission and distribution network is the same as with other power generating technologies.

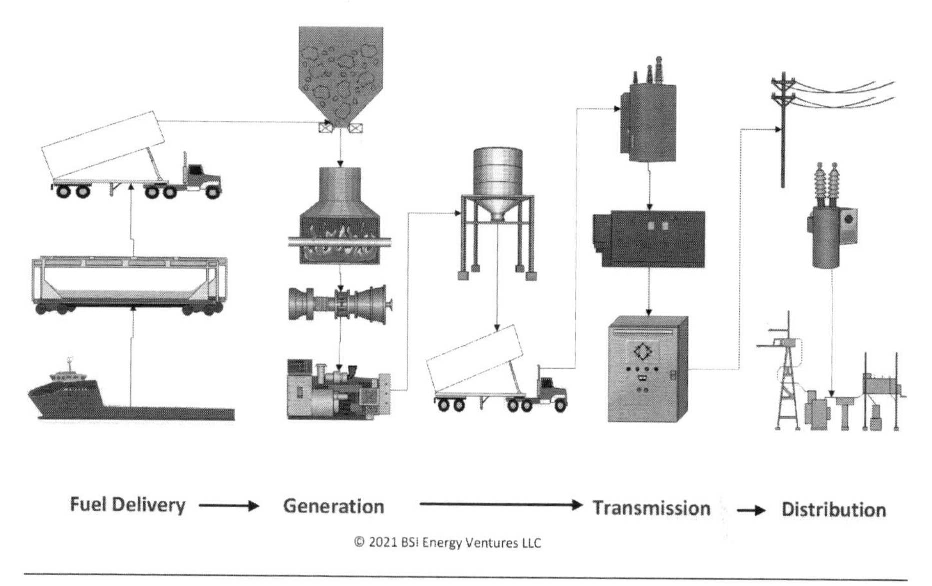

Figure 57. Value Chain for Biomass Power Generation (Illustrative)
(*Source:* BSI Energy Ventures)

Average global supply chain costs for biomass total as much as $0.116 per kWh, based on the author's analysis of data from IRENA, which calculates a Levelized Cost of Electricity (LCOE) of about $0.20 and estimates that 40–50% of the cost is for feedstock including freight. Biomass has the second-highest supply chain cost of all power-generating technologies, according to an analysis by the author. The high biomass power supply chain cost comes from the process of physically gathering the material, hauling from its natural origin such as a forest, and moving the feedstock into and the waste product out of the boiler area. The material is typically bulky and, in some cases, requires specialized vehicles such as logging trucks, and this stage of the supply chain almost always emits tailpipe exhaust that is rarely counted against the clean and renewable qualities of biomass power generation.

Trade-offs and Management Techniques in Capital Project Management

Technology Choice Risk Management: Assessing Multiple Process Routes including "Power to Gas"

One of the most proven biomass technologies is cogeneration, involving burning waste to emit gas that can be burned to produce power and sometimes heat. This model involves much higher efficiencies than burning raw forest debris or wood pellets. However, variations in the exact process route, engineering of the boiler type, and parameters such as temperature are worth studying because they can increase efficiency by a few degrees, which can have critical implications for profitability.

Therefore, Vantaa Energy Ltd of Finland is engaging Wärtsilä, a process technology leader with specific experience in cogeneration, to conduct a feasibility study for a Power-to-Gas facility at Vantaa Energy's waste-to-energy plant in the city of Vantaa in the capital region of Finland.[326] This co-development project leverages Wärtsilä's expertise on processes and technologies, while Vanta brings knowledge of district heating.

The Power-to-Gas facility would produce synthetic biogas using CO_2 emissions to generate electricity. The purpose of the joint study is to assess the optimal size of the project and key factors for project feasibility, along with the cost implications of synthetic biogas for district heating. Vantaa Energy is relying on this project to replace natural gas in district heating with synthetic biogas, thereby reducing company's overall carbon emissions. The Power-to-Gas facility would produce synthetic biogas using CO_2 emissions to generate electricity.

Capital Project Risks and Mitigation: Choosing an EPC with Depth in the Target Process Route

Conventional power plant operators tend to engage EPC firms when embarking on biomass projects to better manage technological and project management risks, especially when the projects are cross-border. By engaging with one or more EPC firms, they are able to harness specific expertise and previous experience with the feedstocks and the conversion process, thereby avoiding learning curves. Here are some examples of conventional operators collaborating with EPC firms for their technical and managerial experience in biomass:

- PowerChina (China) and IBS Energy (Brazil) to build an 80 MW biomass project in Brazil.[327]
- Village Farms Clean Energy (Canada) and Mas Energy (USA) to design, build, finance (including all capital expenditure for construction), own, and operate the biomass-based Delta renewable natural gas project.[328]
- Blue Sphere Corporation (USA), a clean-tech independent power producer (IPP), selected Anaergia (Canada), an organic waste treatment plant operations and technology provider, for the turnkey design, construction, and delivery of a biogas plant in Sterksel, the Netherlands.[329]
- Energy Works (UK) signed an EPC contract with M+W Group (Germany) for a 28 MW waste to energy plant in Hull, UK, which will incorporate fluidised bed gasification and anaerobic digestion.[330]
- Fraddon Bio Gas (UK) engaged FLI Energy (UK) as EPC for the design, construction, and commissioning of the Fraddon biogas plant, which will convert organic agricultural and local food waste into gas and electricity.[331]

Supply Unavailability Risks and Mitigation: Assessing Fuel Supply, Logistics, and Seasonality

Since a biomass energy system depends on fuel, resource assessment is critical in the planning stage, and procurement and inbound logistics are critical in the operations stage. As part of the feasibility assessment, planners must estimate the fuel quantities needed and available, including consideration of factors that may disrupt supply such as weather and seasonality.

In order to mitigate feedstock supply unavailability risk, Pacific BioEnergy Corporation (Canada) undertook two new long-term wood pellet supply contracts for Japanese power producers. Commencing in 2020 and 2022 respectively, PacBio will be supplying 170,000 metric tonnes per annum through 2030 and 2035, respectively, to newly built dedicated biomass power plants.[332]

Outsourcing Risks and Mitigation: Monitoring the Quality and Purity of Incoming Feedstock

As the operational and financial success of biomass plants depends on the availability of high-quality feedstock, operators need to be rigorous about supplier selection, and monitor the quality of incoming fuel supplies. The quality and energy content of fuel is not an uncommon area of dispute with fuel suppliers. In one case, an EPC firm, Babcock & Wilcox Construction Co. (BWCC) sued its client, the operator of the 75 MW biomass power plant in the US, for supplying wood fuel that failed to comply with the contract specifications, and consequently resulted in damage to plant equipment.[333]

Supplier Partnering Risks and Mitigation: Engineering Design and Operating Parameters

Technical specifications related to the boiler and the flow of fuel over surfaces like fluidized beds are critical in ensuring efficiency and reliable output of biomass plants. Most EPC firms rely on partnerships with engineering experts. For example, Phoenix Energy (USA) relies on EQTEC (Ireland) for the gasification technology at multiple 2 to 3MW biomass plants. EQTEC provides its proprietary Gasifier Technology, together with technical design and engineering for the power plants, and Phoenix Energy will secure project funding.[334]

Procurement Bundling Trade-offs: Tapping a Leading EPC Firm to Ensure Overall System Compatibility

The disadvantage of relying on relatively small technical specialists is that working with specialists may sacrifice "bundling" benefits, where one party has end-to-end responsibility and can thereby ensure full integration and compatibility of all systems and components. One way to achieve both technical expertise and bundling benefits is to ally with large proven multinational engineering firms. In one example, Hitachi Zosen hired GE as its EPC for a 50 MW biomass power plant in Japan. GE's Steam Power division will design, manufacture and supply the core components of the project's power block. The package will include the steam turbine generator, as well as the boiler and its air quality control systems.[335]

Contract Term Risk and Mitigation: Aligning Operating and Maintenance Agreements, PPA, and Boiler Longevity

Whereas PPAs for geothermal, hydropower, and nuclear plants tend to be extraordinarily long-term (20–40 years), PPAs for biomass plants tend to be closer to the 20-year mark, and maintenance and operating contracts tend to mirror them. For example, Hokuriku Electric Power Company signed a 20-year PPA with TEPCO Power Grid for its 75 MW Ichihara Yawatafuto biomass power plant in Japan,[336] and French power producer Albioma signed a 26-year inflation-indexed power purchase agreement for annual electricity sales of 120 gigawatt hours (GWh) for its 48 MW bagasse-fired Vale do Paraná Albioma cogeneration plant, located in Suzanápolis, São Paulo, Brazil.[337]

Trade-offs and Management Techniques in Operations & Maintenance

Cybersecurity Risk Management: Ensuring Reasonable Barriers to Hacking

Cybersecurity risk has not been perceived as a serious threat among the biomass power plant operators traditionally. Unlike nuclear or other utility-scale solar or wind power plants, biomass plants operate with much smaller capacity of under 100 MW, with most of them falling under 50 MW rating. Automation in control and monitoring, however, has bolstered in the last few years. It is expected that sophisticated cybersecurity systems will become a necessity as the capacity grade of these plants move towards utility scale. Increasing need to manage more controls and greater monitoring will likely spur greater digitalization and even create digital twins of the physical operations. While these would enhance efficiency, cyber vulnerabilities would also increase. Meanwhile, for a plant with little digital infrastructure, assuring that basic wi-fi and device connections are secure could prevent unforeseen incidents.

Internet of Things (IoT) and Artificial Intelligence (AI) Technology Choices: Establishing a Data Historian to Enable Potential Data Mining for Process Efficiencies

No matter the end product, process economies are critical to stay competitive, and, therefore, so is attaining cost-effective plant maintenance. Condition monitoring plays a vital role in predictive maintenance by allowing a producer to actively prevent breakdowns and optimize production processes. For example, Atlantic Power Corp.'s biomass power plants deployed a real-time enterprise data historian called eDNA, which collects, archives, displays, analyzes, and reports on continuously streaming time-series data. Various tools within the eDNA were developed to help monitor plant efficiencies and optimize various parameters, so that optimal performance could be known, and operations could be adjusted whenever required. Both short-term and long-term trends are monitored to evaluate the impact of changes on plant output and emissions. Operations are fine-tuned accordingly to meet production and environmental targets with the least possible amount of wood burnt.[338]

Peak Capacity Strategies: Designing Inbound Logistics Capacity to Match Plant Size

Determining the optimal system size for a particular application must accurately consider economies of scale, for which a capacity modeling tool should be used, ideally with network modeling features and capabilities to consider feedstock scenarios. Biomass power system cost intensity tends to decrease as the system size increases. For a power-only (not combined heat and power) steam system in the 5 to 25 MW range, costs generally range between $3,000 and $5,000 per kilowatt of electricity. Levelized cost of energy for this system would be $0.08 to $0.15 per kWh, according to data from IRENA. However, building a larger system would require more feedstock, which would require higher inbound transportation costs, and possibly require purchasing feedstock from sources farther away.

Sourcing Trade-offs: Allowing for High and Potentially Growing Transportation Cost Despite Low-Cost Feedstock

Collection and transportation of feedstock is the most critical and costly part of the biomass energy supply chain. In fact, collection, transportation, and handling together occupy more than 60% of the total production cost, whereas conversion cost is less than 30%. The high cost of biomass supply can be surprisingly high, given that agriculture and forest residue seems free. However, transportation costs alone may sometimes occupy up to 30% of total cost. The nature of biomass availability tends to add to its supply risk. Managers must take note of this tradeoff between biomass cost (of availability) and supply risk. For one, robust planning and forecasting of availability can help optimize transportation and save trips.[339]

TCO Trade-offs: Conducting Scenario Analysis to Determine the Optimal Feedstock and Process Route

Since there are so many pathways to energy generation through biomass, the design and engineering phases of biomass plants should evaluate multiple variations of feedstock and conversion technologies, including hybrid and multi-tier scenarios, in order to seek the lowest Total Life-cycle Cost and hence the optimal Levelized Cost of Energy. Effort spent at this stage is the only way to ensure maximum return on investment, since after the plant is designed and built, many more constraints apply, and unplanned upgrades would thus be more costly and potentially infeasible.

Supply Chain Roadmap for Biomass Power

Due to the wide variety of biomass plants, supply chain opportunities need to be evaluated on a case-by-case basis. That being said, in general, as illustrated in Figure 58, 1) Hybrid Technologies, 2) Multi-Stage Process Routes and Pre-Treatments, and 3) Regrowing Natural Resources may present cost savings and environmental footprint benefits. The first two opportunities directly concern supply chain management because production scheduling, management, and control have been a part of supply chain management since its manufacturing roots. The third opportunity relates directly to procurement and the logistics of supply, both of which are directly in the scope of supply chain management.

Although biomass conversion efficiency is generally low, layered, or hybrid, technologies can in many cases increase cost efficiency. This is similar to "value stacking" in energy storage. For example, Combined Heat and Power (CHP) often increases efficiency by orders of magnitude (e.g., from 25% to 75%). Pyrolysis may also substantially increase combustion efficiency.

Multi-stage process routes such as co-firing and anaerobic digestion pre-treatments may also increase efficiency by orders of magnitude that can change investment decisions. Firing multiple different feedstocks, or sequential firing using different treatments or technologies, can often enhance energy production efficiency and thus are worth studying and implementing.

Regrowing natural resources reduces CO_2 footprint. As noted earlier, some plants can consume resources more rapidly than they can be replaced or regrown. Environmentalists have become close observers of the balance between fossil fuel substitution and natural resource depletion. In addition, the burning of some types of biomass generates CO_2 that offsets the benefits of not burning fossil fuels.

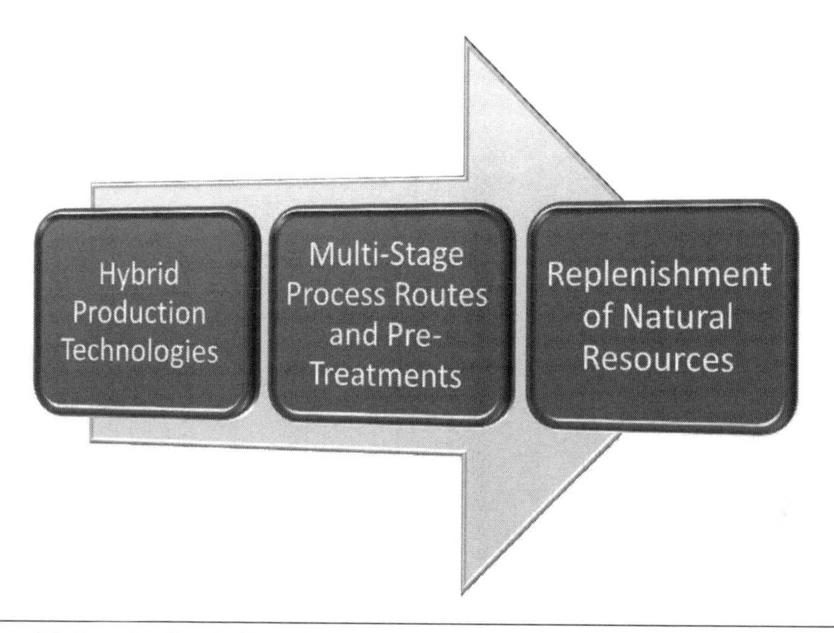

Figure 58. Emerging Supply Chain Opportunities in Biomass Power Generation

Oil & Gas—Upstream

Chapter Highlights

1. Due to the complexity and risks inherent in upstream operations, energy industry decision-makers are typically cautious with technology adoption, relying on suppliers to prove new technologies *in situ* and sometimes at scale before adopting them.
2. Due to the high capital intensity of the industry and the fact that many upstream ventures are funded through consortia of operators, a rigorous risk management framework must be used to decide which risks to accept and which to ask others to share, including the time-phasing of risks throughout the multi-year construction cycle.
3. Because of the scale of most energy capital projects, it is essential to prioritize long lead time and bottleneck items and services, including materials and services in shortage, before other purchases, and to invest in market intelligence and supply market forecasts to ensure reliable assumptions upon which to base procurement plans.
4. Outsourcing should involve a careful decision-making process that makes sure to keep strategic activities in-house, and to incentivize suppliers to achieve continuous gains and share them in some way.
5. Procurement of related products and services should be bundled together to allow suppliers to realize economies of scale and scope.
6. Standardization efforts are worth investing in, especially for very high-value external expenditures. While these may seem like the least likely to be able to standardize, they have the highest leverage—even small gains in standardization can have large payoffs in reducing capital and operating costs.
7. When contracting, long-term agreements provide stability and opportunities to capture efficiencies through partnership programs. Typically, these start off with a multi-year agreement and then provide options to extend for multi-year periods.

8. IoT and AI can generate extraordinary operational and cost benefits and must be implemented concurrently with cybersecurity programs to minimize the risk of hacking.
9. Organizations should try to embed a culture of environmental, social, and good-governance values so that initiatives become part of the corporate culture and are self-funding where possible. Advanced technology, including remote operation, wearable sensors, and similar devices, may increase worker and operational safety.
10. The upstream Oil & Gas supply chain can be rationalized, synchronized, and made more sustainable through supply chain management best practices. Cost can generally be reduced by deploying digital oil fields and converting product procurement to services-as-a-product. Reliability can be ensured and increased by implementing cybersecurity protections. Sustainability can be demonstrated by using renewable energy to power operations.

Introduction: Upstream Supply Chain Characteristics and Cost Drivers

The end-to-end oil & gas value chain is extensive, spanning exploration & production, midstream shipping and separation, and refining, as indicated in Figure 59. Seismic specialists assess geological formations. Drilling teams drill exploratory wells and outfit rigs with equipment to extract the oil & gas. Production teams complete wells and install wellheads to control the pressure of flow and separators to sort out gas, oil, and water. Pipelines and tankers transport crude product to a refinery or gas processing or petrochemicals plant, which converts the crude products to finished petroleum products such as gasoline, diesel, naphtha, and petrochemicals. Further pipelines or vessels carry the refined products to retailing centers for end use.

Exploration and production (E&P) is engineering-intensive and requires specialized technical expertise at every step along the way. For example, geophysicists and petrophysicists evaluate geological formations. Geochemists and petroleum engineers drill exploratory wells and outfit rigs with high-pressure, high-temperature equipment to extract oil & gas. Petroleum engineers and production teams install wellheads and blowout preventers to manage ultra-high pressures and flow rates, and separators to sort out gas, oil, and water. Figure 60 shows a schematic illustration of E&P activities.

Due to the extremely engineering-intensive nature of E&P, it might seem that supply chain management is not very relevant to upstream oil & gas. The formal training of most personnel in exploration is typically scientific (e.g., stratigraphers, paleontologists, and geophysicists) and frequently based in petroleum engineering

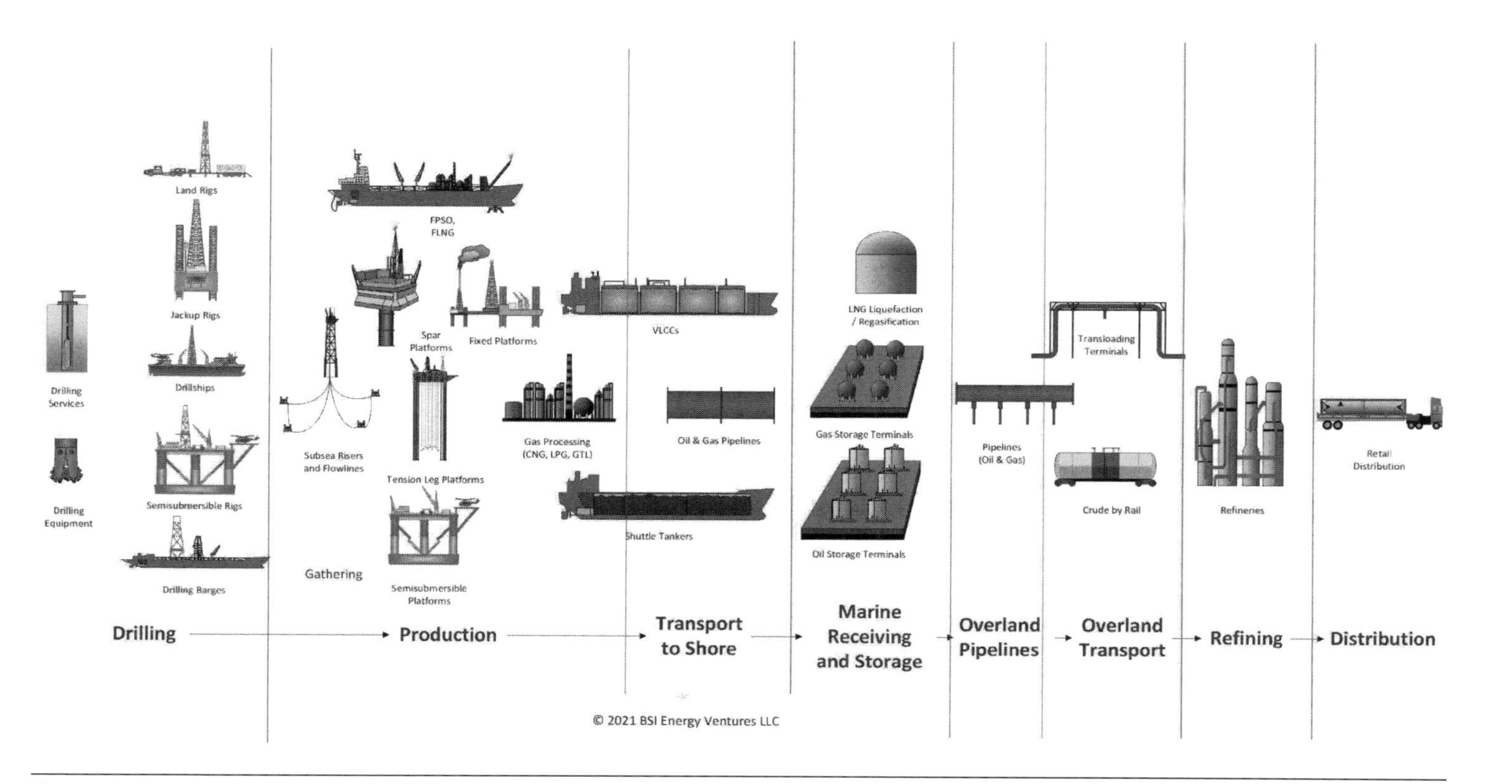

Figure 59. Oil & Gas Value Chain
(*Source:* BSI Energy Ventures)

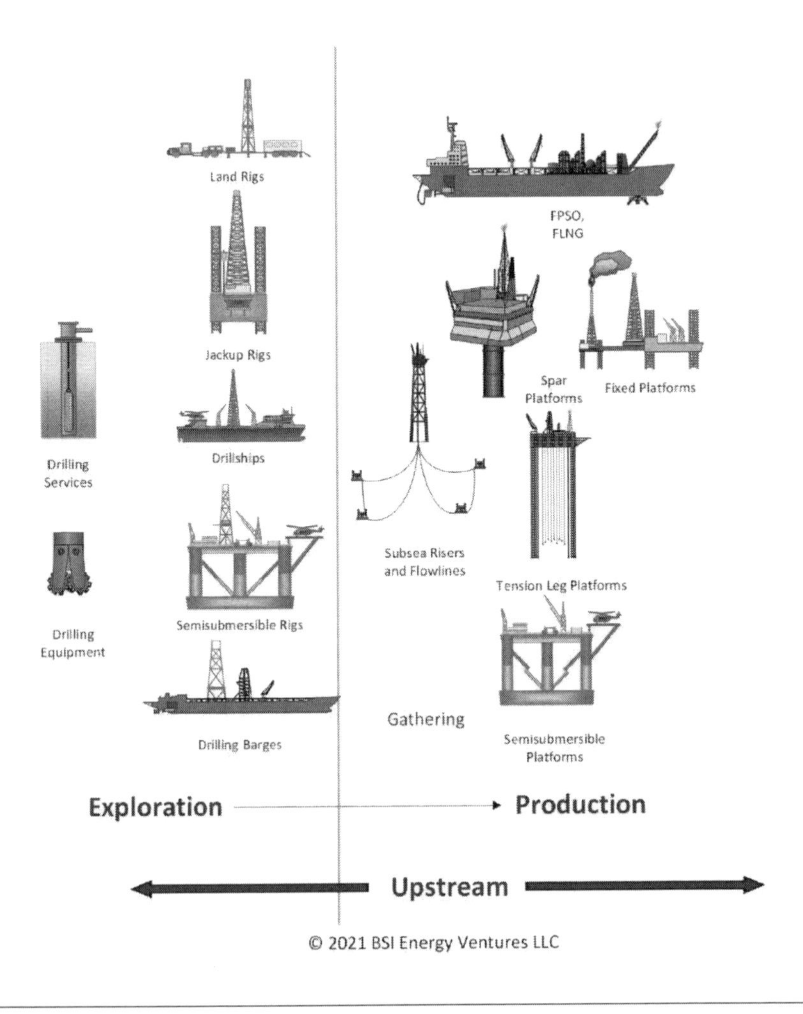

Figure 60. Upstream Oil & Gas Value Chain (Illustrative)
(*Source:* BSI Energy Ventures)

(e.g., reservoir engineers, petrophysical engineers, drilling engineers, production engineers, etc.).[340] Seismology involves more statistics than logistics: the issues and tools are often technical more than business oriented, using Monte Carlo simulation or optimization tools to estimate and probabilistically express reserves trapped in formations.[341]

However, the rigs themselves are often rented rather than owned by the oil companies, and the operation thereof outsourced, which leads to supply chain decisions. Furthermore, many of the personnel operating drilling rigs are subcontracted oilfield service employees, including the drilling crew (e.g., rig manager, driller, derrick man, roughnecks, and roustabouts), production crew (e.g., production foreman and pumpers/gaugers, and roustabouts), and craftspeople (e.g., electricians, mechanics, pipefitters, plumbers, instrument technicians, carpenters, welders, metalworkers, and drivers).[342]

In addition, the work relies on specialized equipment and services from third-party suppliers, creating a supply chain of activities related to "consumable" products and services such as tubulars, drilling fluids (muds), bits, well fracturing & stimulation (including sand control services), Formation evaluation (wireline and mud logging, measuring while drilling and logging while drilling (mwd/lwd, and coring), and well completions (services, liner hangers, packers, perforation, flow control). Procurement managers have other contracts for products and services such as service tools (coiled tubing and drillable, retrievable, and stimulation tools), fishing, oil & gas production chemicals and distribution systems, and artificial lift. In addition, upstream activities involve extensive maintenance schedules for capital equipment, including, for example, compressors (mostly reciprocating for in-field gas gathering or storage, and injection compressors), high pressure pumps for injection and use on production platforms, and turbines and generators for production facility power or drive power. Outsourced services often include well servicing, contract compression services, laboratory services, production chemical services, and water control.

The extraordinary amount of capital at risk in upstream oil & gas makes producers and their suppliers especially vulnerable to market, operational, and financial risk. Some smaller rig operators concentrate regionally (e.g., Diamond Offshore in the Gulf of Mexico and Unique Maritime in the UK) because their familiarity with the area allows them to better manage the risks. Oil companies often structure operational investments to mirror incremental and successive levels of financial commitment, especially during the exploration phase. Planning is extensive and occurs through a multi-stage vetting process.[343] Error rates (wells that are not commercially viable) begin high during early exploration, when it is relatively cheap to fail, but decline as operators reach project stages that require more investment. The main stages of project development are: concept studies, which are typically designed with the expectation of 25–40% error; feasibility studies, which have a range of acceptable error of about 10% and usually amount to 1–2% of the total project cost; FEED studies, which often have error in the 5% range and cost 5–10% of the total project cost; and EPC construction.[344]

Over time, fluctuations in exploration and production activity flow through the supply chain, causing a cycle of inefficiency characterized by "bullwhip" as fluctuations are passed through the supplier capital investment cycle, resulting in oscillating prices, capital investment, and facility utilization. Appendix 1 contains a detailed study of the long-term effects of this chain reaction. Suppliers experience the bullwhip effect more than the producers themselves, for the effects of bullwhip are magnified as it is transmitted through the supply chain. Symptoms of delayed reaction to changes in drilling volume often include rising and falling capacity utilization, extending and contracting lead times, and price volatility. Companies at all levels in the supply chain have used vertical integration, scale, and market dominance to shield against the effects of bullwhip over time, which explains, in large part, why the industry for oilfield products and services is so concentrated.

Therefore, while upstream oil & gas is engineering intensive, it is also supply chain management intensive, and the combination of the two calls for highly experienced and trained supply chain managers to make sound and durable decisions in a complex business and technical environment.

Trade-offs and Management Techniques in Capital Project Management

Technology Choice Risk Management: Carbon Capture & Sequestration, Enhanced Oil Recovery, and More

Increasing technological complexity requires strategic and operational expertise that can access the capacity, the financial resources, and the broad technical skills needed to develop the extraordinary new technology at the edge of science.

Oilfield technology is critical to sustaining production levels while oil & gas reserves are being depleted. The remaining economically viable oil is in harder-to-access places, and much of the oil that remains is of lower quality (typically heavier and harder to process). Drilling increasingly involves high pressure and high temperature situations. Onshore drilling is increasingly reliant on enhanced recovery, while offshore drilling wears equipment out more quickly due to the hot and deep (high pressure) environment, which also exacerbates corrosion and water treatment problems. Pumps must increasingly handle more viscous liquids and slurries. Oil sands and shale bring unique new challenges such as lifting and separation. In the offshore segment, logistics draw heavily on the experiences of the maritime industry, with purpose-built vessels to tap hydrocarbon reserves farther offshore and in deeper waters. Upstream oil & gas production involves a heavy dose of chemistry, physics, fluid mechanics, simulation modeling, and possibly augmented reality and remote-controlled vehicles.

Technologies that can improve productivity are worth a great deal to upstream oil & gas producers. Four areas have proven to be exceptionally fruitful: digital oilfields, deepwater exploration, Enhanced Oil Recovery, and Carbon Capture and Storage (CCS).

- *Digital oilfields.* The application of information technology, including by extension augmented reality and virtual reality, to obtain real-time intelligence about downhole conditions helps operators minimize fluid loss, manage flows within the reservoir, reduce the need for interventions, thereby maximizing production.
- *Offshore and deepwater projects* are unlocking huge reserves, but require multiple times the capital investment of traditional onshore

developments. Unconventional gas projects across the United States, heavy oil and oil sands projects in Canada, Arctic drilling, and ultra deepwater discoveries off the coasts of Brazil and Angola are increasing net available energy resources, but they require unprecedented capital and technology. Major research and development efforts are underway to produce heavy oil economically, and to treat oil with high H_2S content that will allow production to continue in fields across Asia and the Middle East that would have otherwise been plugged. Finally, exploration and production at Australian LNG and coal seam gas deposits such as Icthys are extending the limits of science on a daily basis, but the conditions are so harsh that only the majors need apply.

- *Enhanced Oil Recovery,* the method of increasing oil yield from older oil wells by injecting gases or chemicals to increase the pressure and output, has been around for years (for example, Norway began injecting CO_2 into the North Sea in 1996), and Weatherford is using thermal recovery, water-flooding, and implementing EOR techniques to increase yield of heavy oil deposits.[345]
- Enhanced Oil Recovery is often being combined with *Carbon Capture and Storage (CCS),* which requires ultra-large compressors that only a handful of suppliers have the technology and scale to produce. By connecting existing sources of CO_2 to oilfields, power generators and oil drillers both benefit. Power generators get lower emissions and oil drillers can increase recovery rates by up to 75%.[346] However, the economics of CCS are far from proven, which leaves large companies to manage the initial projects.

Both the costs and the benefits of new technologies evolve rapidly, making long-term technology planning a complex tradeoff of uncertain variables. For example, capturing carbon dioxide emitted from refineries in close proximity to oil production sites and pumping it into nearby oil wells can be an alternative to steam and nitrogen injection. However, the economics of CCS are still being explored. Operations using CO_2 can realize up to a 20% increase in oil yield (many estimates say more) but combining carbon capture and sequestration (CCS) with EOR can be cost-effective only if the infrastructure to transport CO_2 to oilfields is already in place.

For an operator, there are three approaches to managing technology in upstream oil & gas: 1) stay at the cutting edge; 2) delay implementation of technology until it is mature; and 3) conditionally decide case by case or project by project.

One approach to technology is to set up technology investments and partnerships to ensure enough technological strength to develop and implement winning solutions to these challenges by staying at the cutting edge. Shell and Schlumberger formed a research partnership to increase recovery rates from mature reservoirs at lower cost.[347] The consortium is exploring using drill bit sensors to collect more accurate field data, and then leveraging that information to customize the bits to meet the requirements of specific fields. Also, Kuwait Oil Company began its "integrated digital oilfield" program as early as 2010,[348] and is actively seeking technologies to ensure its future production, particularly in 3-D seismic, dynamic pre- and post-processing techniques, and horizontal drilling/multiple lateral technologies.[349] Schlumberger opened a Research and Geoengineering Center in Rio de Janeiro with 300 scientists and engineers to integrate geological science with drilling engineering in the hopes of achieving better deepwater drilling results. This was the first research center dedicated to exploration and production activity in Brazil.[350] Four years later General Electric opened its new multi-disciplinary Global Research and Development Center in Rio de Janeiro for 200 researchers and engineers, and committed to a $500m research program on electrical generation and control technologies as well as oil & gas applications.[351]

A second approach, which could broadly be called "optionality," involves delaying the adoption of unproven technology until it is mature. Many oil & gas majors are slow to implement technology, weighing the cost of failure against the cost of lost production. Measured and incremental investments can limit exposure to large losses. For example, after geological studies determine that the remaining reserves justify the investment, primary production (natural flow, artificial lift, gas injection, hydraulic lift, and plunger lift) is supplemented by secondary recovery (water-flooding, and a "five spot" flooding of multiple "injector" wells near one "producing" well in a repeating pattern). Similarly, operators sometimes opt for tertiary recovery (fire floods, steam floods, and CO_2 injection) if the reservoir characteristics justify it. An incremental approach to technology adoption is the best approach if technology switching costs and the opportunity cost of lost production are both high.

Optionality is not a new concept in oil & gas. The option to lease land for drilling exploration, shown in a flowchart in Figure 61, is an example. The land option cost varies depending on whether the option to lease is coupled with shooting rights (it is expensed if it is not coupled, and capitalized if it is coupled), and whether or not any of the acreage is leased (it becomes capitalized if it is leased and expensed if it is not leased).[352]

A third approach is to optimize technology costs and benefits on a project-specific basis, as indicated in the earlier principles and methods chapter. This strategy would be most useful for new industries such as renewable energy companies, for which the economics of the key variables are not yet clear—variables like future acquisition cost, the rate of progress of productivity improvements and learning curve effects over time, operating and maintenance costs, infant mortality rates, and supplier viability risk.

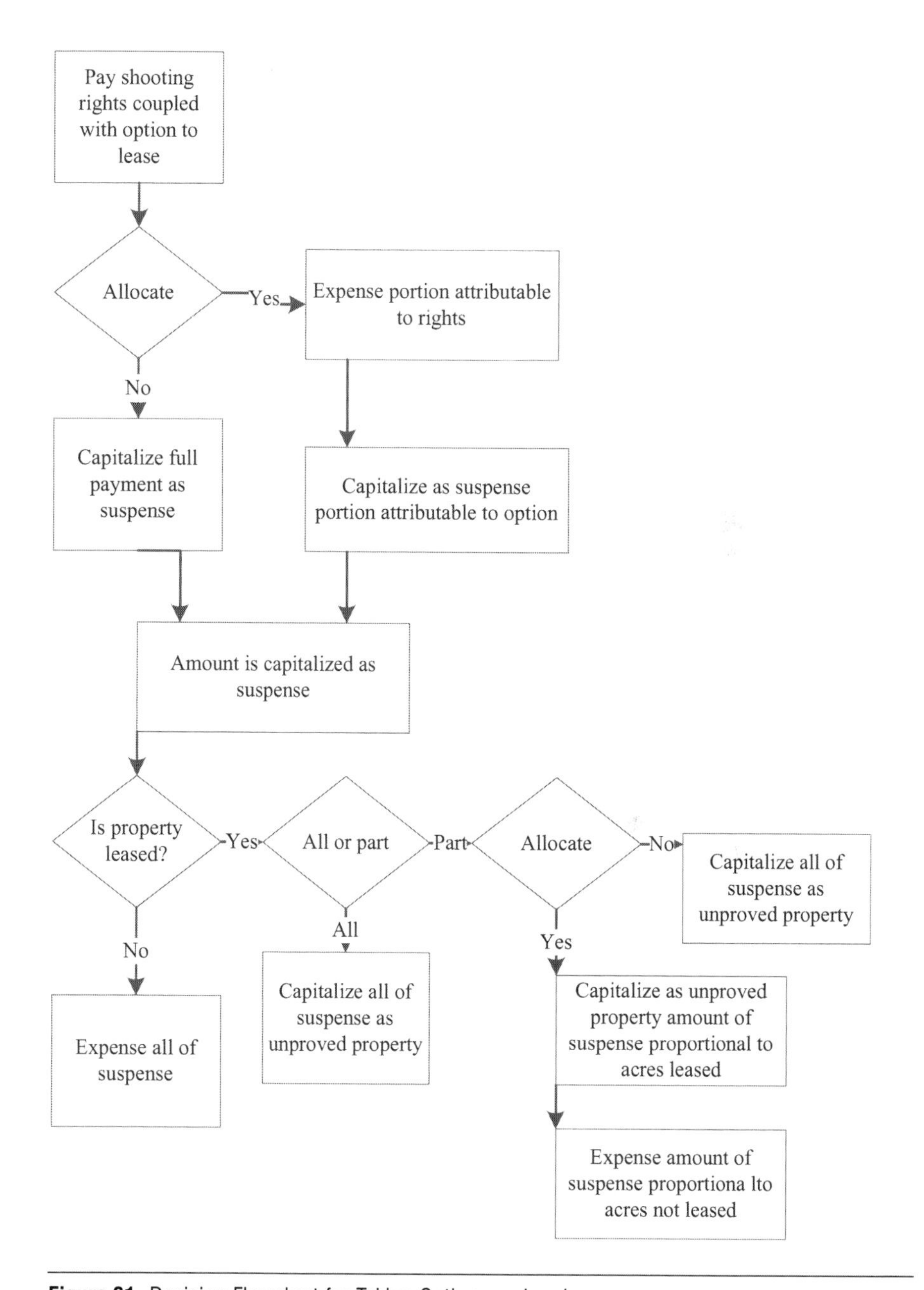

Figure 61. Decision Flowchart for Taking Options on Land
(*Source:* Charlotte J. Wright, *Fundamentals of Oil & Gas Accounting*, Tulsa: PennWell, 2008, p. 101.)

Capital Project Management Structure Choices: Sharing Financial and Operational Risks in Upstream Projects

Upstream oil & gas capital projects often involve massive amounts of capital and come with large risks. Offshore drilling rig rental rates are frequently over $100,000 per day, and one rig rental company off the coast of Brazil sold for 950 days at a daily rate of $320k.[353] Notwithstanding this, financial risk is only one part of the picture. Projects often come with physical risks, environmental risks, and safety risks as well. Just to illustrate the magnitude of risks encountered daily in supply chain management for these projects, consider these examples. The explorations of the Ichthys field off of Australia and the Kashagan field in the Caspian Sea are among the greatest engineering feats of history. Kashagan involved building artificial islands, working through arctic conditions, and drilling at ultra high temperature and pressure. An official with the consortium has called Kashagan "one of the largest and most complex industrial projects currently being developed anywhere in the world."[354] For its Azeri, Chirag, and Gunashli development in Azerbaijan, BP needed to transport over one million tons of equipment and materials, making about 23k individual freight movements.[355] The list goes on. Petrobras converted an ex-tanker into an FPSO vessel for operation in the Cascade and Chinook fields in the Gulf of Mexico. Altus Oil & Gas loaded and shipped the FPSO topsides, as well as a turret and mooring system, for deliveries to Keppel Shipyard in Singapore, and onward to an offshore site in the Gulf of Mexico. The STP buoy & turret weighed 1,200 tons and was about 30 meters tall, so lifting the buoy and finding the right type of vessel to move it involved specialized engineering and planning, especially to maintain a safe and stable center of gravity during the ocean transit. The team's analysis led to a decision to use a slightly adapted, less expensive heavy-lift vessel rather than a specialized vessel.[356]

Traditionally, oil and power companies have sought to share the risk with partners, for example through production sharing contracts in Iraq (BP), Oman (Transco), UAE (ADWEA/GDF Suez), and Qatar (AES–Gulf Investment).[357] In such arrangements, responsibilities are split between the IOC and the NOC. Traditionally, the IOC would fund exploration, appraisal, development, and production, and then transfer ownership of assets to the NOC. The NOC would then provide power and water, and a share of the profit.[358] In recent times more and more risk is being transferred to the service providers. Most LSTK tender guidelines are clear about shifting risk to the supplier for delay-related liquidated damages and defects liability. They also require IOCs to put up performance bonds and agree to indemnities and liens.

Today EPC firms are less eager to enter into LSTK contracts, which allocate nearly all risk to the EPC, leaving them to mop up the cost of less-than-expected or lost money due to escalating labor costs, unrealistic timeline constraints, and penalties imposed by the owner, and owner micromanagement of the process, including stipulations about which technologies and sub-suppliers to use.[359] EPC firms can increase prices to offset the risk, but this makes them less competitive with in-house work and general contractors. Some NOCs and major players in the wind power industry, which is entering a rapid phase of major capital investment, and not looking toward EPCs at all, are preferring traditional methods of construction contracting.

The industry is now using more sophisticated methods to determine how much risk each party should bear. "Real options" analysis can help to better hedge against the risk of failure as well as that of raw material price inflation and equipment unavailability. Texaco and Exxon have used real options for exploration and production decisions. Mobil used real options to decide whether to develop a natural gas field, and Anadarko used them to bid for oil reserves.[360] However, most companies have not used real options analysis for applications involving supplier risk and market risk. Potential applications include deciding on the optimal contract length or choosing between different supply chain risks that could stem from few manufacturing locations (e.g., decentralized vs. centralized production).

Boston Strategies International used risk modeling to determine the optimal contract term for compressors and pumps supply contracts. The analysis tested sensitivities on different price discounts (10, 20, 30, 40%), and found that after securing a 10% price discount for a 1.5-year contract, a buyer should be able to get a 20% price discount for a five-year contract. The relationship between contract term and price discount changes over time with market volatility.

Very long-term contracts can minimize risk-related costs in oil & gas, according to a study of an oil & gas supply chain (see Appendix 1). A simulation model measured the impact of volatility on four types of inefficiencies (high prices, excess inventory, excess capacity investments, and lost orders) at four nodes in the supply chain: refiners, upstream oil producers, equipment OEMs, and component suppliers, over 43 years. It tracked seven economic variables (backlog, orders, production, prices, capacity utilization, and total supply chain cost) under a scenario in which the price of oil was flat, and a scenario in which the price of oil fluctuated. The conclusion was that there are two basic strategies for managing supply chain risk in the face of uncertainty: 1) go short, continually rebidding contracts and tracking the market (this is a type of hedge); or 2) go long. To eradicate the impact of production-inventory-capacity cycles, companies need agreements longer than 20 years. To do this with minimal risk, it said, capital-intensive companies could form contracts that are long enough to ride through both ups and downs, but provide flexibility to share cost risks and leave options for each party to adjust their commitments related to capacity, lead times, and prices as conditions change.

In addition to long-term contracting, operators can assess risk in their CapEx scenarios by including acquisition cost uncertainty and operating and maintenance cost uncertainty in the NPV calculation. Using this approach, three figures should be "hard" numbers: acquisition cost, operating & maintenance cost, and the number of units. Two numbers should be probabilistic: acquisition cost risk (which decreases as a result of learning curve effects), and operating and maintenance cost risk (which decreases with time since model year launch, as a result of a declining rate of infant mortality). Total cost can then be calculated as the sum of: 1) acquisition cost; 2) acquisition cost risk; 3) operating and maintenance cost; and 4) operating and maintenance cost risk.

Sustainability Trade-offs: Reducing Upstream Emissions

Upstream oil & gas exploration and production is heavily energy intensive: about 5% of the energy ultimately produced goes to powering production equipment.[361] In less accessible fields like the Canada oil sands, the energy intensity can be as high as 20%.[362] Usually the power comes from heavy-duty diesel engines and generators or large-scale electric motors (more than 500 horsepower). These engines and generators run on diesel fuel or in some cases natural gas, and the motors run on electricity which is powered by the generators or perhaps a micro-turbine, which itself runs on natural gas. So, the process of producing energy itself consumes a non-negligible amount of hydrocarbons.

Using Renewable Energy to Power Operations

One way that oil & gas companies can reduce environmental footprint and increase energy efficiency at the same time is by using renewable energy to power oil & gas drilling and production equipment such as pumps, compressors, and turbines. The industry has seen many creative hybrid applications such as floating solar and wind apparatuses to power offshore platforms.

ExxonMobil agreed in 2018 to buy 500 megawatts of wind and solar power from Ørsted in Texas; this was at the time the largest renewables deal ever signed by an oil company.

Chevron has committed to build 500 MW of renewable energy plants to power some of its global operations. It contracted with Algonquin Power & Utilities, a Canadian renewable energy developer, to commission power plants running on renewable sources within four years at key sites in the Permian Basin of Texas and New Mexico, Kazakhstan, and Western Australia. Projects will be jointly developed and owned by Chevron and Algonquin, and backed by Chevron's commitment to buy the electricity through power purchase agreements.[363] Chevron already buys 100 MW of power produced by renewable energy power plants (65 MW of wind energy in West Texas and 35 MW of solar power in California) to run its operations.[364]

Increasingly, operators are considering powering the oilfield equipment with renewable energy. For example, Equinor's Hywind Tampen project will get 35% of its power from an 88-megawatt floating wind farm between the Gullfaks and Snorre platforms.[365] Floating solar is also being considered for some applications, but as noted in the Solar chapter, several floating solar installations have caught fire, and fire risk near an offshore oil platform is the worst risk an oil operator could encounter. The concept could be used onshore, however, with little risk.

While using renewables to power oil rigs only makes a small dent in the carbon footprint of oil & gas, many environmentalists would be quite keen to realize a small reduction in the carbon footprint of oil production itself.

Carbon Capture and Storage (CCS)

Capture and sequestration (CCS) has gotten widespread traction recently.

Saudi Aramco has built the capability to capture and process 45 million cubic feet of CO_2 at its plant in Hawiyah, in Makkah province in Saudi Arabia. The captured CO_2 is transported through an 85 km pipeline to the 'Uthmaniyah oil field to be injected into the oil reservoir. The process serves the dual purpose of recovering more oil by maintaining required pressure in the reservoir, while sequestering the unwanted gas. Since the initial injection of CO_2 in 2015, the oil major has doubled oil production rates from four of its wells.[366] The increase in well productivity offset the cost of piping the CO_2 from Makkah to 'Uthmaniyah.

Equinor, Shell, and Total also pursued CCS. They jointly invested in the Northern Lights project in Norway's first exploitation license for CO_2 storage on the Norwegian Continental Shelf, with an initial investment of $800 million, scheduled to be operational in 2024. Phase 1 will witness building capacity to transport, inject, and store up to 1.5 million tonnes of CO_2 per year. Once the CO_2 is captured, Northern Lights will be responsible for transportation by ship, injection, and permanent storage at 2,500 meters below the seabed. The plant will be operated from Equinor's facilities at the Sture terminal in Øygarden and the subsea facilities from the Oseberg A platform in the North Sea.

Enhanced Oil Recovery (EOR)

In addition to CCS, Enhanced Oil Recovery (EOR), which increases the capture rate of existing wells and reservoirs, concurrently reduces the relative energy intensity of the business. One way is through Enhanced Oil Recovery (EOR). Cairn Oil & Gas, which is part of the Vedanta Group, is implementing a more than $1 billion Alkaline Surfactant Polymer (ASP) enhanced oil recovery project to increase crude oil production from the Mangla, Bhagyam, and Aishwarya fields in Barmer, Rajasthan, in India. The ASP EOR technology will help increase the oil recovery factor from 36% to 50%.[367]

Reductions of Gas Flaring and Methane Emissions

Until recently it was common practice to flare the gas that was separated from the oil that came out of the ground. The gas was viewed as an unwelcome byproduct of the oil, and flaring was the easiest and cheapest way of it. Interest in flaring was particularly strong when gas prices were low. There was no economic reason to process it. Flaring created widespread environmental damage over the years. The industry is now committed to reduce gas flaring.

Chevron committed to reduce gas flaring intensity for global upstream operations by 25–30% between 2016 and 2023. Its program has already reduced flaring and associated emissions by 22% since 2013, through various initiatives including tying executive and employee compensation to the program's performance. On average, producers flare 6.1% of production, while Chevron's Permian operations flared just 1.07%.[368]

Shell too aims to keep its methane emissions intensity below 0.2% by 2025 across all its operations through leak detection and repair programs, and through technologies including optical gas imaging cameras. It upgraded valve actuators and well pad designs and replaced older equipment to achieve its goals.

- In Canada, Shell's Groundbirch natural gas project is reducing methane emissions from existing gas wells by updating valve actuators, resulting in a reduction of 13 tonnes of methane emissions in 2019. Around 50 valves were replaced the following year to further reduce methane emissions by 74 tonnes. Furthermore, a newly introduced electrically operated well design called "Gen 4 Multi-Well Pad" was introduced at Groundbirch's newest deployed well. Groundbirch's 25 wells further reduced methane emissions, while increasing production by 40%.
- In Appalachia, Shell's gas operations use pneumatic pumps, assisted by the pressure of the natural gas, to separate water from the natural gas. In 2019, four pumps were replaced with electric versions, which reduced methane emissions by 625 tonnes. Shell plans to introduce four more electric pumps which will further reduce methane emissions by 700 tonnes.[369]

Upgrading to Lower-Emissions Equipment

There are many options for equipment to help reduce methane and CO_2 emissions in upstream oilfield applications, including methane leak detection cameras and devices,[370] variable speed generators, and premium-efficiency motors.

Halliburton has begun fracking using electricity from the power grid instead of turbines and Tier 4 dual-fuel (diesel and natural gas) engines in the Permian Basin.[371] Atlas Copco has a range of variable speed generators for oilfield use, which it claims enables fuel consumption levels to be reduced by up to 40% against traditional fixed speed generators, when combined with an integrated energy storage system. ABB and other motor manufacturers make IE5 ultra-premium energy motors with up to 50% less energy losses and significantly lower energy consumption when compared with commonly used IE2 induction motors.[372]

Supply Unavailability Risks and Mitigation: Contracting & Procurement Strategies

Raw materials and equipment or its major components are prone to becoming bottlenecks, but the supply risk can be managed by passing it through, hedging it, sharing it, and eliminating it (or trying to).

The easiest path is to pass this risk directly and immediately on to customers in the form of a surcharge whenever there is a cost increase, if the market will bear it. Suppliers can either increase prices by the same amount that costs increase, or determine what portion of the increase they think their customers will pay.

The next easiest path is to hedge supply price risk through frequent periodic price adjustments. During the "commodity supercycle" mentioned in the Introduction, the top three global iron ore producers: BHP Billiton, Vale, and Rio Tinto, replaced previous annual benchmarks with quarterly contract prices in 2011. Vale restructured a quarterly agreement with Japanese Nippon Steel that resulted in a 90% increase in ore contract prices. ArcelorMittal estimated that the increase would raise steel costs by a third in the short term in regions where steel makers have traditionally signed long-term contracts, such as China and Japan.[373]

The third option is to share supply price risk. One way to share risk is to commit to a supplier share of business, which may rise or fall with the underlying volume of business. For example, if there are three target suppliers, a buyer can offer a share to each, subject to each supplier's capacity availability, without committing to the dollar value. Depending on the size of the suppliers, this may still be enough to secure "most favored nation" pricing (the lowest that the supplier charges to any of its customers).

The fourth, and usually the most expensive, path is to eliminate supply price and availability risk by buying ahead or planning far in advance or by integrating vertically with suppliers. Cesium, which is used as cesium formate in drilling fluids to prevent blowouts in high-temperature over-pressurized wells, is the most active metal on the earth and is in extremely rare supply. In fact, the supply is almost entirely controlled by China, with the only other new source being found in Canada. Equinor secured its supply by signing a $40 million supply agreement until 2022 with five two-year extension options.[374 375 376]

Many parts suppliers, which used to plan and schedule in increments of weeks, began planning one year or more in advance. One rotating equipment parts and repair company planned bulk steel orders for cast parts up to two years ahead, and postponed smelting and finishing until it had orders against the material. By machining them to specification when orders come in, the company kept lead times at two to three weeks, even for precision machined parts. Ordering long in advance and increasing stock levels goes against the grain of lean principles, but many companies have found it necessary to operate their business without shortages.

Due to strong demand growth, increasing lead times, and scarce raw materials, some suppliers have vertically integrated their supply chains. For example:

- Newkut, a manufacturer of polycrystalline diamond compact (PDC) cutters for drill bits, developed its own diamond synthesis process, and claims that its vertically integrated manufacturing process allows it to achieve high levels of quality.[377]
- Ruhrpumpen (Mexico) owns its own foundry.[378]
- BASF built a plant in Geismar, Louisiana, in 2011, for methylamines, chemicals that are used in about 20 of BASF's specialty amines that it manufactures nearby in order to give it more control over its costs.[379]
- CNOOC bought an 83% controlling stake in Dayukou Chemical Company Limited in 2007 to extend its supply chain to include direct production of sulfuric acid, which is used as a refinery catalyst or pre-catalyst and sometimes for well acidizing (dissolving dirt and clay to improve flow rates).[380]

In order to decide which approach is best in each circumstance, some upstream oil & gas companies have instituted market intelligence programs. One company measures the variability of order volume over time. For each major category of purchased materials and services, it assesses the future demand, order lead times, capacity utilization, and prices, quarter by quarter, with forecasts three years forward. It tracks prices and projections to determine the peak of the cycle and proactively work with suppliers to avoid shortages and price spikes.

Because of the magnitude of the risk and the interconnectedness of many technological and economic variables, many E&P contracts are single or dual sourced, often with solution-based or fixed pricing. This favors large service providers.

Using Substitute materials

Substitution of certain materials, even a slight change in the specifications at the design stage, is another way of avoiding materials shortages.

One example of material substitution trade-offs occurs in drilling fluids. There are many choices, including water-based and oil-based muds, as well as many choices regarding specific ingredients such as barite and bentonite, to achieve the required characteristics and wellbore reactions. Within this framework, the grade of mineral may offer cost vs. volume trade-offs.

One conservation method employed to extend the supply of barite has been to lower the required grade of barite in the fluid mix from 4.2 to 4.1 specific gravity. Essentially this lowers the ratio of barite to water, stretching the supply of barite with few to no adverse affects for drilling under some circumstances.[381]

Alternatives to barite in the drilling mud market include hematite, celestite, ilmenite, and iron ore. However, since they are more abrasive and more expensive, these substitutes have not been widely adopted. Hematite, the primary alternate to barite, is more abrasive, so it wears out equipment (such as pumps) faster. It also has a higher specific gravity (5+), which would require operators to revise mud properties and system designs to compensate if they used it as a substitute.[382]

Outsourcing Risks and Mitigation: Separating Strategic from Transactional

Until outsourcing became a major cost-cutting tool in the 1990s, most companies did more of their activities in-house. In fact, in the 1970s most national and international oil companies were far more vertically integrated than they are today. After the oil crisis, many oil companies began to segment upstream from downstream businesses. This trend converged with the general outsourcing phenomenon that prevailed during the 1990s, resulting in a willingness to consider outsourcing a wide range of activities that had hitherto been performed in-house.

Outsourcing came with risks, however. One major risk was losing the core competence and becoming dependent on the outsourced provider. Another risk was the potential to save money upfront only to pay more later—the classic "rent versus buy" dilemma. India's National Oil Company, ONGC (Oil & Natural Gas Corporation), used incentive-based agreements to capture the benefits of outsourcing while limiting its downside risk. The company signed agreements with Halliburton (for the Kalol field in Gujarat) and Schlumberger (for the Geleki field in Assam) for outsourced production of oil & gas from its ageing fields.

According to the agreements, the two service providers will invest capital and leverage their technical expertise in the aforementioned two stagnant but producing oilfields. A base production level will be decided, exceeding which they will be paid a predetermined fee per barrel. ONGC acknowledged that they are able to avert risk of failure if they brought in a new technology which didn't work. By outsourcing production to oilfield experts, ONGC offloads the responsibility for bringing in whatever technology is needed to Halliburton and Schlumberger, who get paid only if production exceeds the agreed base level.[383]

One type of outsourcing, low-cost country sourcing (LCCS), is receding in importance as globalization has given way to nationalism, and COVID-19 has abruptly made most multinational companies recenter their supply base closer to home. Low-cost country sourcing in upstream oil & gas started several tiers back in the supply chain, invisible to the end customers. Top-tier suppliers subcontract to second-tier suppliers, which lower cost through LCCS. For example, ESAB, a welding equipment provider, moved welding equipment production from the US to Poland to reduce labor costs and develop a strategic base to tap growing markets in Asia, Eastern Europe, and the Middle East.[384] PDC inserts were almost as invisible to the end users of drill bits, but now a Chinese supplier sells its inserts to nearly all the major bit suppliers, which are resold under the market leaders' own brands. LCCS became more visible when Chinese tubular suppliers gained prominence, although Western manufacturers eventually accused Chinese tubular suppliers of dumping at 30–90% below cost.[385] Whether or not dumping occurred, the price competition changed the industry structure and pricing behavior, as well as buyers' expectations.[386] Pump manufacturers also sourced from and set up manufacturing in China, in collaboration with local Chinese companies that in most cases benefited from original equipment supply contracts with those companies. One major pump company's Chinese operation is approved to supply nuclear plants. A top wellhead supplier also went to China, as well as Brazil, Malaysia, Indonesia, and Eastern Europe.

Sometimes unbeknownst to oil & gas customers, suppliers have migrated extensively to China in order to stay competitive. Here are some additional examples in cable and membranes. In cable, Pirelli expanded into the Asian market by acquiring a majority interest in a JV with Nicco Corp. Nexans, which has 12 plants in five Asian countries, bought Olex, the largest cable-maker in Australia and Asia-Pacific. Alcan set up a factory in Tianjin, China,[387] and General Cable acquired Jiangyin Huaming Specialty Cable Co. In addition to such foreign joint ventures and wholly owned companies, there are more than 1,200 domestic Chinese cable manufacturers.[388] In membranes, which are used for water treatment applications in upstream oil & gas, the major suppliers expanded production in Asia to take advantage of lower costs and reduce the cost to serve the Chinese market. For example, GE invested $9m in its Wuxi rolling facility,[389] and Toray set up a joint venture with BlueStar in Beijing.[390] As with cable and almost every other manufactured product, hundreds or even thousands of emerging local Chinese suppliers will make additional ventures with Western companies and become major independent suppliers.

Managing 3PLs and 4PLs

Many operators have concluded that logistics is not a core competency and have consequently turned to integrated logistics providers in order to assure their target levels of service. The arrangements have covered a spectrum from project shipping contracts to full-scale outsourcing of transportation and logistics planning and management. For example, KNPC awarded Agility a contract for door-to-door freight services, customs clearing, and transportation for five years, and Scorpion Offshore outsourced its stockroom to NOV. Scorpion's storerooms are now owned and staffed by NOV and stocked with NOV inventory, which allows Scorpion to focus on its core drilling competencies instead of on inventory management. The services operate on consignment, so the driller pays only for what is used and never deals with obsolete product, slow movers, or write-downs. NOV calculates that the service saves $2.2 million per month for a typical rig.[391]

On a more strategic level, operators hire third-party logistics companies with intentions of achieving certain goals, such as specific cost savings targets.

- Saudi Aramco established an Electrical Stocker Distribution Program on an outsourcing model to rationalize cost and embed innovation in the electrical equipment supply chain. The $275m program also simplified ordering processes, lowered costs, and improved fill rates.[392]

- Petrom, the Romanian oil company, chose Tenaris to manage its inventory in the Black Sea in order to reduce carrying costs and equipment damage.
- Traverse Drilling outsourced its sourcing, purchasing, industrial supply, logistics, and supply chain management activities with the goal of increasing its operational efficiency by streamlining supply and service and allowing it to focus on its core competency (drilling).[393]
- A major oilfield equipment supplier and a major compressor component manufacturer both hired CEVA to manage their spare parts distribution operations.[394] The oilfield equipment supplier estimated savings of $50m from the outsourcing deal.

Partnershipping: Single, Dual, and Local Sourcing

Due to the economic benefits of maintaining few suppliers mentioned above, both operators and suppliers have become adept at forming partnerships. Partnerships can help achieve tight project schedules by engaging collaborative planning before a final investment decision (FID) has been reached. One offshore rig contractor set up raw material purchasing, negotiated an escalation formula for prices and costs at multiple levels of the supply chain, and reserved production slots at a yard in Italy for deployment in Africa, all before the contract was actually awarded. The group was jointly engineering the project simultaneously with price negotiations in order to meet timetables.

Recognizing the value of partnerships, some equipment suppliers form dozens of partnership agreements. A compressor supplier uses these to demonstrate its commitment to meeting service and schedule requirements, even when capacity in the industry is tight. Many of its contracts are for ten years, typically with two renewals of successively shorter terms.

Single and dual sourcing

Most major oil companies employ a widely used category management approach that consists of category planning, category strategy, negotiation, and supplier performance monitoring. They go through a series of steps to "define the buy," select suppliers, structure agreements, and track and improve supplier performance. As with most strategic sourcing efforts, the goal is to get the most value from suppliers and to rationalize the supply base to strategic, partner suppliers.

In upstream, many of the category management 2 x 2 matrix exercises often yield the same answer: the economics of complex projects favor one or two suppliers. Single and dual sourcing minimizes the risks of many technology, process, and project management mistakes by assuring close collaboration and alignment of interests between the owner or operator and the supplier.

Therefore, bid slates often consists of large well-known companies, such as Transocean, Diamond Offshore Drilling, Noble, Ensco, and Seadrill for drilling; Halliburton and Schlumberger for oilfield services; and KBR, Saipem, and Technip for EPC services. While a large number of possible suppliers are considered, a dearth of technically qualified suppliers often results in one or two established suppliers.

A tight project schedule increases the tendency to single source. One EPC firm contracted with a single source to build a gas compression and metering station in Asia. Cost imperatives drove the sourcing behavior. The contractor was responsible for managing cost and was held to performance guarantees, all with almost no slack in the budget. While there were thousands of suppliers interested, the EPC quickly selected preferred suppliers and formed a partnership agreement with a single source. The strategy was driven mostly by the tight schedule (24 months) and the need to deliver long lead time compressors. The project was a success, and the schedule was maintained due to the close working relationship between the EPC firm, its single source, and the compressor supplier.

Another example of single sourcing was motivated by the technical supply market for seismic surveying, which has few providers. Niche services are a common reason to single source because suppliers are generally small and highly technical, making economies of scale unlikely and volume leverage impossible. These reasons explain why a major oil company decided to single source deepsea seismic acquisition and related services.

Another EPC firm, responsible for building an FPSO vessel, dual sourced the job, despite its preference for a single source, due to the sheer size of the project. Long lead time items (huge generators, centrifugal gas compressors, water injection pumps, offloading pumps, and diesel generators) and the "economics of complexity" favored one supplier, but the sheer size and capabilities needed to execute the project required more than any one supplier could provide. One supplier took responsibility for the hull (engineering, procurement, fabrication of modulars, towing, and commissioning). Both jointly shared responsibility for the topsides, with one handling the engineering and commissioning, and the other handling the fabrication and integration. During construction, the two parties subcontracted to hundreds of other companies and met local content requirements.

Situations where construction is geographically distributed frequently call for three to five primary suppliers. For example, one EPC chose three to five suppliers in each of four regions to meet its needs for gas-oil separation plants (GOSPs).

Technip formalized a long-term strategic collaboration with two subsidiaries of Petronas. The oil & gas engineering firm formalized a strategic partnership with MISC Berhad, one Petronas subsidiary, to install wellheads. MISC is a growing player in the offshore industry, offering floating facility solutions mainly for FPSOs.

The partners will work together on onshore and offshore projects, design and build offshore platforms, and exchange expertise. The cooperation with MISC Berhad will enable Technip to deepen its expertise in wellhead and umbilical installation for FPSOs.[395]

FMC Technologies, which makes template manifold systems, wellhead and Xmas tree assemblies, subsea control systems, and tie-in systems for installation on the seabed, uses partnershipping to better leverage its total spending and reduce supplier coordination time. The team's representatives from operations, manufacturing, and engineering share demand forecasts with its suppliers on a quarterly basis, and its innovative frame agreements with key suppliers allow it to secure capacity and quality in exchange for a guaranteed share of its purchasing volume. In addition, the company conducts monthly bottleneck analyses.[396] Partly as a result of these efforts, FMC Kongsberg Subsea won BSI's annual supply chain award for its superior supplier management, order fulfillment reliability, and overall asset productivity.

Joint process improvement typically yields strong benefits. For example, Amoco's partnership with Red Man Pipe yielded continuous, ongoing benefits.[397] Also, Lincoln Electric, a welding equipment supplier, offers to help its customers identify areas of cost savings through a welding audit. If the client addresses the opportunities, which often involves optimizing electrode and power sources, Lincoln guarantees a cost savings on its welding products. The joint process helps reduce welding costs.[398]

Local content

Since 1970, National Oil Companies (NOCs) have increased their control over the world's hydrocarbon reserves[399] from 15% to 90%, while International Oil Companies (IOCs) have shrunk from 90% to 15%.[400] The transition has occurred through the changing nature of production sharing agreements, which are increasingly oriented to provide NOCs title to hydrocarbons, control over operations, upside risk on oil price and volume movements, and a greater share of the profit upside. Meanwhile, as NOCs have taken more control of their countries' resources, their governments have increasingly required foreign contractors to use a greater share of domestically produced goods and services.

Countries actively setting local content levels include, for example, Kazakhstan, China, Brazil, China, and Azerbaijan. Since the movement has affected companies' investments due to the risk of expropriation of assets and other financially and operationally perilous incidents, one company has constructed an index to track risk factors associated with resource nationalism. Reports Verisk Maplecroft: "Regionally, Africa is home to ten countries experiencing a growth in these risks, including Tanzania (third-highest risk), Zambia (17th) Gabon (23rd) and Equatorial Guinea (40th). But host governments are using such measures to wrestle revenues away from oil, gas, and mining operators across the continents. India (15th),

Malaysia (30th), Turkey (36th), Iraq (42nd) and Mexico (68th) are among major producing countries that have also seen their scores in the Resource Nationalism index deteriorate."[401]

Upstream Oil & Gas companies need to plan and budget for local content. Often this involves extended project time frames, more contingency and backup plans, tighter risk management, and more emphasis on human resource management, training, and leadership development.

Procurement Bundling: Solutioning, Should-Cost, and Tier-Skipping

Increasingly, supply chain transactions consist of solutions rather than products or services alone. This is part of an overarching trend in manufacturing. "Yesterday you were buying a product for a price. Today you are buying complete system for a cost," explains one manufacturer.[402] Bundling services helps manufacturers differentiate to increase exit costs, and to maintain or increase their share of the value they generate. Manufacturers that otherwise would have been commoditized by the tsunami of low-cost country sourcing are focusing on a "value creation triangle" strategy of bundling products and services into solutions, embedding technology, and increasing prices to retain margins. Along with the solution comes a new pricing model—sometimes based on productivity or output (for bits, per foot drilled) rather than a unit price per "each." For example, drilling is often contracted on a turnkey, footage basis, incentive basis, or day rate.[403]

Logistics, including component repair and life-cycle maintenance services, is a ripe area for such solutions, and many 3PLs and 4PLs have emerged to make logistics and order fulfillment more automatic and efficient.

- BP named ABB as one of its global main electrical contractors (MEC) responsible for engineering, procurement, and construction of electrical equipment for its major upstream capital projects. The decision was driven by opportunity to benefit from more savings (doing away with overheads due to multiple invoices as opposed to one), lower risk (of delivery of the system by having one accountable unit instead of multiple vendors—tracking one is easier even from a project management perspective and quality control), and overall achieving greater efficiency when designing, procurement, construction, and even installation is done under a single umbrella (through more efficient coordination within ABB).[404]
- Schlumberger's acquisition of Smith, a manufacturer of drill bits, allowed it to bundle products and services, adding higher value than either products or services would on their own. The joint company now offers products that support the entire drillstring, and also take advantage of Smith's Wilson subsidiary's offering in integrated supply service (i.e., supplier management, inventory holding, field audits, warehousing, and other services such as training, testing, and inspections).[405]

- Saudi Aramco bought into the model, selecting Schlumberger to service its new oilfield facility in Al-Khafji. Schlumberger opened a $2m, 20k-square-meter (215k-square-feet) oilfield services facility in Al-Khafji, Saudi Arabia, for repair and maintenance services, in addition to engineering, planning, and operations support. The facility also supports well completion services and equipment, artificial lift, and information solutions including software, information management, and IT infrastructure services to optimize operational processes. It will shorten turnaround times and compete with local third-party maintenance providers.[406]
- Nations Energy Company chose The Wood Group to provide supply chain management solutions under a $10m, two-year contract that covers procurement, inventory, and logistics management services in Aktau, Kazakhstan.[407]
- Scorpion Offshore used NOV's distribution service called Rigstores on its rigs with the aim of reducing maintenance, repair, and operating (MRO) costs. The Rigstore is owned and staffed by NOV and stocked with NOV inventory, allowing Scorpion to focus on its core drilling competencies instead of inventory management. It can leverage NOV's supply chain expertise to ensure the necessary products are available on the rig when needed. The services operate on consignment, so the driller pays only for what is used and never deals with obsolete product, slow movers, or write-downs. NOV calculates that the service saves $2.2 million per month for a typical rig.[408]
- Ingersoll Rand's PackageCare™ program is a fixed-cost maintenance program that offers customers predictable, scheduled maintenance and protection from unexpected repair expenses for the life of the agreement. The company also services third-party equipment under the program.[409]
- KSB provides repair and overhaul services and parts management services (demand planning and warehousing) through Standard Alloys Incorporated.[410]
- SKF signed multiple maintenance and inventory management "solution" contracts.[411]
- Norwegian BTU and PI Intervention announced a new common brand ("Interwell") to market downhole tools and services under one umbrella. The offering will now provide customers with full life-cycle services from design and prototype through operation and maintenance of downhole tools.[412]

One contract term option is to have no contract at all, but instead to "rent" services instead of buying the products. For example, most PDC drill bits are now rented, because the bit body can be refurbished by the manufacturer.

Operators must return the bit body in good condition, however, or they must pay a penalty. Roller-cone bits, in contrast, are mostly worn out during the drilling process and are typically not rented. The decision to rent or buy downhole tools is driven by the recoverability of the asset (whether it is left in the well or not), its price, and its level of exclusivity. For example, intelligent controls, while they are expensive and high-tech, are bought instead of rented because they remain in the well. Conversely, shock jars and reamers are usually rented because they are almost always recovered in good enough condition to be reused. One drill pipe supplier effectively rents its product. It's the buyer's until they don't want to pay for it anymore; then the supplier takes it back and rents it out again. Sometimes drilling contractors buy tubulars and rent them to oil & gas operators.

Rentals are taking on a larger share of the total market as buyers try to keep assets off their balance sheets and operators seek to protect new technologies, driven by the need to report strong financial results and minimize their asset bases. For example, IOCs now rent 90–95% of their needs for bits and downhole tools (NOCs often prefer to own the equipment and typically rent only 40–50%). The 2008–2009 recession accelerated the trend toward renting vs. buying as crude oil prices plummeted from a peak of $147/bbl in 2008 to an average of $43/bbl in 2009 Q1. Faced with lower returns on their drilling investments, operators sought to cut costs and reduce risk by renting instead of purchasing. Instead of paying $40–50k for a bit, the operator can pay $10–20k to rent one for only as long as it needs the equipment.

Suppliers generally prefer to rent because it allows them to protect their intellectual property, compared to selling tools and bits. Another driver for the suppliers is profitability, for the firm gets repeat revenues on the same asset. If the supplier can refurbish a bit body three to four times, it profits more from the asset than it would if it had simply sold one bit.

Standardization and modularization

Traditionally, it has been difficult to achieve standardization or economies of scale on very large installations. Rig costs do not exhibit economies of scale, as shown in Figure 62 below. In addition, complex machinery and equipment such as compressors and turbines require customization to each application to function at peak performance in each operating environment. Given the operating cost and the opportunity cost of lost production, such customization is usually considered a small price to pay for high operating performance and production rates.

Nonetheless, oil & gas companies and their contractors have achieved major cost reductions through standardization of processes and modularization of large installations during the engineering, construction, and installation phases.

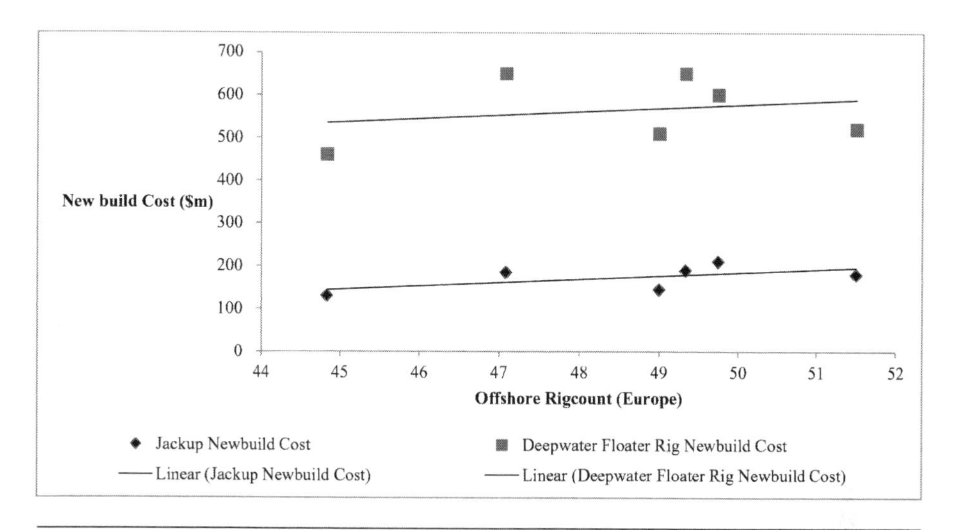

Figure 62. Jackup and Deepwater Floater Rig Economies of Scale
(*Source:* Boston Strategies International estimates and analyses)

Statoil, FMC Kongsberg Subsea, and Aker Solutions cooperate in standardizing specifications for subsea workover systems. Multifunctionality for workover systems means the same system can be used independently of supplier of subsea infrastructure. This results in cost savings and simplified well maintenance. In addition, one supplier can work on another supplier's equipment, which provides extra safety.

BP used standardized processes to build the colossal offshore platforms in the Azeri-Chirag-Gunashli field in the Caspian Sea. It set up a production line and standardized offshore jackets and float-over topsides. By building onshore and transporting to its offshore resting place, BP was able to take advantage of both lower construction costs and economies of scale through standardization. It plans to use the concept as a standard practice in the future.[413] The topsides—wellbay, process, utilities, drilling, and quarters—were identical and modular. Each jacket substructure was built on the same concept. The first phase was six weeks ahead of schedule, and the second was 4–5 months early. The design man-hours per tonne of topsides decreased from 42 in the first phase to 31, and then to 11 in the second phase. The design man-hours per tonne for drilling decreased from 50 to 10 between phase 1 and phase 2. The cycle time from cutting first steel to a completed deck went from 33 months to 31 months and is targeted at 29 months for phase 3. Completion and testing dropped from 900k to 800k early in phase 2, and subsequently to 500k later in phase 2. Jacket building time dropped from 18 months to nine months, and float-over deck installation fell from 56 hours to 35 hours, while jacket installation time fell from 37 days to 29 days.[414]

Shell was able to reduce drilling days from 60 days to 25 days and reduce drilling costs by 30% on the Pinedale Wyoming shale gas field, which involved hundreds of multi-stage, fractured wells. It reduced completion costs by 60% and cycle time for completion from one well every 60 days to five wells in 20 days. On the Groundbirch, Canada, shale gas field, its first well took 40 days, compared to 15 days three years later. In the Groningen, Netherlands, gas field, where it built seven compression stations to maintain adequate pressure for drilling, it was able to use the same construction team, design, and methodology to achieve 20% lower costs and cut 10–15% from the project cycle time by the third station.[415] The company estimates savings up to 35–40% in project cost and a 15–40% reduction in project cycle time through standardization.[416]

Should-Cost Analysis

Sometimes market prices become significantly divergent from competitive levels, and structured initiatives to realign them such as strategic sourcing programs and e-sourcing platforms can yield big benefits for buyers. For example, Petronect, the e-procurement portal for Petrobras, implemented the SAP® Supplier Relationship Management application and the SAP NetWeaver® Portal component, which cut the average closing price for bids by 22% and reduced operational costs as a percent of sales by 10%.[417]

In a more targeted way for specific bids, "should-cost" analysis can provide a useful reference point for bids when only one or two suppliers are qualified and full market transparency of prices and costs cannot be assured.

The analysis steps are:

- Tailor to specific orders
- Start with a base period price using cost inflation factors
- Update to current prices
- Subtract excess profit
- Subtract cost of over-engineering
- Subtract diseconomies of scale
- Subtract diseconomies of scope
- Subtract cost of coordination; go with LTA

Our sample nine-step "should-cost" analysis is based on an API 610 two-stage horizontal split case pump. The price is normalized, if necessary, to the current price using an inflation factor. After inflating for the category, the cost factors for the unique specifications of the equipment can be modeled. The resulting analysis produces a time-series of capacity utilization, lead time, and prices for the specific equipment. By applying algorithms based on the differences in the shape of the capacity utilization, lead time, and price curves, it is possible to create a profit margin adjustment factor. If the market is in an "up" mode, the profit margin adjustment factor inflates the prices. If the market is in a "down" mode, the dominant price index adjusts downward (and possibly deflates) the historical price. In the model, prices are upwardly flexible and downwardly sticky.

The "should-cost" methodology should also take into account possible opportunities for standardization or product simplification, economies of scale from volume purchasing, economies of scope from purchasing multiple types of products or services from the same supplier, and potential savings from more coordinated production and inventory planning.

- Standardization. Highly engineered rotating equipment such as compressors and turbines often must be customized for each application in order to be cost-effective over their lifespans. However, some equipment, like some types of pumps, may be able to be standardized, for example on an ANSI specification, at significant cost savings.
- Economies of scale. Unit costs and prices can decrease as the number of units increases. For example, Capstone's turbines became less costly the more the company produced. At one point the company produced turbines with a total capacity of 22 MW at an average cost of $3 million/MW, whereas at another point the company produced 52.8 MW of microturbines at an average cost of $2.4 million/MW. The 18% reduction in cost per MW was due to economies of scale in production.
- Economies of scope. Economies of scope represent savings that can be realized not through higher volumes, but by amortizing volumes of disparate purchases over a more concentrated fixed overhead cost base (this is our supply chain adaptation of a theoretical economics definition). For example, Siemens supplies motors, electrical distribution and control equipment, and lighting, but few buyers think to leverage their purchasing power by consolidating their purchases of transformers and lighting. In this example, consolidating transformer with lighting purchases can save contracting, purchasing, and IT costs.
- Potential savings from more coordinated production and inventory planning. By collaborating with suppliers on production planning, inventory stocking and warehousing, and logistics, further savings may be possible.

The layered analysis, which results in a "should-cost" price, is illustrated in Figure 63.

Furthermore, patterns between prices, lead times, and capacity utilization can drive price forecasts. For example, if prices stay steady while capacity utilization falls, this could indicate a period of "softness" during which sellers could successfully make additional profits, and buyers could successfully negotiate prices down. Boston Strategies International has been monitoring the relationships among prices, lead times, and capacity utilization for decades.

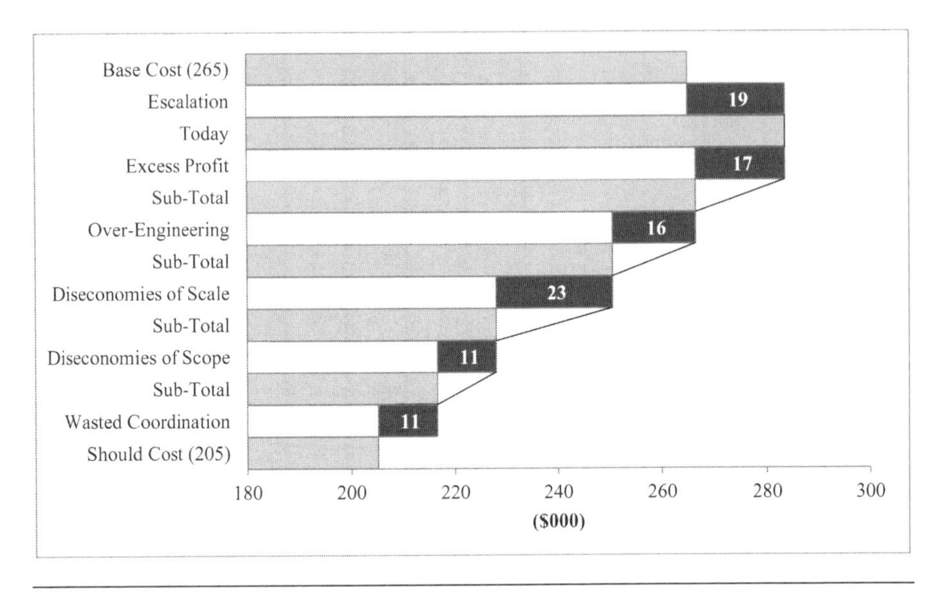

Figure 63. "Should-Cost" Waterfall Chart
(*Source:* Boston Strategies International)

Tier-skipping (buying direct from sub-suppliers)

Tier-skipping—cutting out the middleman—is another avenue to cost reduction. At the procurement and construction phase, operators must evaluate the pros and cons of opportunities to source directly from sub-suppliers. For example, wind turbine suppliers also install foundations, but buyers can sign independent contracts with the foundation suppliers and the wind turbine suppliers if they want to weigh the cost complications savings against the risk of subsequent warranty erosion and possible operating and maintenance.

For example, some operators are now buying fracturing sands (proppant) from a materials science company instead of through distributors. The company's technologically engineered proppant lasts longer and requires less additive compared with sand for shale (or tight sands) wells. Given rapidly rising prices and the perceived lack of value-add by distributors, the company, and its customers are giving the middleman the boot.

Also, a minerals supplier that once used to sell only to drilling mud companies is now selling directly to operators who want to mix their own drilling and completion fluids. The company has been able to achieve economies of scale because it supplies minerals to many different industries. Those economies of scale allow it to offer advantageous terms to operators.

Contract Term Risk Management: Achieving Economies of Scale through Contract Extensions

Negotiating a longer-term contract allows suppliers to realize economies of scale, which can be a win-win proposition if they pass some or all of the cost savings back to the buyer in the form of lower prices. However, suppliers may become complacent if a long-term contract shelters them from competition. Procurement managers often cope with the paradox by offering an initial fixed-term agreement with one or two options to renew. For example, Abu Dhabi National Oil Company (ADNOC) awarded Tenaris a long-term contract, worth $1.9 billion, for the provision of tubulars and Rig Direct® services for five years, with possibility of a two-year extension.[418] Also, Aker BP awarded Archer a five-year contract with an additional three-year extension option for the provision of platform drilling operations and maintenance services on the Ula and Valhall installations in the Norwegian sector of the North Sea.[419]

Trade-offs and Management Techniques in Operations & Maintenance Management

Upstream oil & gas is experiencing digitalization and margin pressure at the same time, with the result that operations and maintenance programs represent a unique blend of old-fashioned lean management principles and practices, alongside some of the most advanced digitalization and automation tools and techniques.

Internet of Things (IoT) and Artificial Intelligence (AI) Technology Choices: Digitalizing Intelligent Oilfields

Predictive maintenance

The number and value of complex assets, combined with their criticality to production, makes upstream maintenance operations a ripe field for predictive maintenance optimization. Furthermore, predictive maintenance can be optimized through artificial intelligence (AI), since a trained expert can often detect telltale warning signs of impending equipment failure. Therefore, predictive maintenance and artificial intelligence are often intertwined in upstream oil & gas operations and maintenance.

Abu Dhabi National Oil Company (ADNOC) completed the first phase of its large-scale multi-year predictive maintenance project to maximize asset efficiency and integrity across its upstream and downstream operations. The predictive maintenance platform is an integral part of ADNOC's Panorama Digital Command Center at its headquarters. Using machine learning and digital twins, ADNOC's maintenance platform can predict equipment performance, enabling planned maintenance. This significantly reduces downtime and even the maintenance cost by up to 20%. In its first of the four planned phases, the project covers the modeling and monitoring of 160 major turbines, motors, centrifugal pumps and compressors across six ADNOC Group companies. Altogether, the project aims to enable the central monitoring of up to 2,500 critical machines across all operations by 2022.[420]

ExxonMobil is using predictive maintenance at its Permian shale assets in West Texas and southeast New Mexico, optimizing well performance and AI algorithms to improve operating performance. An additional layer of intelligent applications will provide a complete, end-to-end view of the Permian operations and will be able to improve capital efficiency and support production growth by as much as 50,000 oil-equivalent barrels per day by 2025.[421]

Total is also deploying AI-driven software for the analysis of subsurface geophysical data and equipment monitoring to interpret subsurface images, particularly from seismicstudies using computer vision, automating the analysis of technical documents using Natural Language Processing. These programs allow Total's geologists, geophysicists, reservoir, and geo-information engineers to improve efficiency and accuracy, besides reduce time in exploring and assessing oil & gas fields.[422]

Intelligent Oilfields

Saudi Aramco's Khurais facility is equipped with IIoT devices and digital twin technology that enables remote operation using real-time 3D Virtual Reality (VR) software. Khurais is the first oil field in the world with Advanced Process Control (APC) for smart production rate adjustments, with monitoring application that uses predictive and prescriptive analytics to optimize fuel gas in boilers, contributing to fuel gas savings and reducing CO_2 emissions. These technologies help reduce overall power consumption at Khurais by 18%, slash maintenance costs by 30% and cut inspection times by 40%. Operators can now make decisions based on information obtained through real-time IIoT controlling systems, replacing physical human intervention, from assessment to determining enhancements/modification.[423]

Optical technology is often critical to the sensing component of advanced control systems. Optical gas imaging (OGI) is a thermal imaging technology that uses high-sensitivity infrared (IR) cameras to detect very small fugitive emissions of industrial gases. Shell's Appalachia gas operations have been using an optical gas imaging (OGI) camera to identify methane emission leaks since 2012. More than 400 wells and 143,000 individual components were inspected during 800 trips to individual wells between 2012 and 2019. In 2019 alone, more than 600 tonnes of methane emissions were reduced.[424][425][426] ExxonMobil is conducting field trials of eight emerging methane detection technologies, including satellite and aerial surveillance monitoring, at nearly 1,000 sites in Texas and New Mexico to further reduce methane emissions.[427]

Cybersecurity: Bridging the IT/OT Gap

Upstream oil & gas operations have been the target of numerous cyberattacks, for example on Saudi Aramco in 2017. Malicious actors, whether state-sponsored or not, can cause widespread disruption and financial damage by shutting down oil flow to the world at the source, which seems to be more effective at achieving their goals than shutting down more localized infrastructure. This leaves upstream oil & gas operations seeking a way to reap the efficiency and productivity benefits of the digital era while minimizing the risk of attack.

For cybersecurity, operators usually turn to cybersecurity specialists and their solutions. Chevron uses SecurityGate.io's risk management platform, which enables it to complete more cyber assessments and assess risks faster and without incurring the cost of hiring permanent staff. Chevron benefits high-visibility dashboards and micro-level reports that bridge the IT/OT gap, besides decentralizing processes to enable swifter troubleshooting.[428]

Peak Capacity Management Strategies: Optimizing Production, Inventory, and Asset Control

Some of the unique and differentiating aspects of upstream oil & gas operations and maintenance activities pertain to: production, logistics and inventory management for volatile and stochastic demand, asset tracking, outsourcing, managing 3PLs and 4PLs and transporting ultra-large oversized cargo, supplier quality management, and managing exceptionally stringent health and safety priorities as they extend throughout the supply chain.

Lean and Flexible Production Management

Since upstream operations are asset-intensive—rigs can cost several hundreds of thousands of dollars per day, and subsea and deepwater operations further exacerbate asset intensity—shortfalls in productivity can rapidly erode the financial return on those assets. Worn drill bits reduce productivity, eventually to the point where they are costing more in rig time than it would take to buy a new bit.

Production scheduling optimization can increase the efficiency of upstream assets, simultaneously helping to reduce costs and increase throughput (revenue and profit). Thailand's PTT Group, which has operations in upstream oil & gas, deployed a software application to simulate its gas separation plant number five (GSP5) so it could increase production capacity. AspenONE Engineering and Aspen HYSYS created a process simulation model of the ethane recovery unit (ERU). The initiative optimized the ERU, increasing plant capacity to 106% of design capacity. This led to an improvement in daily profitability of $60k/day, according to AspenTech.[429]

Flexible manufacturing and assembly is a powerful tool for dealing with demand volatility. As noted in the constraints management section of Part 1, reducing cycle time in every process along the value chain makes the entire system more able to adapt to changes in demand. One way to operate "Lean" is to make rapid changeovers. An example might be deploying drilling rigs or installation vessels to where they are needed quickly. Statoil reduced lead time for subsea tie-backs by 50%, decreased the capital expenditure by 30%[7], and reduced the time from discovery to production from seven to three or four years.[8] The company is using standardization as a way to make smaller reserves profitable. Exxon Angola's Kizomba floating production project experienced a five-month shorter cycle time for the second project (B) than for the first one (A): Kizomba A took 36 months, while Kizomba came onstream after 31 months of construction.

At the component level, pump suppliers are implementing mechanisms so they can adjust to changing demand more economically and responsively. Flowserve added second and third shifts during the economic turmoil that followed the financial crisis in order to increase its ability to flex capacity quickly without adding real estate or selling plant assets.[430] The company also shifts production globally to balance its capacity utilization across plants. ITT's modular assembly operations help it scale production capacity to actual demand, as well as reduce its order cycle times. ITT builds its modules on skids that can be easily transferred across product lines. This enables it to move the product to whichever production operation has the lowest utilization rate, thereby balancing capacity. Balanced capacity allows it to lower the average capacity level of all of its operations since there are fewer and lower peaks.[431]

Logistics and Inventory Management for Stochastic Demand

How do you formulate a stocking policy for a product that is only needed in enormous quantities at unpredictable times? During the Gulf oil spill, BP ruled out the majority of dispersant suppliers simply because the suppliers could not offer enough capacity to produce the volumes it needed. The dispersant market is small, and Nalco was chosen as the sole supplier because of its ability to quickly reallocate resources to produce mass quantities of Corexit.[432]

Specialized inventory management solutions exist for intermittent demand. The underlying replenishment statistics rely more on the probability of equipment failure, as is common in service parts, than on replenishment, as is common in typical fast-moving goods applications. Analyses based on "mortality rates" are used by analyzing Weibull distributions rather than analyses based on "z" statistics, which assume a distribution around a normal curve.

Other, process-based, approaches can also deal with "unpredictable" demand. For example, lean manufacturing, lean distribution, and other derivatives of lean thinking reduce buffers, and along with them the chance of getting stuck with a lot of inventory that isn't needed, or conversely coming up short when demand spikes.

Refurbishment has become a popular alternative to replacing aging equipment due to its shorter lead times. When supply was tight in recent years, many operators turned to refurbished or used equipment rather than ordering new equipment in order to reduce delivery lead times.

- Hitachi Cable developed a continuous processing technology that allows the recycling of silane cross-linked polyethylene waste that is generated while manufacturing power cable. The polyethylene waste can be reused as insulating material in new power cables.[433]
- In anticipation of continued demand that is overburdening rolling mills for new tubes, one OCTG manufacturer increased its capacity to refurbish OCTG by re-cutting the threading and recoating the pipe if the pipe is sound enough to be reused. Refurbishing pipe takes half as long as it does to produce as new pipe.
- NSK reconditions and refurbishes bearings rather than replacing them in order to save its customers money, and to diversify its revenue stream. It recently helped a paper mill avoid $75k in capital expenditure this way.[434]

Asset Tracking

An asset tracking initiative or technology investment can pay off if the cost of error and subsequent redeployment (manufacturing, transportation, warehousing, handling, etc.) is high.

The tagging of drill pipe at Petrobras is an example. In order to ensure that drill pipe is in the right place at the right time and in the right quantity, it required Weatherford to track all the items that it ships to its offshore platforms. Weatherford tagged 7,500 segments of drill pipe, with the help of Trac ID Systems, a Norwegian RFID tag provider. Workers tag pipe sections in the field and key in data such as the location and whether or not there were any defects or damage to the pipe. With the information gathered via the RFID solution, Weatherford can set proper stocking levels, track inspection histories, and monitor condition in real time. In addition, Weatherford can "peg" (identify and trace to a certain job or lot number) the location of pipe while it is en route from its stocking point to the offshore rigs. For this application, Weatherford is using ISO 18000-2 compliant one-inch wide 134.5 kHz tags.[435]

Asset tagging can also pay off by communicating with remote devices to increase production and yield. Weatherford and Marathon Oil are using RFID to remotely open and close the cutter blocks on reamers, thereby increasing well productivity. The tag is positioned in the middle of the inside of the drill stem, while the reader is positioned on the drill stem itself. The result is more reliable than hydraulically controlled reamers, which have a higher risk of damage and failure because of their manual connection from the wellhead to the reamer. Weatherford named the system RipTide. Marathon Oil, which bought RFID technology provider In-Depth Systems in 2001, also uses RFID in coiled tubing, packer setting, wellbore cleanup tools, zonal isolation, perforation, and cementing. MI-SWACO also uses RFID for drilling and wellbore cleanup applications.[436]

Asset tagging can also improve compliance, safety, and inventory availability. Halo LLC, a manufacturer of ropes, chains, and similar fastening equipment, is using RFID to warn of impending safety inspection requirements on safety-critical slings.[437] Using wireless monitoring, GE and Baker Hughes are offering ancillary monitoring and control products that allow wireless remote monitoring and web portal access to product inventory status.[438 439]

Sourcing Trade-offs: Managing Distributors and Integrators

One of the decisions most often faced by procurement staff at energy companies is whether to buy direct from manufacturers or through a distributor. Usually, distributors offer low prices on some but not all the items, and each distributor has different "loss leaders," so many buyers use a "market basket" of commonly purchased items to determine whether it is less expensive to buy them directly or through the distributor.

Purchase price is not the only factor, however, and sometimes it is not even a major factor. The total cost equation must account for not only the cost of stocking, but the service level—availability and speed—to field support operations, which is where distributors sometimes have an added advantage due to their field network. The choice often trifurcates to buy direct, buy through the distributor, or engage the distributor as an in-house third-party stockist.

Considering the total costs involved, Saudi Aramco developed a stocker-distributor model for electrical maintenance, repair, and operating (MRO) supplies. The transition from in-house stocking to managed inventory services improved service levels, reduced cost, and allowed management to focus on making decisions related to their core competency instead of managing myriad small inventory and replenishment details.

TCO Trade-offs: Including Throughput in "Total Cost"

Large energy companies often prefer to engineer each major piece of rotating and static equipment to the in-situ process conditions in order to maximize throughput, flow rate, or other operating parameters to maximize profit. Whether or not to standardize complex static and rotating equipment involves a particularly complex analysis because it needs to be based on reliability analysis, which is engineering intensive and usually requires extensive data over a life-cycle of time for any specific piece of equipment.

The important aspect of this TCO analysis is that it should take into account not only the TCO of the equipment in question, but also the incremental revenue and cost associated with incremental changes to the process. For example, the TCO analysis for a pump should not only consider the lifetime cost of the pump, but also the financial impact of any incremental throughput associated with the change in specifications. One Latin American national oil company conducted such throughput-driven analyses of the optimal sizing and custom specifications using parametric cost analyses on a wide range of equipment from pumps and compressors to oil rigs and storage tanks.

HSE Considerations: Using Remote Controls and Robotics for Improved Safety

In addition to safety-related blowback from the Gulf of Mexico spill, upstream oil & gas companies have used remote control technology to reduce the instances involving risk to human life. For example, Aker BP plans to use Remotely Operated Controls System (ROCS) at its Deepsea Nordkapp semisubmersible rig. The tubing hanger installation had been done through a dedicated umbilical controlled from a large topside hydraulic unit, an operation that resulted in hours of additional high-risk operations. The new technology offers the same controls functionality to the tubing hanger, without the topside hydraulic unit and without a large, heavy, and costly umbilical and controls system. Using the ROCS significantly reduces in-person time spent on this risky operation.[440]

Supply Chain Roadmap for Upstream Oil & Gas

The upstream Oil & Gas supply chain can be rationalized, synchronized, and made more sustainable through supply chain management best practices. Even though many companies have been working on supply chain initiatives for a long time, emerging technologies and business models make certain best practices high priorities today.

Carbon Capture and Storage, which has the potential to prevent CO_2 from ever entering the atmosphere, should be implemented wherever it can be shown to be technically and economically viable.

"Digital oilfields" encompass many technologies, of which optical readers, sensors, virtual reality, artificial intelligence, pattern recognition, and digital twins are a few. More broadly, smart IoT devices and drone inspection are enabling significant, structural, and permanent cost reduction with relatively modest investment. Digitalization is a smart investment for the upstream oil & gas industry, in that it lowers the cost of production, making it harder for renewables to compete.

Converting *product* purchases to *product-as-a-service* contracts often simplifies procurement and bookkeeping, makes suppliers more accountable, and lowers cost. This relatively new business model tests vendors' knowledge of the system in which their products operate and allows them to achieve far greater savings than they would have by just reducing the price of their product or solution.

Reliability can be ensured by increasing and implementing cybersecurity protections. The many high-profile cyberattacks on oil & gas facilities are early warning signs of an ever-increasing digital threat facing upstream oil & gas. If companies ignore the threat, they will inevitably pay dearly for it. Cybersecurity investments are not optional anymore.

Sustainability can be enhanced by using renewable energy to power operations. Against the backdrop of pressure from investors and governments to reduce carbon footprint, upstream oil & gas operators have a high-visibility opportunity to test renewables by using solar or wind, likely with battery storage, to power drilling operations or production platforms.

The safety profile of operations can in certain cases be improved by the use of remote equipment such as drones and robots to conduct operations that may be safety risks for people. For example, using drones to make parts deliveries to offshore platforms eliminates the need for people to fly helicopters in sometimes dangerous weather conditions, and inspections of hazardous areas by drones reduces the risk of human exposure.

Figure 64 summarizes the supply chain value creation opportunities in upstream oil & gas.

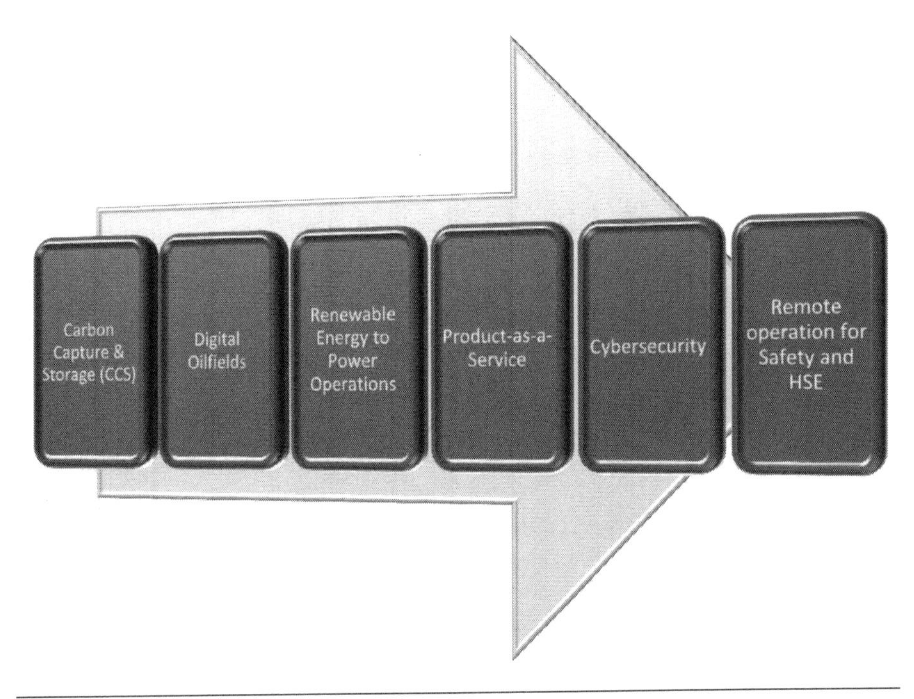

Figure 64. Emerging Supply Chain Opportunities in Upstream Oil & Gas

Oil & Gas—Midstream

Chapter Highlights

1. Logistics costs are driven by the location, extent, and type of processing, so changing the temperature or form, e.g., Liquefied Natural Gas (LNG), Gas to Solids (GTS), or Gas to Liquids (GTL), may reduce shipping costs.
2. Depending on a variety of parameters, it may be possible to minimize midstream transport costs by configuring production of power or refined products adjacent to the wellhead, in so-called Gas to Power (GTP) or Gas to Commodity (GTC) procedures.
3. The pervasive nature of regulatory risks at the midstream project planning stage make it essential to align all commitments to trading partners with final approvals.
4. The length of project timelines and the consequent potential for economic conditions can change before Financial Investment Decisions (FIDs) make contractual and financial hedging "best practices."
5. Strong investments in physical and cybersecurity are needed to ensure safety and profitability. Technologies such as drone surveillance, ultraviolet corrosion detection, and predictive maintenance should be evaluated.
6. Supply chain improvement opportunities include: 1) the expansion of LNG, 2) small-scale modular gas processing technologies, 3) drone monitoring, and 4) smarter process automation solutions.

Introduction—Supply Chain Economic Cost Drivers and Design Constructs

Midstream value chains vary widely depending on the type of hydrocarbon concerned, broadly represented in Figure 61. After the crude flows through the wellhead and undergoes initial separation in a temporary or permanent Gas Oil Separation Plant (GOSP), the oil and the gas are routed separately to any number of pipelines and shipping/receiving/storage terminals. For large facilities, especially export junctions, large tank farms provide buffer capacity that aligns the marine or pipeline transport capacity with the rate of field production. Depending on the hydrocarbon viscosity, the terrain, and the length of the pipeline, booster stations may maintain pressure in the pipe either by compression (for gas) or by pumping (for oil).

Midstream logistics flows and capacity design should minimize total value chain costs, including logistics (transportation plus inventory) costs, and to buffer against potential fluctuations in demand. This often involves considering scenarios that advance or postpone processing. Petroleum can be stored and transported as crude oil or refined oil, and natural gas can be compressed and liquefied to reduce cubic volume and shipping cost—liquefication reduces the volume of natural gas by a factor of 600.[441] Furthermore, natural gas can be transported, stored, and directly consumed in multiple forms, for example as:

- Natural gas
- Liquefied natural gas (LNG)
- Compressed natural gas (CNG)
- Gas to Solids (GTS)
- Gas to Power—transmitted to where it will be used as electricity (GTP)
- Gas to Liquids (GTL)—converted to methanol or ammonia for transportation fuel
- Gas to Commodity—uses natural gas to produce aluminum or some other commodity that is energy-intensive, rather than ship it (GTC).[442]

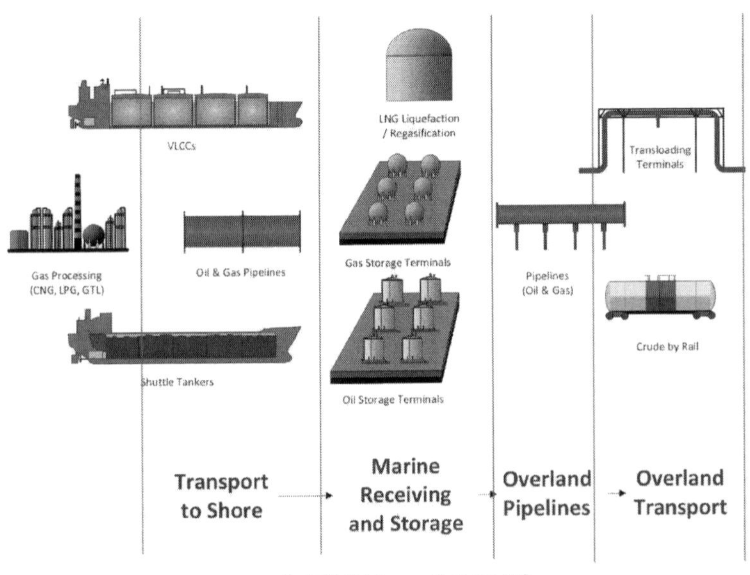

Figure 65. Midstream Oil & Gas Value Chain from an Offshore Production Platform (Illustrative) (*Source:* BSI Energy Ventures)

Midstream scenarios may need to consider the delivered energy content as well as the logistics costs. This may require specialized skills in areas such as fluid mechanics and cryogenics, or at least enough background to understand the physical constraints. For example, transport in methanol form allows shipment in a conventional tanker but decreases the energy (BTU) potential of the shipment substantially, since methanol contains less than half as much energy per shipload as gasoline.[443]

Oil & Gas Pipelines

Oil & gas pipelines consist of gathering lines, transmission lines, and distribution lines. Gathering lines collect the oil or gas from wells and field processing stations. Transmission lines carry the product over long distances. Distribution lines, mostly for natural gas, feed the product to many smaller points of consumption such as residential, commercial, and industrial customers.

Before natural gas flows through transmission lines, the "wet gas" is processed to remove heavier hydrocarbon gases and strip out water, hydrogen sulfide (H_2S), carbon dioxide (CO_2), nitrogen, and helium. Pre-processing typically consists of four steps: oil and condensate removal, water removal, separation of NGLs (typically ethane, propane, and butane), and removal of non-hydrocarbon gases. What travels through the pipelines should consist of nearly pure methane, although ethane, CO_2, and propane may exist in trace amounts.

Oil pipeline operators schedule sequences of discrete shipments either as batches or in a mode that allows products of different grades to intermix inside the pipeline. If they are to be kept separate, separators called "pigs" keep products like gasoline and kerosene from intermixing. Operators at the receiving facility process any "transmix," the blended contents from inside the pipeline, before sending the product on to further shipment or to storage.

While flowing through the pipeline, operators use Supervisory Control and Data Acquisition (SCADA) systems, which act through a series of in-line valves through a series of sensors, relays, and controls, to regulate the pressure, speed, and direction of the flow.

Globally, the oil and natural gas pipeline network tends to attract less capital investment than upstream projects, but while upstream investment is volatile, fluctuating with oil & gas prices, midstream CapEx, and consequently pipeline mileage, tends to be more stable. There is volatility in pipeline construction, however, both regionally and over time. Regionally, major pipeline projects can drive extraordinary growth while pipelines are in active construction. Over time, many pipeline projects are cancelled due to environmental clearance roadblocks or financing constraints, or both, as they move through the phases of capital project evolution.

Oil & gas pipeline capital investment grew by about $250 billion between 2013 and 2020,[444] as partially illustrated by the map below, which shows only the

Eurasian pipeline network (the global gas pipeline network is far too extensive to be shown on any one map). This figure does not even include the totality of the investment and upcoming mileage to be added in the Eurasian region through megaprojects such as the Turkmenistan-Afghanistan-Pakistan-India (TAPI) Pipeline or the Trans-Anatolian Natural Gas Pipeline (TANAP),[445] which are under construction, and the Trans Caspian Pipeline (TCP), which has not yet passed the Final Investment Decision (FID) stage of approval. Nor does it include the NordStream I or NordStream II pipelines from Russia to Europe, which are near to the point of completion after much political wrangling as of the date of writing. If this were a density map, it would show an intense concentration of gas coming from Russia towards continental Europe.[446]

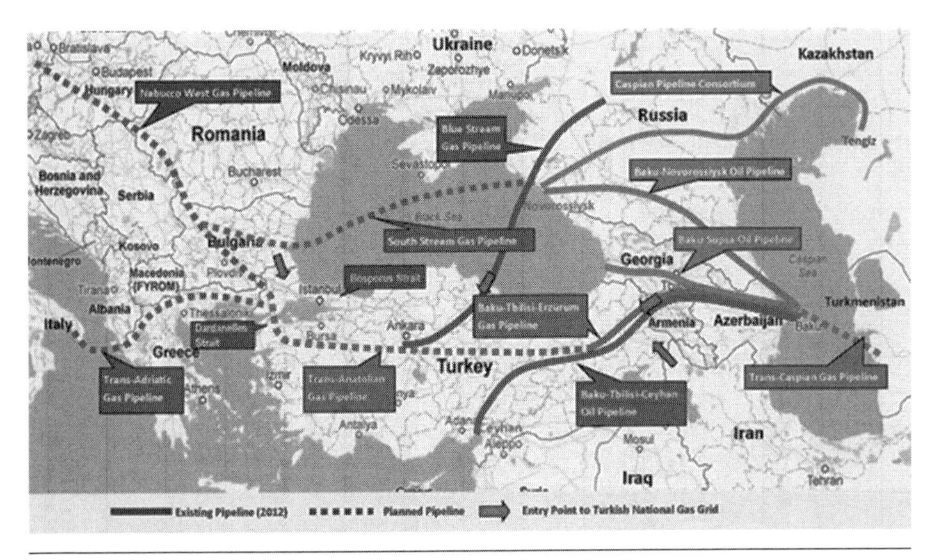

Figure 66. Eurasian Gas Pipeline Network
(*Source:* Dr. Omer Esen, Security of the Energy Supply in Turkey: Prospects, Challenges and Opportunities. International Journal of Energy Economics and Policy, Vol 6, Issue 2, 2016.pP. 288).

The United States also saw an explosion of pipeline construction between 2012 and 2017, with the preponderance of these pipeslines originating or terminating in Texas and Louisiana. For example, MarkWest Liberty Midstream and Sunoco Logistics built a 50k bpd ethane pipeline from the US Gulf Coast to refrigerated storage facilities along the East Coast, moving products via ship to Gulf Coast markets.[447] Also, Koch Pipeline and NuStar Logistics developed a pipeline connection and capacity lease agreement that will give Koch Pipeline future additional capacity to transport shale oil from the Eagle Ford shale to Corpus Christi, Texas.[448]

During the engineering, procurement, and construction phases of midstream projects, management involved in construction of pipelines needs to be familiar with international and national standards that apply to both the construction and the operation of oil & gas pipelines, such as ISO 21809, ISO 13623, and ISO/TC67. ISO 21809-5:2010 specifies requirements for qualification, application, testing, and handling of materials for reinforced concrete coating for oil & gas pipelines of concrete thicknesses of 25 mm or greater.[449] ISO/TC67 attempts to form an internationally accepted standard that incorporates the best aspects of many countries' individual standards. While the standard for pipelines (ISO 13623, "Pipeline Transportation Systems") is the governing standard of this committee, the group is working on standardizing sub-standards related to line pipe, coatings, cathodic protection, mechanical fittings, pipeline valves, and actuators. Working group committees are composed of country-nominated "experts" in each sub-specialty.

Oil Tankers

Figure 67 represents oil transportation via water by six classes of tankers depending on their size: Ultra-Large Crude Carriers (ULCCs) are the largest, ranging from 320–550k deadweight tonnes (DWT); Very Large Crude Carriers (VLCC) are next, ranging from 160–320k DWT. SuezMax tankers run from 80–160k DWT; AfraMax tankers run from 45–80k; and PanaMax tankers are between 25k and 45k DWT. Tankers below 25k DWT are simply called Product Tankers. There are over 2,200 oil tankers in service today. Over 90% of them are VLCCs, SuezMax, and AfraMax sizes.

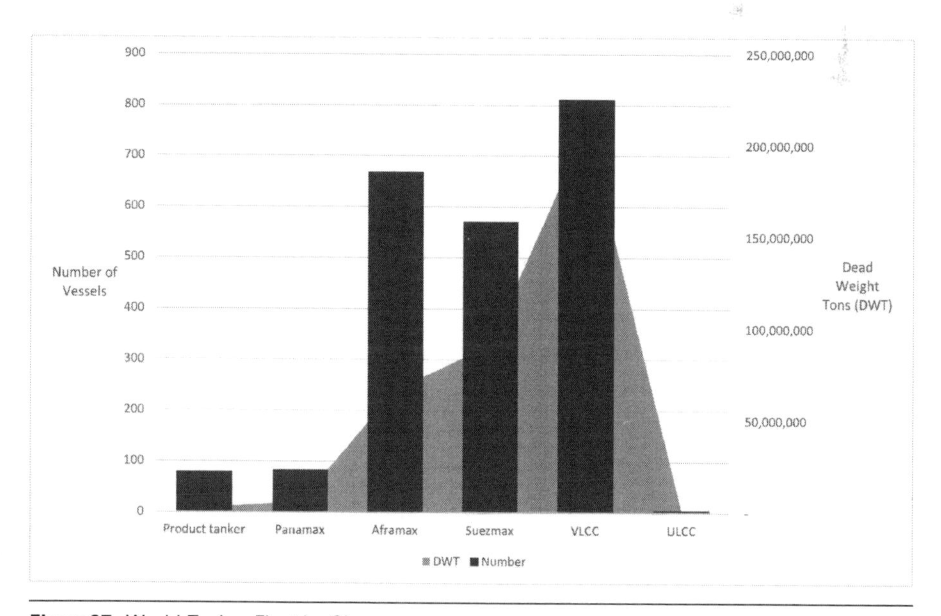

Figure67. World Tanker Fleet by Class
(*Source:* Boston Strategies International analysis)

New builds cost tens of millions of dollars to over a hundred million dollars, depending on the vessel size. Charter rates are volatile. As of this writing, time charter rates for VLCCs were ranging from $20,000 to $42,000 per day.[450]

Environmental and safety regulations and considerations can significantly affect the tanker market. Environmental requirements to burn low-sulfur fuel, for example, the International Maritime Organization's (IMO) Sulphur Emission Control Areas (SECA) 0.1% (2015) and 0.5% (2020) cap on sulfur content, have raised fuel costs. Environmental requirements to discharge ballast water farther offshore in accordance with the 2017 Ballast Water Management Convention, and the decision of some carriers to install ballast water treatment systems to avoid disrupting marine life when ballast water is discharged, are raising operating costs. Also, piracy is an ongoing threat and cost, and there is often not much choice on routing, so shipping lines are forced to use patrol boats to escort the ships. Carriers must buy war risk insurance and pay crews bonuses due to security hazards. Finally, due to negative publicity, major oil companies are increasingly reluctant to operate old ships that could be perceived as unsafe (even if they are safe); the drive to replace them with newer ships will raise costs.

Liquefied Natural Gas (LNG) Infrastructure and Shipping

Numerous phase change technologies facilitate the more efficient transport of natural gas. These include Liquefied Natural Gas (LNG), Gas to Solids (GTS), Natural Gas Hydrates (NGH), Gas to Liquids (GTL), and other process technologies. These have the potential to change the economics of production and make the product more transportable.[451]

LNG has attracted the most investment and commercialization so far. The transformation of the US from an LNG importer to a net exporter shaped the LNG landscape. Cheniere's Sabine Pass terminal in Louisiana was originally intended to be an import (regasification) terminal until 2010, when US shale gas and shale oil production rose to levels that allowed the country to export LNG instead of importing it. Sabine Pass secured $8 billion of investment as an export (liquefaction) terminal and won approval to export starting in 2011. This, coupled with the US government's 2015 decision to end a decades-long ban on crude oil exports, which was later extended to natural gas, contributed to a major growth in global LNG trade. Kuwait National Petroleum Corporation's Mina Al-Ahmadi floating, dockside GasPort became the Middle East's first LNG receiving facility (its offshore location reduced capital costs compared to onshore LNG facilities due to the absence of most conventional onshore real estate issues). LNG export projects subsequently developed in Qatar, Egypt, Trinidad, Oman, Israel, Angola, Nigeria, Canada, Mozambique, and Russia.[452]

Daniel Gergin explains in *The New Map*: "Europe now has more than thirty receiving terminals for LNG, which can be ramped up on short notice. They are also part of an increasingly dense global network. Worldwide, over forty countries now import LNG, compared to just eleven in 2000. Exporting countries have increased from twelve to twenty. Overall global LNG demand in 2019 was almost four times larger than in 2000, and liquefaction capacity is expected to increase by another 30 percent over the next half decade. Methane molecules from a growing number of countries now jostle and compete with one another for customers across the globe."

Novatek's $30 billion Yamal LNG liquefaction and export facility in the Arctic permafrost, 300 miles from the North Pole, may represent the extent to which LNG has passed the tipping point and is set to be a permanent fixture in the energy economy.[453] The project took seven years to complete—begun in 2013, Yamal's fourth train became operational in February of 2021.[454] Meanwhile, the US continues to build more LNG export capacity, such as the Venture Global terminal in Louisiana.[455] The buoyant price of LNG time charter rates, which exceeded $120k per day during peak season in 2020, is indicative of demand that appears to be strong and relatively stable after years of extreme volatility.[456]

LNG currently makes up about 10% of global gas supply, and the most-trafficked LNG routes today are intra-Asian, especially between Japan, Australia, the Philippines, Indonesia, and China, and between Asia and the Middle East, with transshipping to Africa, Europe, and South America. The largest importing countries are Japan, China, South Korea, India, Taiwan, and Spain. Currently, about two-thirds of global exports come from Qatar, Australia, Malaysia, Nigeria, and Indonesia. The US will soon emerge as a large LNG exporter as capacity ramps up there.[457]

The public visibility as well as the long timeframe of these projects brings huge regulatory and NIMBY ("not in my backyard") risks, with the implication that owners would be wise to secure a final investment decision (FID) before committing to suppliers, and to be aware of pending regulatory changes or upcoming revisions to standards. For example, the US enacted a pipeline safety bill after a 2010 pipeline explosion in California killed eight people, requiring operators to install automatic shut-off valves on new or replaced pipelines, and to test the pressure of pipelines in populated areas. Also, ISO instituted the 28460:2010 standard, which specifies requirements for onloading and offloading procedures and safe transit of LNG carriers through port areas.[458] It applies to ship, terminal, and port service providers.

Tradeoffs and Management Techniques in Capital Project Management

Since pipeline capital costs generally far outweigh pipeline operating costs, proper management of pipeline construction is essential to overall project profitability. Depending on the region, terrain, labor, and steel costs, pipelines may cost in the range of $300k to over $2 million per mile, which in expensive cases can encourage siting of refineries and processing facilities close to the wellhead. In contrast, pipeline operating costs are only about 2–5% of capital cost.[459 460]

Due to generally long project time frames and high market risk, midstream projects are often co-funded by multiple consortium partners. The major shareholder often plays a key role in project management and/or selection of the EPC firm that manages sub-suppliers. For example, a gas processing plant in Qatar leveraged a shareholder (Mitsui) as an EPC, which subcontracted to Hyundai for gas turbines, steam turbines, heat recovery steam generators, and desalination units. Another LNG processing project was established as a joint venture between a US EPC firm and a Japanese EPC firm and involved two local joint venture partners to satisfy local content requirements. The contract was set up as a management contract with reimbursement for material costs plus an adder, with a cap. While 250 suppliers were considered, only suppliers that demonstrated quality were allowed to bid.

Capacity constraints for key equipment can cause project bottlenecks. For LNG, compressors require significant lead time. Most suppliers follow a FIFO or first-come-first-serve prioritization of their capacity. Some suppliers have offered capacity reservations in exchange for firm orders or a certain volume of business. Some have sold production slots for cash.

Technology Choice Risk Management: Selecting a SCADA System

LNG shipping is by nature a technical domain. Natural gas flow and storage is affected by temperature and pressure changes (which depend on heat transfer, changes in viscosity and surface tension, and erosion and corrosion problems). Natural gas flow is assured by minimizing corrosion, preventing hydrate formation conditions, predicting the effectiveness of inhibitors, preventing wax deposition that may impact pressure and flow, avoiding slugging (liquid buildup that prevents gas from flowing), adjusting the number and pressure of the producing wells, and closely monitoring shutdowns and restarts.[461]

One of the more important choices faced in the development phase of pipelines is the choice of a Supervisory Control and Data Acquisition (SCADA) system. The system specification and vendor support significantly influence the degree of operational flexibility and asset performance of the pipeline.

SCADA systems may be particularly important when there is a high mix of products and batches flowing, and delivery timeliness and purity is critical to downstream customer operations. To make matters more complex, the ability to connect multiple field devices on the one hand, and sensors and relays on the other, through cloud architecture is rapidly becoming essential.

A high-performing SCADA system can increase service availability by 2%, throughput by 1%, and life-cycle costs by 5%, according to Honeywell, one of the larger providers, which offers central control room cockpits, compressor and pumps station controls, line information and metering instrumentation and controls, and integrated terminal management including fire and gas leak detection and control systems.[462]

PetroChina engaged in a $12 million contract with Honeywell for the SCADA on Phase II of its 9,600km west-to-east pipeline, the longest in the world.

Capital Project Risk Mitigation: Standardizing Materials and Modularizing Facility Design

Some buyers are willing to pay a commitment fee for an option on capacity. Figure 68 shows how one owner used real options to decide whether or not to invest in an FGSO (Floating Gas Storage Offloading) unit.[463] The company evaluated four substantive choices: 1) acquire options on steel to be able to build the ship, 2) acquire the ship itself, 3) refit a shipyard to construct the ship and possibly others, and 4) construct an offshore development in addition to the ship. Each option had a cost of acquisition and a cost of disposal or decommissioning if the choice turned out to be below expectations.

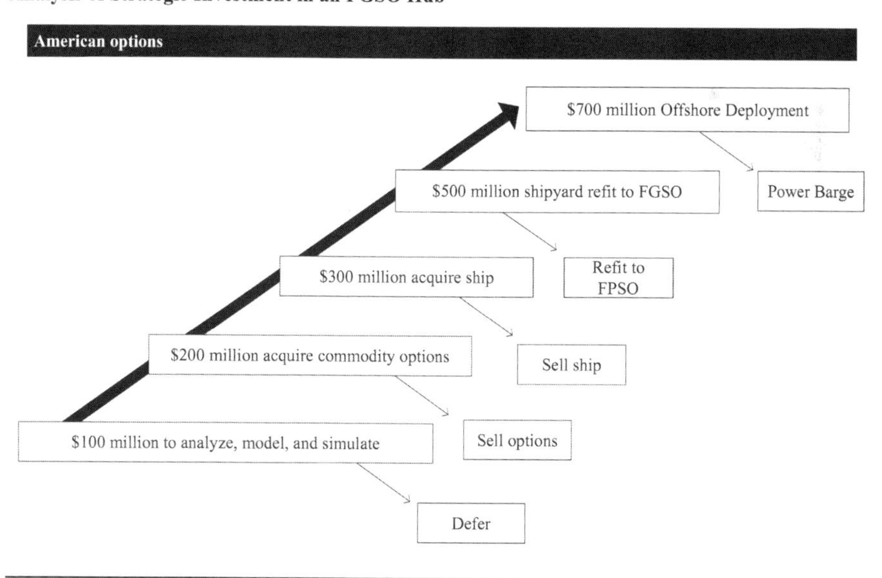

Figure 68. Analysis of Strategic Investment in an FGSO Hub Using Real Options
(*Source:* Roger N. Anderson, Computer-Aided Lean Management for the Energy Industry, PennWell, 2008, p. 275.)

After summing the NPV with the option value, the lowest-cost option turned out to be to refit the FGSO shipyard, as shown in Figure 69.

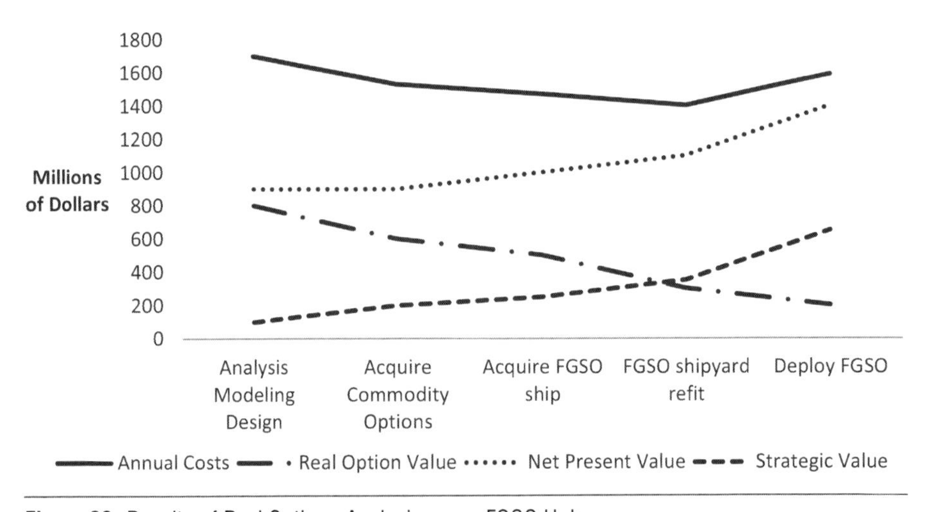

Figure 69. Results of Real Options Analysis on an FGSO Hub
(*Source:* Roger N. Anderson, Computer-Aided Lean Management for the Energy Industry, PennWell, 2008, p. 275.)

Project Management Tools

Disciplined project management is a critical success factor for large midstream projects, since delivery and operational logistics need to be thoroughly planned in advance. In addition, at the equipment and component level, global sourcing is a key lever for keeping cost down.

The Nord Stream 1 pipeline provides an example of rigorous management of a physically challenging and complex project. The size of the project made it a particular logistics challenge. Twin pipes carry 55 billion cubic meters of natural gas per year over 1,224 km, from Vyborg, Russia (near Russia's border with Finland) to Greifswald, Germany, via the Baltic Sea. Laying the pipe cost about $9.6 billion[464] and involved 202k concrete coated pipes, each of which is 12 meters long and weighs about 23 tonnes. Thirty pipe-laying vessels worked simultaneously on the project, including Saipem's Castoro Sei and Allseas' 300-meter-long Solitaire.[465] Galvotec Alloys Inc., a supplier of anodes, used a complex, project-specific database to monitor the production and delivery of its anodes to the pipeline (delivery took up to eight weeks). The manufacturer produced over 13,000 pages of records to accompany the shipment.[466]

Careful logistical planning helped to avoid surprises. The developer operated out of two logistics hubs, one in Kotka in the Gulf of Finland and one in Mukran on the German Island of Rügen, which helped to coordinate the physical delivery and materials management aspects. The supply chain plan allowed time buffers to minimize the risk of bottleneck activities, which has kept the project on schedule. Critical path sections, such as the German landfall section, were executed four weeks before stipulated by the project plan, which enabled the rest of the project to continue according to schedule. Finally, external auditing and certification ensured confidence and reliability. DNV certified Galvotec's manufacturing and quality process.

At the equipment and component level in the supply chain, global sourcing is keeping costs of compressors, blowers, and pumps down. To cite just two examples, Kobelco bought 44% of China-based Wuxi Kobe Steel and provide its process gas compressors intellectual property to Wuxi so Wuxi could manufacture compressor systems in China at lower cost than Kobelco could elsewhere.[467] Almost ten years later, Kobelco purchased the rest of Wuxi Kobe Steel to make it a wholly owned subsidiary.[468] Many other turbocompressor manufacturers have also established joint ventures and made acquisitions in Asia to keep their costs down and tap growing local markets.

Design Standardization

Standardization of design across multiple terminals can reduce CapEx significantly. At first it may seem that each project is unique—that there would be no learning gleaned from project to project—due to their average size and large capital expenditure. Contracts are long (for example, 20 years), and consortium models add complexity to contracting. However, there is a strong learning curve benefit as contractors construct similar projects, so repeating the supply chain structures (suppliers, specifications, and systems) can take advantage of learning and reduce cost.

BG Group, which handles natural gas distribution, claims to operate the lowest-cost LNG terminals among its peers due to the use of standardized designs. BG's Atlantic and Egyptian LNG terminals took only 4–6 years to build, including the tender to an EPC firm and the EPC tender to subcontractors. In contrast, competing LNG terminals at Rasgas, Qatargas, Oman, and Nigeria took 6–16 years, with the difference in speed due to the lack of a standardized design. BG's terminals operate at $175–x275/tpa (excluding financing and based on a two-train build), whereas the facilities at Rasgas, Nigeria, and Qatargas operate at approximately $300, $350, and $425/tpa. The difference in cost is also due to BG's standardized approach. The Egyptian LNG project is owned by BG, Petronas, EGPC, EGAS, and GdF. Its standardized design allows the consortium to lower CapEx to levels achieved only with smaller trains (3.6 MTPA), minimize design changes, and achieve more flexibility by having multiple smaller units rather than a single large unit. The group claims lower purchase cost, shorter construction schedule, less financial and technical risk, and lower operating costs. On top of all those benefits, the project team claims higher reliability (96%) and availability (98.5%) than on other similar installations.[469]

The findings give hope for the many other offshore projects that are underway. No matter how large or deep, each one consists of components like wet and dry trees, jumpers, manifolds, risers, and SCADA systems, which should be eligible for economies of scale and learning curve benefits.

Asset Standardization

Prior to finalizing design specifications, engineering can have a major influence over project price by standardizing materials and equipment. Shell's high-performing standardization program allowed it to reduce purchase prices by 30% for valves and cut variety by 50% through the use of its extensive Materials and Equipment Standards and Code (MESC) catalog. The catalog, which is based on ISO and IEC standards to ensure interoperability, integrates 370 Design and Engineering Practices that standardize tools and facilities, reducing recurring engineering and design work and consolidating spending on standard items. In addition to reducing purchase cost, the practice also reduces delays due to supplier confusion and costs related to unnecessary rework. Continuous feedback from users and participation from external standards bodies keep specifications up to date.

Modularization

Modularization can lower the break-even point of LNG projects. Edge LNG converts stranded and flared natural gas to LNG using its Cryobox technology, which can produce as much as 15 metric tonnes (10,000 gallons) of LNG per day. The rig can be moved between wells and can be up and running within an hour of arriving on site. Peak production is attained wihin 10 minutes, which is much faster than the 18 hours needed for some big conventional liquefaction plants. The daily output of one unit is a tiny fraction—about one-3,000th—of large-scale plants in Qatar and Russia, but the benefit is the unit's portability. The small scale makes the system economical for isolated pockets of reserves that don't justify building gas-gathering pipelines.[470] To date, EDGE has delivered over 30,000 gallons of LNG to a delivery point at a New England gas utility over 300 miles away from the Marcellus production site.[471]

Sustainability Trade-offs: Preventing & Controlling Pipeline Leaks

Multiple technologies have advanced significantly in recent years, reducing the likelihood of pipeline leaks. Corrosion detection has become much more accurate, enabling early detection and preventive correction. Innovations in pipe healing and sealants have enabled rapid fixes for what used to require excavation. Gas detection technology has become much less expensive and easy to deploy because of improvements in device size, accuracy, and portability. Also, drones can be used to monitor pipelines and send images for analysis, thereby identifying pipeline leaks earlier.

Supply Unavailability Risks and Mitigation: Planning for Pipe, Compressors, and Other Long Lead Time Equipment

Lead times for pipeline pipe can extend for months and must be scheduled in advance if the order and/or the diameter is large and if pipe mills are full (80%

or higher capacity utilization). Excluding delivery, lead times can be as low as 4–6 weeks or can extend for a year or more for a large order during peak. Steel buyers and their EPC firms should place orders as early as possible to avoid experiencing project delays due to bottlenecks at the steel plants and rolling mills. In addition, engineered compression systems can bottleneck production schedules and should be ordered as early as possible.

Outsourcing Risks and Mitigation: Focusing on the Core Midstream Business

In new pipeline construction projects, many, even most, aspects of project engineering and pipe-laying are typically outsourced. During pipeline operation, many aspects of operations and maintenance are also typically outsourced, such as gas gathering and treatment, pipeline operation, SCADA system maintenance, pigging, and gas compression.[472] In LNG, gas marketing is commonly outsourced, as are procurement and administrative information systems.[473] Vessel brokerage usually goes to specialist ship brokers.[474]

Supplier Partnering Risks: Maintaining Flexibility in LNG Supply Chain Infrastructure Commitments

The financial and regulatory risks of LNG projects are large, and the time frames are long, so it is necessary to regularly update projections and hedge risks as necessary. Commitments to vendors—for example, for pipelines, liquefaction and regasification terminal infrastructure, and for tankers, vessel capacity—are aligned with the appropriate degree of flexibility until final approved decisions are reached and permits are secured. Contracts and financial commitments should include escape clauses where possible to minimize the downside risk of stumbling blocks in the project plan.

Procurement Bundling Tradeoffs: Contracting for Process Control and Metering Systems

For a large midstream operation, process control is multifaceted, involving hardware such as physical valves and meters, electronic Programmable Logic Controllers (PLCs), Remote Terminal Units (RTUs), safety shutdowns, and flow computers and software systems, all of which may need to be supported by an experienced technical customer service team.[475] Buying the whole package along with maintenance and support services from a leading vendor can assure end-to-end compatibility and alignment of processes and systems during upgrades and extensions.

Contract Term Risk and Mitigation: Calibrating Term with Stability and Liquidity

Actual LNG fuel contracts, as well as LNG shipping contracts, many of which historically used to exceed four years in length, have shortened over time as the industry has matured and a robust short-term (under four years) and spot market has emerged.[476] The same may be said of natural gas pipeline arrangements, which now follow demand set on well-established natural gas spot and futures markets, to the point where long-term supply is more stable (even if gas prices remain very low).

On the other hand, supply contracts for operations and maintenance services and systems tend to have longer time horizons, to the extent that liquidity in the market de-risks their long-term future. Of course, the term of each contract needs to be evaluated in light of the service being provided (operations, maintenance, equipment servicing, software support, etc.) and any future plans for expansion or upgrade that would form natural contract break points.

Tradeoffs and Management Techniques in Operations & Maintenance Management

Key drivers of midstream operating and maintenance costs include production planning and control, maintenance of storage tanks and pipelines, and routing efficiency. These cost drivers recur across the range of IoT/AI, cybersecurity, capacity management, sourcing, TCO, and HSE aspects of supply chain management, discussed below.

Internet of Things (IoT) and Artificial Intelligence (AI) Technology Choices: Automating Batch Scheduling & Dispatch

Careful planning and scheduling characteristic of pipeline operations can over time be learned through artificial intelligence and embedded in algorithms for more effective and efficient dispatching. A scheduler sends batch orders in a repeating sequence to avoid commingling fluids and to separate the ones that have less tolerance for contamination. Based on orders ("nominations"), the scheduler assigns batches to available pipeline capacity—for example, a repeating cycle of diesel, gasoline, and jet fuel. Careful scheduling optimizes the delivery of the most profitable product to the customers in the time frame that they want, especially if there is a wide variety of products and grades.[477] This can be learned by a machine.

In addition, AI has an important role in reliability-centered maintenance and Six Sigma, which are covered in Part 1. Corrosion and leaks in tank storage can be predicted with sensors, instrumentation, telemetry, and AI, thereby reducing unplanned repair that increases operating costs. Corrosion is a perennial problem that must be addressed in order to operate safely. Tank storage is sensitive to safety and environmental concerns, whether the tanks are atmospheric, low pressure, or high pressure, and whether they are single-walled or double-walled. Choice of materials (carbon steel, aluminum, stainless steel, or fiberglass reinforced plastic [FRP]) can significantly reduce the corrosion problem. However, in most cases corrosion still causes some degree of stress cracking, fatigue, galvanic corrosion, crevice corrosion, cavitation damage, and hydrogen embrittlement, just to name a few potential causes. Intelligent, AI-driven reliability centered maintenance can eventually embed the standards and practices that prevent corrosion, vapor escape and faulty seal welds. Various API standards can be used for machine learning development, such as API RP 570.[478]

Cybersecurity Risk Management: Protecting Flow Control Systems

Cybersecurity is critical to midstream security as operators use more control systems, surveillance apparatus, and tracking systems.

Supervisory control and data acquisition (SCADA) systems help to monitor pressure and warn of abnormal conditions. PetroChina uses OASyS, a SCADA system to monitor the pressure within the 2,100 km Lan-Zheng-Chang oil pipeline, which has a total elevation change of 2,500m and passes through complex, varied terrain that could impact flow rates. The system provides early indications of leaks and potential disruption, thereby minimizing the negative impact of disruptions. The field of SCADA cybersecurity is being serviced by an emerging industry of IT leaders and specialists.

Unmanned aerial vehicles (UAVs) have become cost-effective for pipeline surveillance in recent years, and one Gulf gas producer signed a five-year agreement with a drone surveillance operator, as noted previously in this book. Drones are both a potential target and a potential source of cyberattacks, and there are large communities issuing technical guides, policy studies, laws, standards, and best practices regarding drone cybersecurity. A major state-owned LNG producer in the Middle East uses drone inspection, surveying, and data visualization services. The provider collects engineering-grade inspection data from onshore and offshore oil & gas assets and delivers detailed inspection reports via data visualization in a GIS-enabled, cloud-based, IoT-based visualization interface data management software. The result is faster inspection and surveying than in the current manual land-based approach, especially for pipelines in very remote and difficult to navigate terrains.[479]

Radio frequency identification (RFID) tags were at one time problematic in liquid environments because the signal did not transmit through liquid. However, advances have made RFID an option for some applications. One vendor, Container Technology, is testing passive RFID tags for use in liquid and metal environments where RFID technologies did not previously work.[480] The tags, which comply with EPC Class1 Gen2 standards, can be placed on 55-gallon drums with a read range of about 20 feet. RFID systems can be sniffed, tracked, hacked, and counterfeited,[481] so RFID vendors and IT consultants are offering a wide variety of advice and solutions. Blockchain applications are also being developed in some cases.

Peak Capacity Strategies: Managing Interruptions & Disruptions

Shipping, pipeline, and tank storage markets are highly regional and dynamic over time, fluctuating with both supply and demand for oil and shipping capacity.

Demand for midstream storage and transport capacity rises and falls with overall economic growth, the price of oil, and the inventory of crude and refined oil. Over recent years, global economic growth has been rocked by the global financial crisis and the COVID-19 pandemic. US GDP fell by 31% in the second quarter of 2020 and then rose by 33% in the fourth quarter.[482] The daily price of oil has ranged from nearly $150/barrel in 2008 to $11/barrel in 2020.[483] US inventories of crude oil ranged from 26 days of supply to 42 days of supply and back to 28 days of supply between February 2020 and February 2021.[484] The US went from being an oil importer to being an oil exporter in 2019, creating a permanent rebalancing of oil trade flows.[485] Bunker fuel (ship fuel) prices, in addition to fluctuating with the price of oil itself, were influenced by the IMO's low-sulfur standards that came into effect in 2015 and 2020.

Demand and supply factors rarely align, resulting in gaps, regional and temporal imbalances, and consequently rate peaks and troughs. Pipeline and storage capacity is essentially fixed, and rates are locked in by contract, but ship capacity swings from year to year depending on shipbuilding cycles. The LNG tanker market had cyclical overcapacity crises in 1977 and 1981. Anglo-Dutch ships went unused and were sold and resold at successively steeper discounts. At one point El Paso Energy ended up writing off its entire investment in LNG, completely scrapping three vessels.[486]

As a result, charter rates can and do fluctuate widely in response to these demand factors. In March 2020, COVID-19 lockdowns resulted in a sudden surge of crude inventory. With nowhere to put it all, shipboard storage became the answer. Tanker rates quadrupled.[487]

Parcel tankers can manage these imbalances better because they can handle up to 52 different types of cargo on the same ship (one per hold). Thus, in event of a trade imbalance, they are more likely to be able to maintain full loads in both directions by handling smaller lots of more cargoes.

Capacity availability information systems can mitigate capacity crunches during peak periods. Boston Strategies International helped one oil major build a multi-regional database of available terminal and storage spots in response to a shortage of carspots.

Sourcing Tradeoffs: Owning versus Leasing Vessels

The emergence of a consistent LNG market has in some cases presented large producers with the option of continuing to lease LNG ships on a time-charter basis or buying and owning LNG carriers. The decision should consider many variables, including the difference (or spread) between the cost of purchasing an LNG vessel versus the daily time charter LNG shipping rate, interest rates, the volume and duration of expected shipments, and operational assumptions such as terminal utilization rates and boil off ratios. Scenarios should be constructed based on a variety of future LNG price trends.

TCO Tradeoffs: Getting More Performance from Compression Systems

Since fuel is a major operating cost, the choice of driver for pipeline compressors can significantly influence operating costs. Moreover, the rapid pace of technological change is increasing the efficiency of all of the available technologies over time. The choice of which driver to use—gas turbine, steam turbine, electric motor, or even diesel generator—determines the cost performance at the desired pressure and flow characteristics, given the types and cost of fuel available at the locations where it is needed.

Next, upgrade and retrofit options must be periodically evaluated to increase the efficiency of older units as much as is practical. Motors made to comply with relatively new efficiency standards (e.g., NEMA Premium and IE3/4) can make for economically attractive upgrades.[488]

For turbines, materials used in rotors and blades, and the machining processes used to make them, have greatly increased heat tolerance, and with it, power-generating efficiency. More examples of power efficiency improvements through technology enhancements and upgrades are discussed in the Power Generation chapters.

HSE Considerations: Conforming to Regulatory Standards and Laws

Midstream transportation of hydrocarbons is regulated by various government agencies, depending on the nature of the safety risks. In the US, while the Federal Energy Regulatory Commission (FERC) regulates the interstate transmission of natural gas, oil, and electricity from a commercial point of view, the Pipeline Hazardous Safety Administration issues highly detailed codes and specifications for safety during such transmission, which are specific to the liquid or gas being shipped and the structure supporting it (pipeline, tank truck, etc.).

Supply Chain Roadmap for Midstream Oil & Gas

Supply chain improvement opportunities include: 1) the expansion of LNG, 2) small-scale modular gas processing technologies, 3) drone monitoring, and 4) smarter process automation solutions. Seen through a supply chain lens, LNG is a supply chain *innovation*; Gas processing technologies are supply chain *customizations*; drone monitoring is a *rationalizing* (cost-cutting) strategy; and smarter process automation solutions are *synchronization* (supply-demand matching) improvements. Therefore, Midstream Oil & Gas is fertile ground for supply chain enhancements across all four supply chain strategies (*rationalization*, *synchronization*, *customization*, and *innovation*).

Now that LNG has achieved mainstream and worldwide adoption, it continues to chalk up new milestones such as new countries and routes. It has passed the tipping point and can achieve major benefits by continuing to scale more quickly than before. Supply chain management can facilitate the growth by sizing terminal capacity, optimizing physical flows, engineering ship logistics and schedules, and building ship-to-shore infrastructure including liquefaction and degasification facilities and the needed bridges, tunnels, and roadways to enable high-volume access.

At the other end of the quantity spectrum, small-scale gas processing solutions are offering targeted and localized supply chain solutions. For example, Edge LNG converts stranded and flared natural gas to LNG using its Cryobox technology, which can produce as much as 15 metric tonnes (10,000 gallons) of LNG per day. Outside of LNG, Calvert GTL offers technologies for converting flare, biogas, landfill, and methane gas into diesel, kerosene, jet fuel, methanol, and ethanol at small scale. This type of modular solution fills a need where large-scale LNG investments can't be justified.

Drone monitoring is now used routinely to surveil and inspect pipelines worldwide. This saves time and labor, and often provides earlier warning of security, maintenance and safety incidents. This being said, it still has room to demonstrate these benefits at many facilities that have not yet adopted it.

Although SCADA systems have been around for a long time, the increasing length and flow of pipelines, coupled with rapidly mounting cybersecurity risks, are making smarter and more secure process automation solutions an ongoing challenge and opportunity for supply chain management.

These significant value-creating supply chain opportunities for Midstream Oil & Gas are shown graphically in Figure 70.

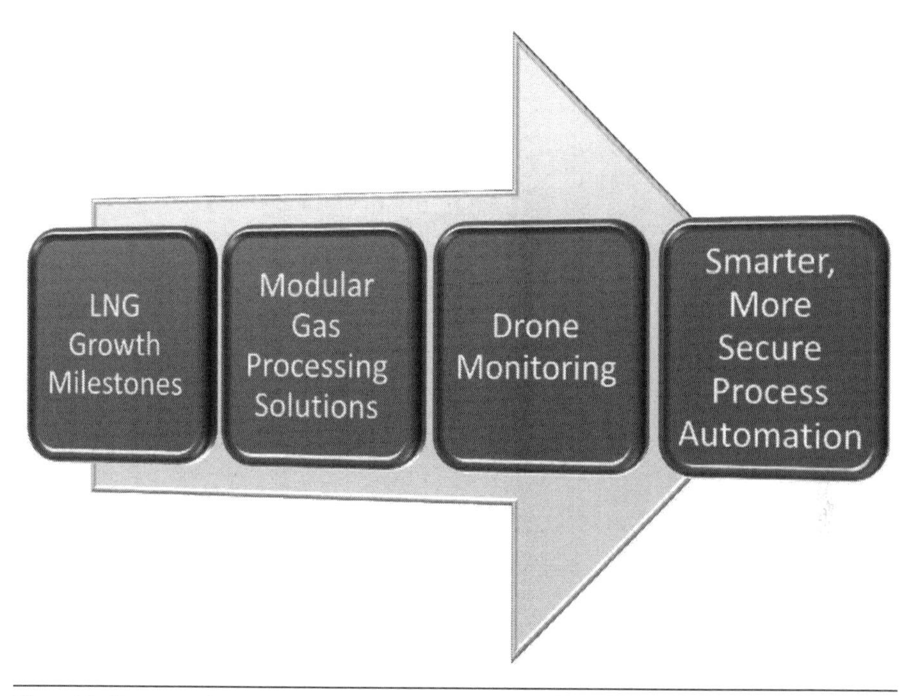

Figure70. Emerging Supply Chain Opportunities for Midstream Oil & Gas

Oil & Gas—Downstream

Chapter Highlights

1. Since total life-cycle cost of refineries depends heavily on upfront capital investment decisions, the capital project supply chain is an especially important supply chain responsibility.
2. When managing capital project construction projects, supplier adherence to construction schedules may be optimized by bundling materials and components orders that need to come together as a system in a narrow time window.
3. Due to the high capital intensity of the plant and equipment, advanced planning tools should be used for single- and multi-echelon network planning.
4. Process control and manufacturing execution systems should optimize profitability based on constraints including capacity, the grades and quantities of inputs and catalysts, capture of intermediates, and recovery of energy and waste.
5. Due to extreme temperatures and pressures, HSE is paramount and must be embedded in culture and standard operating procedures. Management should adhere to international and recognized standards and include partners and contractors in HSE planning.
6. Supply chain improvements can often be realized in four Downstream areas: 1) More Robust Risk Governance; 2) Cybersecurity for IoT Devices and Systems; 3) Smarter Predictive Maintenance; and 4) Low Sulfur Fuel.

Introduction—Supply Chain Economic Cost Drivers

Refining economics are largely driven by choices made during initial design, especially the choice of crude, processing routes (the number and type of processing units), and facility design. The choice of crude (crude oil evaluation relative to a reference crude) account for 70–80% of the operating costs, which means that traditional operating cost reduction initiatives will have relatively little impact. Furthermore, investment costs are often roughly ten times annual operating costs. One example might be a plant with an investment cost of $2.8b and annual operating costs (excluding raw materials and depreciation) of $314m. The main operating cost is usually maintenance, accounting for well over half of total annual operating costs. Other significant operating costs include staff, energy (including fuels like natural gas, electric power, and internally consumed components such as gas from light ends), and insurance.[489]

Operational productivity is largely constrained by the physical design of the facility. Due to the aforementioned economic parameters, project risk mitigation depends mostly on controlling upfront capital investment decisions rather than reducing ongoing operating costs. The wide variety of refineries and crude oils (every crude oil is different, with some being paraffinic, while others are napthenic, aromatic, or asphaltic) makes it difficult to generalize about supply chain management applications in refineries.

The basic refining process has not changed in about 60 years. A high-level conceptual flow focusing on flows into and out of the refinery is shown in Figure 71. Core processes include distillation (separation of crude); conversion (decomposition including cracking or breaking down large molecules, unification or aggregating small molecules, or re-forming—rearranging molecules through isomerization or catalytic re-forming); and treatment (desalting, hydrodesulfurization, solvent refining, sweetening, solvent extraction, and dewaxing).

About 94% of US refineries have a hydrocracker, and about two-thirds of those have a coking unit as well. The rest are simple refineries that perform only atmospheric distillation, catalytic reforming, and refining with no further cracking or processing steps.[490] Most investment projects center around residuum processing options such as delayed coking and visbreaking, or fluid catalytic cracking.

Production planning and execution is more complex in downstream oil & gas and in petrochemicals than in process industries such as paper milling, sugar refining, and minerals processing. First, an oil refinery can produce dozens of different products out of benzene, whereas in the minerals industry a processing plant might be able to convert limestone only into two or three calcium carbonate products. Second, some hydrocarbon byproducts can be reused as intermediates in other processes (for example, an olefins plant can produce aromatic byproducts that can be recycled for use in fuels blending). Third, due to the toxicity of the chemicals involved, shutdowns, turnarounds, and maintenance are more frequent and more intense than in other process industries. Fourth, some processes like blending and sulfur recovery are batch processes, which require different replenishment signaling, production scheduling, and materials management from the continuous processes.[491]

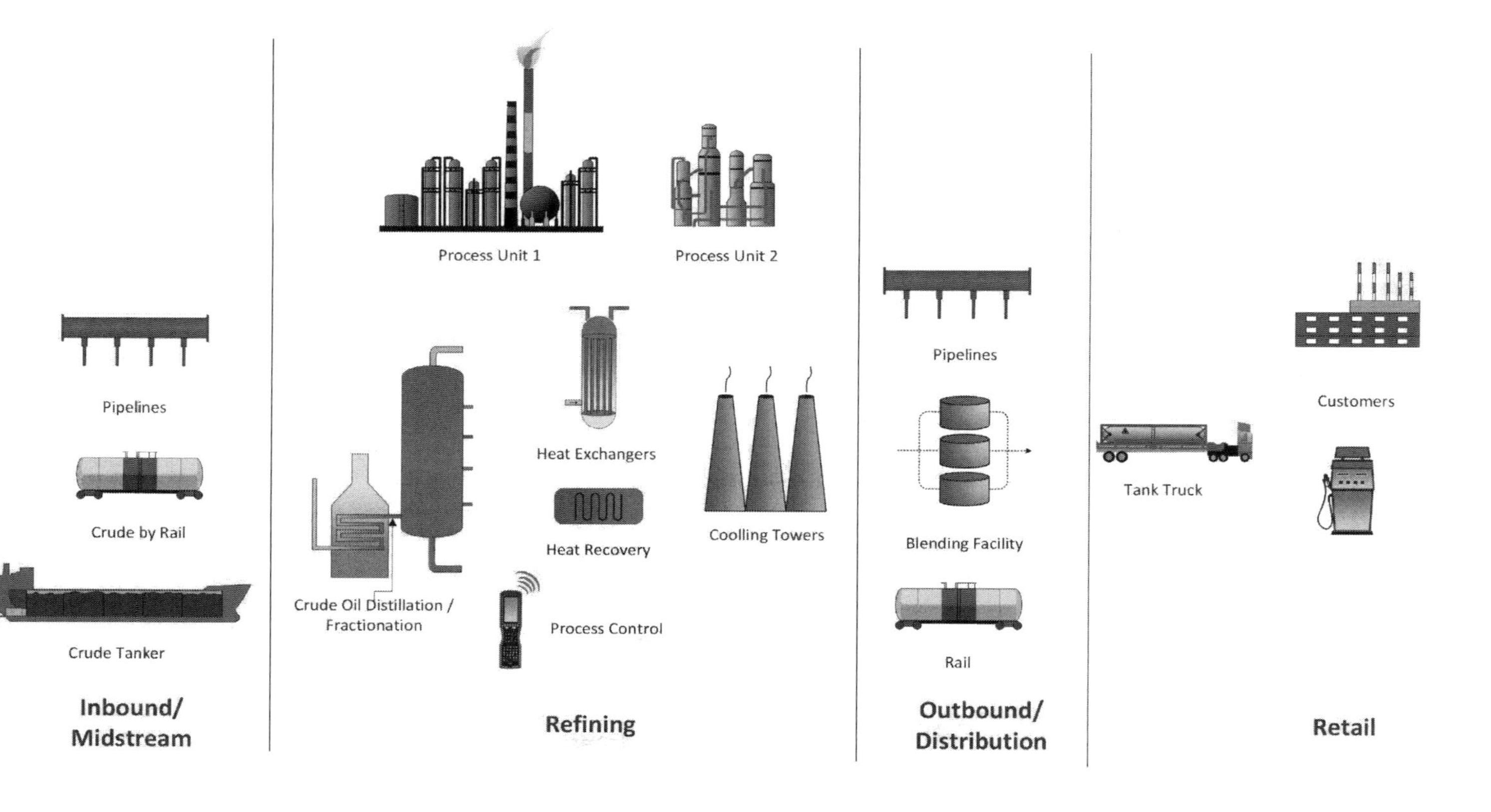

Figure 71. Downstream Petroleum Value Chain (Illustrative)
(*Source:* BSI Energy Ventures)

In recent years debottlenecking and incremental expansions have almost entirely replaced greenfield capacity. For example, ExxonMobil Chemical increased capacity for its Synesstic brand alkylated naphthalene base oil at its Edison, New Jersey, refinery by about 40% through a debottlenecking program.[492] The need for refinery capital investment may also be limited by the deployment of new process technologies and catalysts that effectively increase the capacity of current plants. One industry-leading refiner recently conducted a multi-year study of new process technologies and refinery catalysts in order to stay abreast of the new technologies.

Trade-offs and Management Techniques in Capital Project Management

Optimizing a network of refineries may reduce the need for overall capacity by ensuring the maximum economical shipping range from each plant. When designing supply chain networks, many applications exist for multi-echelon production optimization modeling. When considering where to place a new refinery, PEMEX uses a wide range of criteria, including: type of technology project, configuration of plant, size of the new refinery, distance to sources of oil and consumer areas, availability of raw material, quality of crude available, operational efficiency with which the company plans to operate the infrastructure, geographic analysis, environmental impact, job creation and economic impact, existing infrastructure (roads, pipelines, refining), utilization of existing wastewater, and cost of the land, among others.[493]

Technology Choice Risk Management: Digitalizing Process Control Systems

As with midstream operations, one of the most impactful technology choices that refiners need to make is the choice of a process control system. Increasingly, refinery efficiency is determined by automation and controls. Automation may be mechanical, electrical, or informational in the context of robotic process automation, basically converting repeatable tasks into hands-off automatically recurring events. Some may refer to that type of automation as *digitization*. More and more, these devices are smart and connected, and production optimization relies on digital twins from which operators control the mechanical, electrical, and IT substructures. This smart, interconnected, and sometimes self-teaching neural network is referred to here as *digitalization*.

Chevron used Aspen Technology's aspenONE Advanced Process Control (APC) applications to quickly return APC controllers to service after unit turnarounds. The application reduced the effort required to maintain controller models and saved thousands of dollars for each day the controller would have been offline, according to AspenTech.[494]

Capital Project Management Risk Mitigation: Contracting and Managing Construction Schedules

Just as single sourcing focuses on one supplier to deliver operating materials or services, Design-Build-Own-Operate (DBOO) is its analog in refinery capital project management. DBOO is a supplier-partnering configuration that counts on a single partner to deliver a whole project. Limiting the number of suppliers involved in managing the capital project can keep the project on schedule by reducing complexity and interfaces. Furthermore, structuring project management so one party has control all or most of the way through often improves project economics and timing. DBOO can keep projects on schedule by reducing complexity. For example, Air Products designed, built, and owns and operates a hydrogen plant at Marathon Oil's Catlettsburg, Kentucky, refinery.[X] The value proposition to refiners is the solution's cost-effectiveness, the state-of-the-art technology, and DuPont's expertise in safe and reliable operation.

Lump Sum Turnkey Projects can also ensure continuity throughout the construction process and ensure tight project management control. In one case, an operator awarded an LSTK contract to a sole source supplier. The operator set the contract budget and negotiated the agreed contractor price in order to keep contractor profits reasonable. Execution was thorough and on time (24 months), and the project met safety and HSE goals. The close relationship between the operator and the EPC helped iron out problems with sub-suppliers and transportation logistics. In another, the operator engaged an EPC firm to build an ethylene complex on a contract that provided a fixed price for labor but a cost-plus arrangement for materials. At the operator's direction, the EPC engaged many suppliers (there were about 6,000 prequalified bidders and 200 Tier 1 suppliers) and did not execute any alliance or partnering agreements. It also strongly encouraged low-cost country sourcing. In contrast to the LSTK project above, this project lasted 48 months (twice as long) and was not as profitable for the contractor.

Construction schedule delays can sometimes be avoided by ordering materials and components at the optimal level of bundling, since many materials, components, and whole systems need to come together at just the right time. Refinery construction involves a complex production plan comprising materials, components and whole systems that must arrive just in time—not early and not late. Major equipment, storage facilities, steam systems, cooling water systems, and other systems must all be procured and delivered, built, or assembled onsite. Rule-of-thumb estimates or cost curves for similar units constructed elsewhere with a similar capacity are used to predict costs. If all goes well, the actual construction is not eventful, but getting all the components in place at the right time can be stressful.

[X] Catlettsburg Refinery. A Barrel Full. http://abarrelfull.wikidot.com/catlettsburg-refinery Accessed January 29, 2021.

Appendix 2 contains a taxonomy of major and some minor equipment and major systems that may need to be procured or constructed. For major processes such as Fluid Catalytic Cracker (FCC), Hydrocracking Unit, Coker, Naphtha Hydro Treater (HT), Reformer, Isomerization Unit, Alkylation Unit, Polymerization Unit, H2 Unit, Vapor Recovery Unit, Gas Plant, Amine Treater, and Shell Claus Off-Gas Treating (SCOT) Unit, major equipment typically includes pressure vessels, fractionation columns, storage tanks, heat exchangers, piping, valves, pumps, compressors, and turbines.

Sustainability Trade-offs: Moving Toward Desulfurization and Ethanol Blends

SK Energy invested more than $840 million in vacuum residue desulfurization (VRDS) facilities in Ulsan, South Korea, to produce 40,000 barrels of low-sulfur fuel oil per day in anticipation of an upsurge in demand from new IMO rules to reduce sulfur content in marine fuel to 0.5% (down from 3.5%).

China's Jinxi Petrochemical Corp, a refinery unit of state-owned PetroChina, loaded its first shipment of 3,000 tonnes of very low-sulfur (0.5%) fuel oil (VLSFO) in March 2020, for its marine bunker hub at Zhoushan in Zhejiang province to supply international vessels. One of PetroChina's nine designated producers of VLSFO, Jinxi will be able to make 700,000 tonnes a year of the cleaner shipping fuel. Petrochina is one of the first producers of VLSFO in China, where the government also announced a new tax regime to incentivize low-sulfur fuel production and sale.[495]

Bharat Petroleum, which is the national coordinator in India for ethanol blending by state-owned refiners, is working towards the new Indian government's 10% biofuel target by 2022 by adding more feedstock options. In order to improve its current 5% blending rate, it is widening its feedstock options to include rice in addition to corn.[496]

Supply Unavailability Risks and Mitigation: Managing Shutdowns and Restarts Through the Extended Supply Chain

Shutdowns are a normal part of the maintenance cycle, but they can also be precipitated by hurricanes and other natural disasters. While operations personnel often manage shutdowns at the refinery itself, production scheduling, maintenance, and related departments are responsible for effectuating the procedures at suppliers and coordinating the impact, if any, with customers. Comprehensive, safe, and practiced shutdown procedures are essential to economic operation in the case of planned shutdowns, and safety in the case of emergency shutdowns.

Outsourcing Risks and Mitigation: Hiring Safety and Process Auditors

While there are certain risks to outsourcing operations and maintenance activities, there are on the other hand certain activities that should be done only by third parties. For example, safety and efficiency audits should be conducted by outside experts at triggered milestones identified in standard operating procedures. Hazard Identification (HAZID) studies and Hazard and Operability Study (HAZOP) studies are examples of when external involvement can be particularly useful. Also, some consultants offer Refinery Operations Critiques that audit process efficiency in major refinery process units including alkylation, isomerization, delayed coking, FCC, and hydrocracking, and the optimization of the interactions between process units.

Supplier Partnering Risks and Mitigation: Negotiating with Retail Channel Partners

For many years downstream operations, and especially retail gas stations, earned lower profit margins than upstream activities, and they are hard to manage, requiring a different core competence than upstream or midstream operations, so oil & gas companies have spun them off, divested them, or formed partnerships to mitigate the undesirable financial impact of being present in the downstream business at all. The third option, partnering, is alluring in that it allows the major to leverage its brand, while offloading some of the risk and potentially less attractive financial performance.

Formulating the value proposition and identifying and negotiating with suitable downstream suppliers is essential to success of a partnering strategy. When partnering in downstream businesses, energy company management must develop a plan for joint funding of capital outlays and shared operating costs, discounts and rebates, branding and advertising, engineering procurement and construction of retail sites, and merchandising, including selection and procurement.[497]

Procurement Bundling Trade-offs: Consolidating Procurement Volumes

Integrated oil & gas companies may benefit from consolidating refinery tenders with upstream procurement bid packages. For capital projects this could mean using the same EPC firms and buying pipe from the same suppliers. For operational purchases it could mean consolidating chemical purchases such as catalysts with upstream chemical purchases such as oil & gas production chemicals. Although suppliers generally view the contracts as separate, it may be possible to use the volume and price leverage at the opportune moment in the respective contracting cycles.

Contract Term Risk and Mitigation: Contracting for Oversized Cargo Shipments

As with Upstream operations, transport of immense project cargo is characteristic of one type of logistics management challenge for Downstream. Here are some illustrative examples. Petrobras transported large pieces of equipment for the Abreu e Lima Refinery. Each piece weighed 119 tonnes and shipped from the Port of Houston.[498] Qatar Petroleum and Shell's Pearl Gas-to-Liquids (GTL) Project shipped one million tonnes of construction supplies to its joint venture with Transcar and Transoceanic. And last, but not least, for a single shipment, ConocoPhillips hauled two 300-ton coker drums from Kobe, Japan to its refinery in Billings, Montana (US). The logistics provider shipped the drums in two pieces each multimodally—first by ocean freight to Portland, Oregon, then by barge for 300 miles, then by 96-wheel tractor-trailer over 700 miles of back roads. The speed was limited to 35 mph, and the truck haul was restricted to night hours, due to local regulations. The planning took three years and resulted in a 700-page planning document.[499]

Trade-offs and Management Techniques in Operations & Maintenance Management

Internet of Things (IoT) and Artificial Intelligence (AI) Technology Choices: Optimizing Production and Predictive Maintenance Processes

Real-time optimization of input grades, quantities, timing, capture of intermediates, and recovery of energy can substantially improve profitability. This is often accomplished through a process control sequence. For example, a Supervisory Control and Data Acquisition (SCADA) system might instruct control valves, which change the volume picked up by flow meters and sensors, which send real-time information via transmitters to a control center, which interprets the data and a programmable logic controller (PLC) that initiates another information and control cycle.

What was previously automated in an analog sense is becoming smart in a digital sense. Most major refiners have fully implemented real-time advanced process automation systems, and many are integrating predictive maintenance and machine learning.

- Petrobras implemented AspenTech's process optimization solution at its REVAP refinery in an equation-oriented real-time optimization initiative. Running nine times per day, the solution optimized feed selection, which effectively increased heavy crude processing capacity.

Petrobras is evaluating moving to a multi-refinery version of real-time optimization.

- A major refiner optimized its fuel gas system to burn more efficiently and eliminate flaring. The solution involved a predictive controller embedded in Aspen DMCplus®, along with multi-refinery optimization and artificial intelligence. The project achieved significant energy savings while allowing the unit to maintain stable pressure and product quality.
- Braskem implemented aspenONE™ APC (Advanced Process Control) software to identify operational changes that led to real-time optimization, thereby increasing ethylene production and maximizing product mix and overall profit margin. The system helped to reduce energy costs and steam import rates.

Smart production management pays big dividends in refinery operations. A Gulf coast refinery realized $5 million in energy cost savings by using real-time process control that modulated the activation and speed of compressors, reactors, heat exchangers, and furnaces. The savings equated to 5% to 10% of energy reduction per operating unit. The deployment of variable speed drives (VSD) instead of the use of on/off switches or control valves helped reduce energy consumption up to 50% compared to across the line motor starters. In addition to energy conservation, built-in condition-based monitoring capabilities reduced downtime by 20% and the plant saved nearly $5 million (approximately 10% of the total maintenance budget) in maintenance costs by remotely monitoring large rotating equipment, valves, transformers, and motors via instruments, panels, and switchboards.[500]

Blending, mixing, and storage have historically been relatively manual activities, albeit automated, but these too are becoming more information intensive. Some suppliers have revamped their distribution networks and associated services to become and to remain competitive regional sources of logistics services with the help of information systems. For example, ExxonMobil implemented Invensys' InFusion-based system for inventory, order processing, packaging, and shipping operations, and has since expanded the application to other lubricant facilities.[501]

Cybersecurity Risk Management: Assessing Vulnerability Throughout the Plant

The wave of cyberattacks, including Dragonfly and Stuxnet, that has hit refineries shows that firewalls don't always provide adequate protection, especially if the attacker enters through back doors or acquires stolen credentials. Moreover, IoT devices such as connected valves and sensors connected to Wi-Fi networks, and handheld gas detectors (so-called Operational Technology, or OT) can make easy points of entry for hackers.

Even with all the technical precautions, one of the hardest-to-control aspects of cybersecurity is managing human behavior, especially involving field personnel and small businesses which are sometimes the least prepared to defend against a well-organized attacker. One attacker found a shortcut around corporate cyber protections by infecting a popular Chinese menu that was posted on an oil company's server.[502] Employees must be vigilant about secure management of their access credentials and wary about phishing attacks, USB use, and other day-to-day activities.

The American Petroleum Institute (API) offers a security risk assessment methodology (API 780) that contains tools such as risk registers and heat maps. In addition, software vendors such as VirusTotal, Dragon, Claroty, and aDolus offer cybersecurity solutions for downstream oil & gas environments that address some of these information technology (IT) and OT dangers.

Capacity Strategies: Single- and Multi-Echelon Production Planning

Refinery optimization problems are more complex than traditional warehousing or distribution networking optimization problems for discrete manufacturers. In refinery planning, there are many different processing routes to get to the same output, and each step of each route presents a different set of choices, so the modeling exercise requires extensive modeling preparation and computing power. In contrast, a network design problem for a discrete products manufacturer may involve many SKUs, but each SKU usually stays the same as it moves through the supply chain. Furthermore, in classic distribution warehousing problems, the demand for each product is independent, whereas in multi-refinery production planning, demand for most intermediate materials and products is dependent on the next processing step.

Production planners often divide planning and scheduling into three time frames, reflecting the degree of control over capacity at each time horizon. Strategic planning involves planning over the time frame when the capacity is fixed, which is usually several years. Master scheduling involves planning over the time frame during which demand is visible, which is often several months to a year or more. Production scheduling usually involves the time frame over which customers expect their orders to be delivered, which may be a few weeks. Whether for planning or execution, however, the problem is the same, so we treat all of the above requirements as "production planning."

Depending on the scale of the refinery, the number of grades and intermediate products, and various other factors, production planning may involve intensive operations research models that have the ability to consider a large number of variables simultaneously (such as grade, cost, profit margin, batch size, byproducts, intermediate processing, interplant transfer options, etc.)

Single Echelon Production Planning

Refinery production planning quickly gets complex if varying grades of output have different profit margins and involve different processes, some of which are continuous and some of which are batch (like blending and finishing). In these cases, single-echelon refining production planning consists mostly of optimizing production sequencing within the constraint of a fixed plant capacity.

Large plants plan production with linear programs that consider the different blends of crude that are scheduled to arrive at different times to maximize profit based on the characteristics of each crude (heavy, sour crude is more expensive to process), the profit margin of each blend, and the available capacity by unit.[503] Analysis can result in creative problem-solving and extraordinary productivity gains. One lubricants refinery operates above nameplate capacity by vacuum pumping (increasing volume throughput without degrading quality), reducing viscosity, and shifting dewaxing between two refineries to achieve operational improvements in hydrocracking productivity.

Still, dealing with planned variations (such as maintenance turnarounds and shutdowns to change product mix) and unplanned day-to-day exceptions complicates planning and scheduling.

Multi-Echelon Network Planning

In a multi-refinery configuration, the combinations of choices become so complex as to quickly require computational support and intelligence, especially if the refineries produce intermediates that can be alternately produced at one or the other facility, and even just accounting for the usual constraints of product quality, variable demand by finished product, and the cost of moving materials and finished product between facilities can be overwhelming. While crude oil is the primary input, intermediate outputs can include ethylene, benzene, chlorabenzene, and ethylbenzene, and final outputs can include products such as acrylonitrile, acetic acid, pvc, vinyl acetate, phenol, acetone, and ABS.[504 505]

Many refiners have adopted mixed integer linear programs to solve this application of a network design problem. For example, Unipetrol, a Czech refinery and petrochemical group, deployed the Advanced Process Control (APC) solution from aspenONE® APC to improve profitability and reduce fuel consumption at its ethylene production plant. The solution increased the effective capacity of existing assets and cut $2.6 million a year in energy costs. Shell's lubricants marketing division also adopted aspenONE® to optimize procurement of feedstock and maximize refinery profitability. The solution was integrated with a transportation and distribution network management application, which helped it respond more flexibly to market opportunities, and increase margin in a multi-site environment.

Multi-plant and multi-level refinery optimization require advanced modeling techniques. Since optimization is considered a trade secret, most companies don't publicize much about their algorithms or tools, especially if they are homegrown and have involved substantial investment. In one publication, Khalid Al-Qahtani from Saudi Aramco summarized the types of models that can be used to optimize networks of refineries: linear vs. nonlinear, deterministic vs. stochastic, static vs.

dynamic, and mechanistic vs. empirical (empirical models are input-output models). Neural networks are a type of model that adjusts to patterns and relationships, which can be useful for modeling with variables related to mass and heat transfer, fluid mechanics, thermodynamics, and kinetics.[506]

Sourcing Trade-offs: Procuring Site Security Systems & Solutions

Refineries spend large amounts of money on site security due to the ever-present risk of mal-intentioned intruders and the potential severity of security breaches. In this regard, there is almost always a trade-off between low-tech and high-tech security methods. Consider for example the question of whether to buy and set up intrusion barriers, or to invest in pattern recognition and facial recognition image analysis at entry points, or both. Similarly, whether to patrol perimeters with trucks and drivers, or to use automated and remote drone surveillance, or both. The costs always need to be traded off against the additional benefits.

Rather than reinvent the wheel each time such an investment or purchasing decision needs to be made, refinery management should adopt risk management frameworks and standards that offer a quantitative way of reaching such decisions. For the Society of Petroleum Engineers, this author presented a management checklist and concrete steps that managers should take and frameworks they should consider adopting, to ease and improve decision-making about such risk-return trade-offs. It includes steps about Governance & Organization that ensure organizational accountability for governance of risk management across the supply chain, international standards that cover safety and security, a comprehensive framework for supply chain risk management including metrics and measurement, and supply chain roll-out recommendations.[507]

TCO Trade-offs: Moving to Predictive Maintenance for Rotating Equipment

Process equipment maintenance can be done in four ways: reactive, planned, preventive, and predictive. It is generally acknowledged that as one progresses along that path from reactive to predictive maintenance, long-run costs will trend downward. Reactive maintenance saves money in the short run but can cause big headaches later when multiple systems fail badly. Planned maintenance, at fixed intervals, avoids the worst of those consequences, but may spend more than needed by maintaining and replacing equipment before it needs to be replaced. Preventive maintenance, which is conducted when equipment shows physical signs of wear and tear, reduces premature repair waste but does not eliminate it.

Predictive maintenance, by harnessing the value of analytics, simulations, and artificial intelligence, can predict potential upcoming failure with greater accuracy than the usual manual inspections that are done in most preventive maintenance, thereby minimizing maintenance costs as well as downtime. Predictive maintenance can help lower operating costs over the active life of rotating equipment. For pumps, the initial cost often represents 15% of TCO. Energy and maintenance costs are a much larger portion of total cost, but are not always considered, often for lack of time and lack of data. According to the IEA, 35% of lifetime costs are for energy, 22% are for maintenance, operating and installation costs are each about 10%, and environmental costs are 8%.[508]

HSE Considerations: Ensuring Supply Chain Safety During Shutdowns and Restarts

Safe work areas, safe environmental practices, and air quality monitoring for refinery shutdown and restart processes should be assured by detailed project management Gantt chart and by Responsible, Accountable, Consulted, Informed (RACI) matrices that clearly assign responsibilities for emergency and non-emergency shutdowns and restarts. The American Fuel & Petrochemical Manufacturers (AFPM) and other organizations provide networking and resources that could be useful.[509]

Supply Chain Roadmap for Downstream Oil & Gas

Supply chain improvements can often be realized in four Downstream areas: 1) More Robust Risk Governance; 2) Cybersecurity for IoT Devices and Systems; and 3) Smarter Predictive Maintenance; and 4) Low Sulfur Fuel. Risk governance and cybersecurity assure production reliability, so are *synchronization* strategies. "Smarter predictive maintenance" lowers operating costs, so is a *rationalization* strategy. Low sulfur fuel is a sales opportunity (hence an *innovation* supply chain strategy) since it is required by IMO standards; it is listed here because low-sulfur marine fuel is a fundamental supply chain concern for any business that ships product via oceangoing vessels, including the oil & gas midstream business.

Risk governance has attracted increasing attention especially since the 2010 BP blow-out in the Gulf of Mexico. The discipline is facing a new challenge now that Internet-connected devices and embedded software contain hidden vulnerabilities everywhere. Many new standards are emerging. Staff capabilities need to be augmented. Risk management needs to be embedded much more widely than before. Therefore, it is in this way a relatively new opportunity and obligation for most downstream oil & gas producers.

Cybersecurity for connected devices will be critical to enabling Downstream to be a reliable link in the oil & gas supply chain. The scope of this challenge is expanding rapidly as the number of connected devices increases to include handheld and wearable gas detectors, maintenance and inspection drones, and inline process instrumentation.

Predictive maintenance allows earlier visibility of worn equipment and potential failures, allowing more accurate detection and less costly replacement or repair options. The availability of sensors and optical instruments to record the precise movements, including vibrations, or rotating equipment greatly enhances maintenance and operating effectiveness.

Low-sulfur fuel is both an emerging supply chain requirement, for vessel operation, and a sales opportunity, for trading and bulk orders, for downstream producers.

Figure 72 illustrates these value-creating supply chain opportunities.

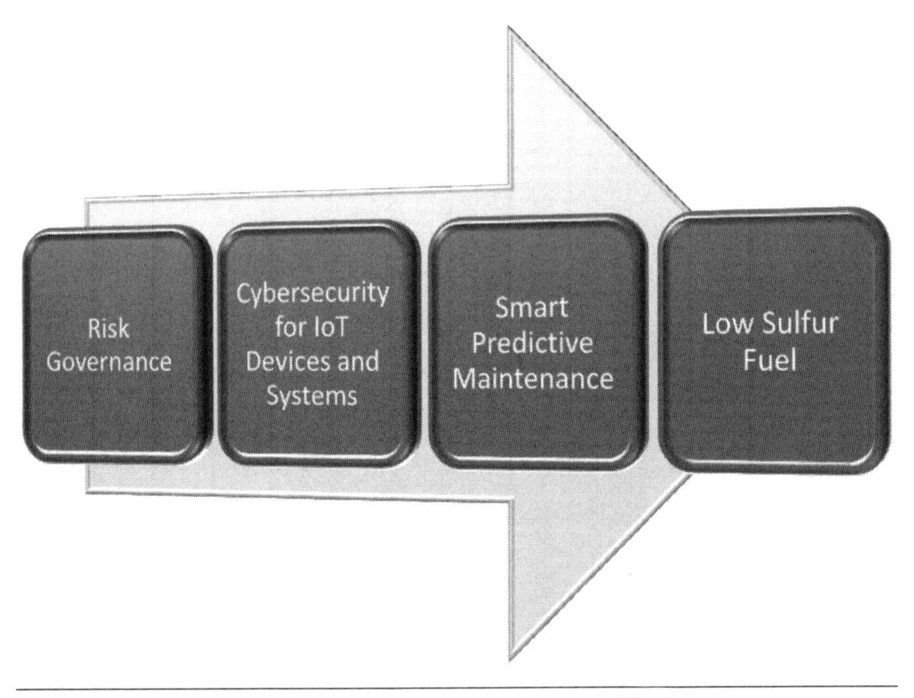

Figure 72. Emerging Supply Chain Opportunities in Downstream Oil & Gas

Geothermal

Chapter Highlights

1. Geothermal plants are most frequently economically viable where surface pressure makes the steam easily accessible, which especially tends to occur near tectonic plate junctions and in areas characterized by consistent volcanic eruptions and earthquakes.
2. Geothermal capital costs are relatively high on a per MW basis, while operating and supply chain costs are relatively low. This makes geothermal exploitations useful for very long-term PPAs, and also makes them interesting objects of study for synergies with other energy and thermal technologies involving heat and power, and energy storage, for example.
3. Upgrading to binary plants, which increases pressure, may in some cases be an attractive substitute for adding new flash plants.
4. AI can be an effective tool for increasing the effectiveness of predictive maintenance, which can squeeze out additional profit margins by ratcheting down operating costs.
5. Local content is often a consideration in geothermal plants since their topographical presence near coastlines and on islands frequently finds them situated in developing economies desirous of skills and knowledge transfer.
6. Geothermal power supply chains may be more cost-efficient and better aligned with demand by: 1) integrating hybrid energy technology formats such as district heating with Combined Heat and Power (CHP), 2) adopting energy storage, and 3) adding binary plants to existing plants instead of building new ones.

Introduction

Geothermal power utilizes the thermal energy that is stored deep within the earth's crust that vents out in the form of high temperature and high-pressure steam. This is used to turn a turbine to generate electricity. In order to proceed with the commissioning of a geothermal plant, the first crucial step is to conduct seismic studies and drill down into the earth, similar to oil & gas exploration and drilling. Drilling is a costly and a complex activity involving a wide range of equipment and downhole tools. The hot water, passing through heat exchanger in many cases, generates steam to turn a steam turbine, which produces electricity from the attached generator. The geothermal power plant also has a cooling tower to reduce steam exiting the turbine into water to be injected back into the ground. The power thus produced feeds the grid via a high-voltage transformer. Figure 67 presents a schematic representation of the geothermal value chain.

Although geothermal steam pressure can be tapped anywhere in the world with enough drilling, it is most efficiently harvested near the junctions of tectonic plates, and where there is active magma flow underground but close to the earth's surface. In the United States, California and Hawaii have substantial geothermal resources. The "Ring of Fire" is a perimeter of plate junctions surrounding the Pacific Ocean that is rich in geothermal energy. Not surprisingly, then, Southeast Asia produces a greater share of geothermal power than any other area, at 2% of total power production. Sub-Saharan Africa is next, at 1.2%, and all other world areas generate less than 1% of their electricity from geothermal resources. Of course, the proportion varies widely among individual countries and states. For example, Costa Rica produces about 15% of its power from geothermal reservoirs.

Where steam is readily available at the surface, "flash" plants separate steam coming from "hot dry rocks" from the associated liquid, then convert the steam to power through a steam turbine and electrical generator. For lower-pressure situations, "binary" plants also include heat exchangers which produce cooled water for reinjection below ground to accelerate the production of more steam. The equipment involves cyclonic separators, condensers, steam turbines, and piping, as well as cooling towers (for Flash plants) and gas accumulation systems.[510] The illustrative value chain in Figure 73 shows upstreamexploration, drilling, and completion, a flash plant, a steam turbine, an electrical generator, cooling towers, transformers, substations and local power distribution.

Geothermal power plant capital costs tend to be high when measured on a per MW basis—about $2–$7 million per MW.[511] Preparation and drilling is by far the largest capital expenditure line item, making up 54% of the capital. The turbine-generator and auxiliary systems represent 13%, EPC fees 11%, the steam supply 10%, buildings and ancillary systems 7%, roads and local infrastructure 3%, and electrical systems 2%.[512]

Geothermal power plant *supply chain* costs, on the other hand, are relatively small—only about two cents per kwh, about the same as for solar power. Nearly two-thirds of the supply chain cost (63%) is for engineering services such as "exploration and resource assessment costs, drilling costs for production and re-injection costs, as well as additional working capital given that the success rate for well could vary between 60–90%, field infrastructure, the geothermal fluid collection

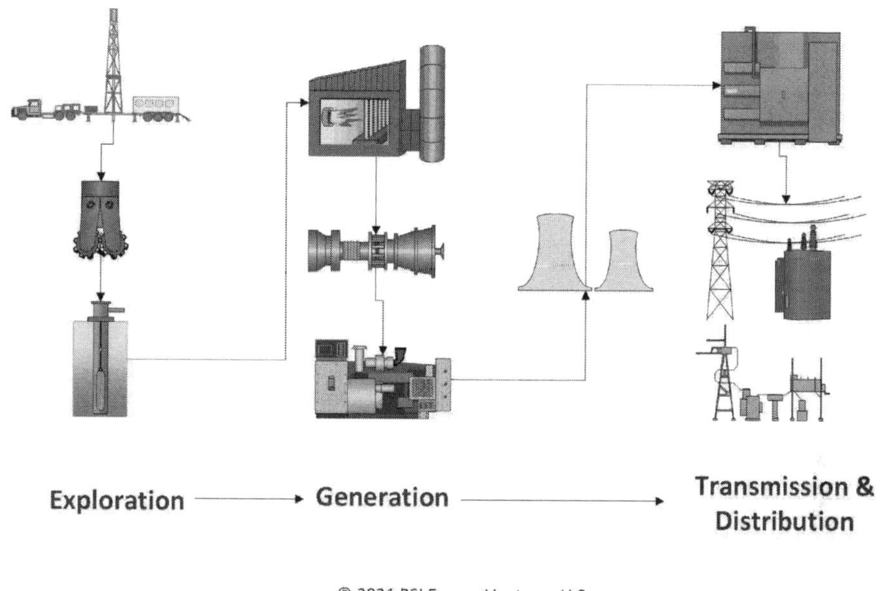

Figure 73. Geothermal Power Value Chain (Illustrative)
(*Source:* BSI Energy Ventures)

and disposal system, and other surface installations; and project development and grid connection costs."[513] The balance (37%) of the Total Invested Cost (TIC) is for equipment.

Trade-offs and Management Techniques in Capital Project Management

Technology Choice Risk Management: Deciding on Binary, Flash, or Both

Adding binary plants to flash geothermal operations may be an acceptable substitute for building new plants in some cases. As an example, the US Energy Development Corporation (EDC), a privately held oil & gas exploration and production company, wanted to expand its generating capacity in the Philippines, but was limited because most of the remaining expansion areas for geothermal energy were inside protected areas.

To avoid project development in the protected quarters and keep the investments low, EDC added binary plants in each of its existing geothermal plants to boost power-generation capacity without needing to build new power plants outside existing geothermal concessions.[514] The new binary plants are low-capacity plants with 5 to 20 MW capacity and easy to set up compared to big-capacity conventional power plants with a capacity of 100 MW or more. These binary plants maximize steam from the existing reservoirs, which obviates the need to build plants.

In doing so, EDC will be pioneering the re-injection technology, a heat recovery process wherein excess heat from wet steam will be harvested and re-injected in the reservoir. Over time, the excess heat adds to the sustainability of power-generation capacity of the plants. Steam in the Philippines is wet steam. Before it gets to the turbine, the water, called brine, is usually thrown away. Here, EDC will harvest it and reinject it into reservoir and to produce more heat.

Capital Project Risks and Mitigation: Lowering Financial Exposure Through Feed-In Tariffs and PPAs

Geothermal projects have a lower project risk profile than most other renewable capital projects such as solar, wind, and nuclear power. Nuclear projects have very smooth and predictable returns once they are up and running, but they have a notoriously high risk of running far over budget during the construction phase. Solar and wind projects have more predictable initial capital costs, but are susceptible to extended periods of fewer-than-expected sunny days and lower-than-expected wind speeds. Topographical weather patterns just can't compete with the reliability of steam rising from the Earth's crust. As a result, project owners spend relatively more of their project setup and management time working out feed-in tariffs, PPAs, and maintenance cost reduction schemes than worrying about revenue volatility.

Supply Unavailability Risks and Mitigation: Avoiding Steel Supply Bottlenecks

Steel plays an important role in the construction of a geothermal plant. The condenser, for example, uses stainless steel in the shell and hotwell plates, and in other internal parts including nozzles. Even the brine, condensate, or steam pipelines are made of carbon steel.[515] Consequently, unavailability or delays in procuring steel and the long lead time equipment that is made of steel can delay project schedules and commissioning dates. Unfortunately, steel shortages happen with periodic regularity. For example, in 2020, a significant increase in China's demand for steel led to a spike in India's exports, resulting in shortages of steel, which were exacerbated by an iron ore shortage due to large lease ownership transitions.[516] Also, floods in Australian coal mines in 2011 impacted steel production and supply globally.[517]

The Olkaria II geothermal project exemplifies how the Kenya Electricity Generating Company Limited (KenGen) conducts its steel procurement to minimize any kind of potential supply risk. Firstly, it uses local procurement for steel as much as possible. Carbon steel for piping and supports were mostly sourced through Kenyan firms, resulting in cost and schedule savings. Where local availability could have been a challenge, items such as specific grades of stainless steel, certain structural components, and pipe support specialty components were identified with enough lead time incorporated in project schedules to procure from overseas suppliers.[518]

Outsourcing Risks and Mitigation: Avoiding Drilling Contractor Delays

Regardless of the power plant technology, outsourcing can present risks. In nuclear, the focus of risks presented in this book are primarily safety oriented. In fuel cells, the focus of risks discussed in that chapter are intellectual property oriented. In geothermal power, outsourcing risks are performance oriented. At least that was the experience of Daldrup & Söhne, a German geothermal specialist, when it could not spare a resource to cater to a scheduled drilling assignment in Belgium. It made a decision to outsource the drilling operation, which unfortunately led to delays and downtime costs for drilling rigs and power plants with corresponding negative financial results.[519]

Supplier Partnering Risks and Mitigation: Realizing Synergies Across Geothermal and Other Power Projects

Since some of the equipment used for geothermal power generation is the same as that used for combined cycle (and other fossil-fuel fired power plant types) and hydropower, such as steam turbines and generators, Turkey's Zorlu Energy Group signed a package deal with Toshiba that extended the terms of its thermal and hydro power plants to its new geothermal plants.[520] Under the terms of the agreement, Zorlu Energy Group engineered steam flow and pressure levels in three locations where Zorlu Energy Group planned to construct plants, and, pending successful exploration, assigned Toshiba the first right of refusal for the supply of steam turbines and generators for all three geothermal plants.[521] The agreement increased both parties' confidence in supply assurance, and helped them achieve creative solutions that might not have been possible with a new non-incumbent supplier.

Procurement Bundling Trade-offs: Leveraging EPC Expertise from OEMs

Geothermal projects are opportunities to hire EPC services from Original Equipment Manufacturer (OEM) supply contracts because the suppliers know their equipment, are highly qualified in applications engineering, and frequently have experience implementing power plant solutions around the world. In one case, Turboden, a subsidiary of Mitsubishi Heavy Industries, served as the engineering, procurement and construction (EPC) contractor for its 9MW binary cycle geothermal project in Nevis, *and* the supplier of high efficiency Organic Rankine Cycle ORC (a technology that convert slow-temperature heat sources into a mechanical energy) turbines and binary-cycle generation equipment. The Nevis project is developed by Nevis Renewable Energy International (NREI, which is an affiliate of Texas based Thermal Energy Partners LLC.[522]

Contract Term Risk and Mitigation: Planning for Generations

As previously mentioned, geothermal projects have predictable long "tails" of power output and revenue. Italy has a long tradition of geothermal energy utilization, with nearly 1,103 MWe installed in two areas of the Tuscany region (Larderello/Travale and Monte Amiata) operated by Enel Green Power. Some plants in the Larderello/Travale region of Italy totaling 700 MWe have been in industrial operation for more than 60 years.[523] Therefore, they provide an unusually strong basis for longer-term power purchase agreements than might be expected on other power technology projects. For example, TransAlta Corporation's PPA for its 50 MW renewable geothermal power plant for Salt River Project (SRP), an Arizona utility, was 24 years long,[524] and it concluded another purchase contract for 86 MW for the similar duration.[525]

Trade-offs and Management Techniques in Operations & Maintenance

Cybersecurity Risk Management: Ensuring Security of Automation Controllers

Even though geothermal power generation does not use exotic raw materials like fuel cells do, or burn radioactive fuel like nuclear plants do, susceptibility to cyberattacks could still cause power outages or shut down power from the plant to the grid.

Gürmat Elektrik, which operates the biggest geothermal power plant in Turkey, recognized that its existing data and networking infrastructure was incapable of delivering contemporary levels of performance and lacked robust cybersecurity. According to Rockwell Automation, Gürmat had an unsecured, unpatched network, coupled to obsolete controllers, servers, and software with very little access control. The company was seeing issues with a SCADA system that was slow; software was outdated; and the mean time to repair (MTTR) for any network-related breakdown was high. And there was a large amount of legacy hardware ("legacy" is a term used to describe old software or hardware that's outdated but still in use). All of these factors made the entire system highly vulnerable to threats and failures.[526]

Rockwell Automation helped Gürmat Elektrik migrate the existing infrastructure over to an entirely new Connected Enterprise network, which integrates unique, isolated standalone systems into one standardized, higher-reliability solution. This solution not only offered the required security, but also delivered detailed information and diagnostics to avoid downtime, while giving Gürmat the ability to collect, collate, and deliver the information in a format that allowed it to make real-time operational decisions.

At the controller level, Rockwell replaced older Allen Bradley CompactLogix programmable automation controllers (PAC), which were located remotely at the end of kilometre-long fiber lines, with newer models, while the larger Allen Bradley ControlLogix controllers were enhanced with redundancy and up-to-date network cards, including Modbus+ cards (which are used to collect data from third-party systems, including controllers, vibration monitors and motor control centers). At the server level, an ftServer model 4800 was deployed in conjunction with FactoryTalk View SCADA and a more up-to-date version of FactoryTalk Historian, which has resulted in performance increases and continuous availability of real-time data.

Internet of Things (IoT) and Artificial Intelligence (AI) Technology Choices: Predicting Maintenance Requirements to Reduce Repairs

Predictive maintenance applies to geothermal as it does to hydropower and gas-fired plants, and where there is predictive maintenance there is artificial intelligence (AI).

Toshiba Energy Systems & Solutions (ESS) conducted a demonstration project of anomaly-predictive diagnosis to improve geothermal power plant utilization factor using Internet of Things (IoT) and artificial intelligence (AI) technologies at the Patuha geothermal power plant of PT Geo Dipa Energi in Indonesia. The project, funded by NEDO (New Energy and Industrial Technology Development Organization of Japan) since 2018, was conducted on the 60 MW Patuha Geothermal Power Plant in West Java, Indonesia, owned by PT Geo Dipa Energi (Persero) (GDE), Indonesia's state-owned geothermal energy company. The application analyzes the plant's past and present operating data using big data analytics, identifies equipment anomalies, and predicts impending problems. This project is expected to achieve a 20% reduction in the rate of problem occurrences at geothermal power plants, thereby making geothermal more attractive and increasing adoption relative to other power plant technologies.[527]

Peak Capacity Strategies: Adding Energy Storage and District Heating to Geothermal Power Generation

Although geothermal plants are capital-intensive, they are relatively inexpensive to operate.

Construction costs include well-drilling, pipeline construction, resource analysis, and power plant design. Capital investment can be split into three phases: exploration and drilling, construction of power facilities, and discounted future re-drilling/well stimulation. The capital cost for a geothermal power plant is about $2,500 per kilowatt (kW).

Operating costs range from $0.01 to $0.03 per kilowatt-hour, and plants can operate at 90% availability or more. Maintenance costs increase if a geothermal power plant operates at greater than 90% availability. This high price is justified because the maintenance costs are covered by the increased production of the plant.

Theoretically, then, geothermal plants would be good candidates for hybrid energy technology formats, such as geothermal + solar PV supplementation, or geothermal + district heating in cold climates, or geothermal + long-term energy storage, such as hydrogen storage in salt caverns or compressed air energy storage. That way they could be built to lower design capacities (at lower cost) and still satisfy demand peaks (by drawing on storage).

The National Renewable Energy Laboratory (NREL) is funding an ongoing research program aimed at developing hybrid energy technology formats based on geothermal energy that also improve residential building efficiency.[528]

Sourcing Trade-offs: Building Local Skills Alongside Geothermal Power Capacity

As the most robust geothermal resources are located at continental plate junctions, many excellent geothermal reservoirs exist on islands and in less developed coastal areas. Hence, procurement needs to take local content into consideration when building plants.

Malawi lies on the southern end of the western branch of the East African Rift System. As one of the poorest countries in Africa, the country faces major electricity generating difficulties due to limited conventional fossil fuel resources and over-reliance on hydroelectric power, with severe shortages during the dry season. As the country lies to the west of the Great Rift Valley, it has rich geothermal potential, which has not been exploited. Adding geothermal to the energy mix will contribute to the stable supply of clean energy and in Malawi's development.

The partnership of Malawi's Ministry of Natural Resources, Energy and Mining (MNREM) with Toshiba Energy Systems & Solutions Corporation includes capacity-building as a key element of its technology deployment.[529] The contract includes the engineering and manufacture of major equipment, operation and management guidelines, and capacity building programs.

TCO Trade-offs: Reducing Lifetime Cost in the Capital Construction Phase

As previously mentioned, the investment cost structure of geothermal power plants is heavily weighted toward upfront construction costs; once a plant is built, operational availability often exceeds 90%. The higher the operational availability, the lower operations and maintenance costs can go, and the more profit can be squeezed from the investment—up to a point. Above a certain production level, operations and maintenance costs begin to increase. Initial equipment engineering and purchase decisions, as well as ongoing operating levels, need to be based not on the initial acquisition cost, nor the operating cost, but the net present value of the sum of both.[530]

Supply Chain Roadmap for Geothermal Power

Geothermal power supply chains can become more cost-efficient (Rationalized) by integrating hybrid energy technology formats such as district heating Combined Heat and Power (CHP) and adding binary plants to existing plants instead of building new ones. Furthermore, supply chains can be better aligned with demand (Synchronized) by adding energy storage. These three paths to supply chain value are shown in Figure 74.

Because of their high uptime rates, geothermal plants complement more intermittent power generation technologies such solar PV and secondary applications such as district heating in cold climates. The combination of technologies and applications may increase the return on investment in geothermal installations.

Retrofitting low-capacity binary geothermal plants to existing plants is generally more profitable than building greenfield conventional power plants with a capacity of 100 MW or more. This may in some cases be economically attractive.

Since geothermal plants are generally available more than 90% of the time while demand nearly always fluctuates, energy storage can buffer the gap between supply and demand, potentially increasing revenue and decreasing costs.

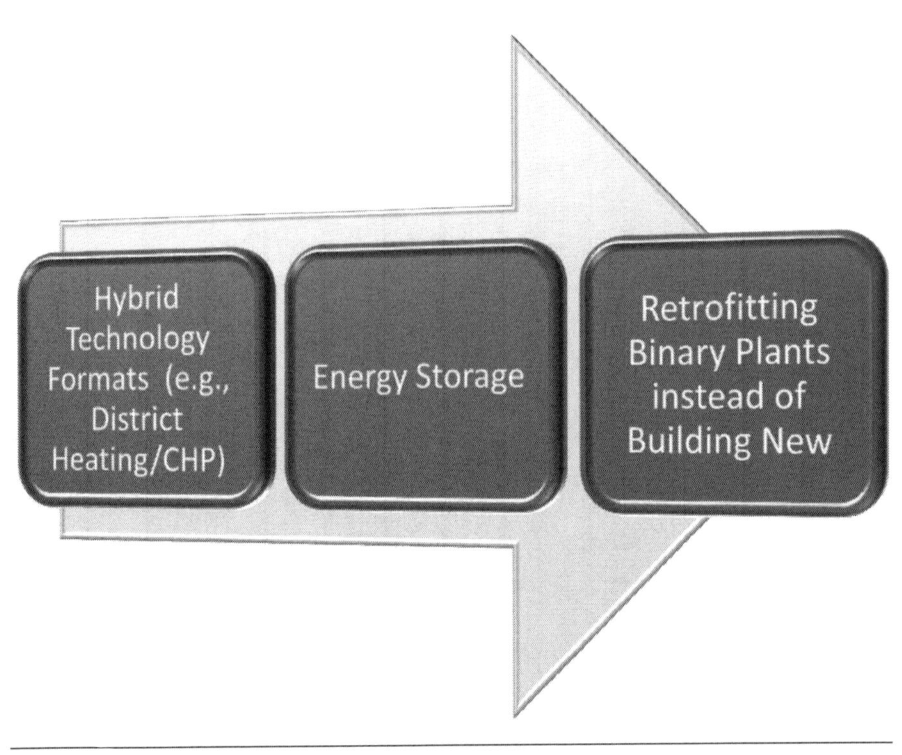

Figure 74. Emerging Supply Chain Opportunities for Geothermal Power

Gas- and Coal-Fired Power

Chapter Highlights

1. Gas- and coal-fired power plants typically involve large capital investments and many interrelated decisions pertaining to critical supply agreements for engineering, procurement construction, commissioning, process control systems, operations, maintenance, and repair.
2. Scenario analysis and real options are decision analysis tools that can help navigate the interrelationships and arrive at optimal long-term decisions.
3. Construction contract management is critical in managing supply chain costs. Liquidated damages, bid bonds, performance bonds, and payment bonds clauses can have significant financial implications.
4. For brownfield coal or natural gas–burning power plants, it may be possible to add renewable co-production to reduce CO_2 and particulate emissions. While such conversions do not always make economic sense, they can be beneficial in certain cases.
5. At the construction planning phase, long lead times for gas and steam turbine generators can bottleneck the plant delivery schedule.
6. When considering strategic supplier partnerships, suppliers' capacity and ability to sustain activity for a long period of time, typically 20 to 25 years, should be evaluated, for thermal power plants based on mature technologies like gas- and coal-fueled combined cycle operations that are usually built for long useful lives.
7. The North American Electric Reliability Company (NERC) provides useful and easy-to-follow cybersecurity standards and guidance.
8. Predictive maintenance improvements offered by sensors, relays, and connected devices, as well as machine learning (artificial intelligence) can optimize plant performance.
9. Combined purchase and operating agreements can take advantage of opportunities to achieve lowest total cost over the lifetime of major equipment.

10. Standardized equipment can help reduce maintenance contracts and service parts inventory management costs.
11. A safety culture and periodic safety audits help to identify and correct practices that could otherwise lead to incidents or accidents.
12. High-potential supply chain improvements can come from: 1) Predictive Maintenance; 2) Performance-Based Agreements; 3) Reliability Engineering / Life-Cycle Cost Management; 4) Cybersecurity; 5) Co-Production; and 6) Alternative Uses for Coal.

Introduction—Supply Chain Cost Drivers and Design Constructs

The value chain for gas- and coal-fired power begins with bulk inbound shipments of coal or natural gas for fuel. The fuel drives gas turbines, or it fires boilers which generate steam that drives steam turbines, or both in the case of combined cycle plants. Generators convert the mechanical output from the turbines into electricity. Switchgear, transformers, capacitors, and other electrical distribution and control equipment relay the electricity from the generator across transmission lines to substations and on to individual customers. Figure 75 represents this value chain, from fuel procurement through power distribution.

The complex interrelationships and trade-offs involved in setting up a large power generating capability pose a conundrum for those in charge of major projects or programs. Supplier partnering strategies, technology choices, regulatory constraints, environmental objectives, and outsourcing decisions each involve uncertain costs and high risks. Moreover, these factors are mostly interdependent (e.g., changing the technology might change the outsourcing strategy, which could change the supplier partnering strategy), presenting a dizzying array of possible supply chain configurations for a major project. Figure 76 shows a mindmap of these interrelated factors.

Financial and contractual exposure in large power projects is large. Decisions must often be made before a final investment decision due to long lead times for capital equipment. Formal government consent frequently does not occur until the project is long underway.

Financial options can be created without physical assets, too. The first way is simply as an analytical construct—"real options analysis" can function like an NPV analysis. The Tennessee Valley Authority used real options analysis to decide on power purchase options, and Enron used real options analysis to make new product development decisions, switching options for gas-fired turbines.[531]

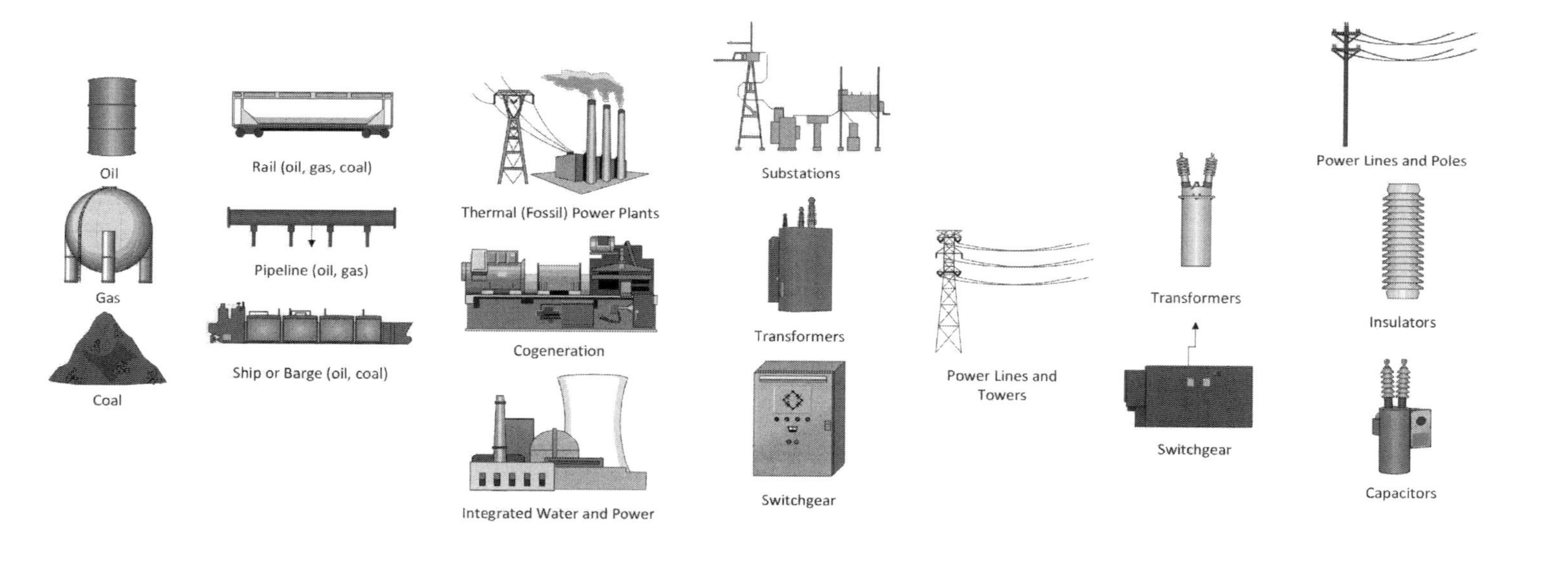

Figure 75. Gas- and Coal-Fired Power Value Chain
(*Source:* BSI Energy Ventures)

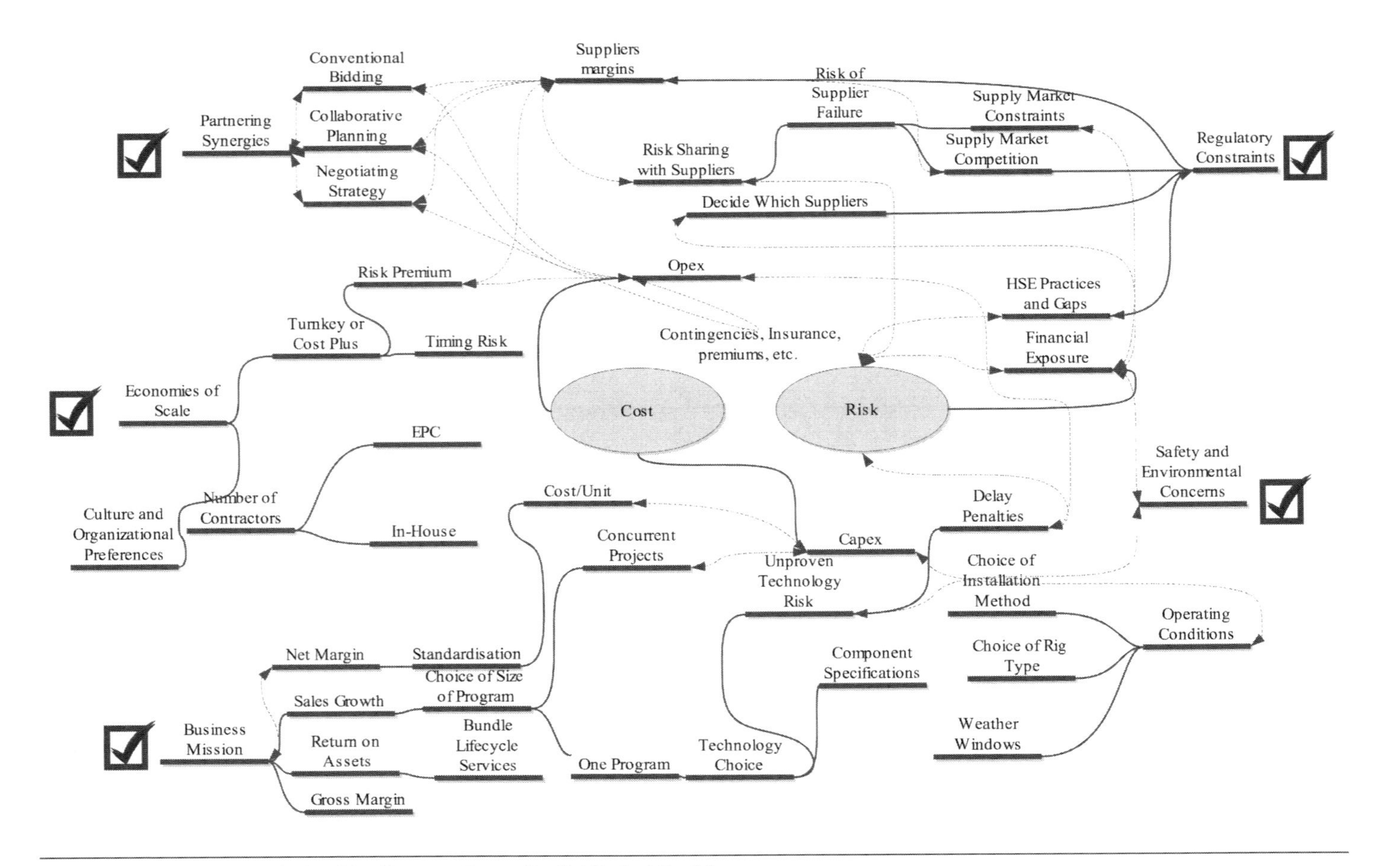

Figure 76. Complex Interrelationships Between Power Capital Project Decisions
(*Source:* Boston Strategies International)

Trade-offs and Management Techniques in Capital Project Management

Technology Choice Risk Management: Determining the Optimal Power Portfolio with Simulation Tools

The cost of adopting the wrong technology in the power industry is large, in terms of opportunity cost due to high capital requirements, potential mismanagement during the construction phase (which can result in rework, penalties for not meeting delivery schedules, or poor operating efficiency on commissioning), and operating errors that can damage equipment or endanger personnel.

- The upfront investment required to build a power station sometimes starts in the billions of dollars. Making the wrong bet on technology, scale, or cost vs. quality trade-offs can represent a major long-term headache.
- Generating capacity typically comes in large chunks, with major gaps between increments. For example, US utility Duke Energy once faced the choice of whether to build two 800 MW coal-fired plants or one, for a total project cost of either $1.5b or $3b, respectively.[532]
- The wrong choice of specifications can reduce the plant's operating efficiency for its entire lifespan, increasing operating or maintenance costs.
- The need for peak power further accentuates the investment and project management risk, magnifying any errors in investment decisions, especially if the plant operates at a low-capacity utilization.

The most common approach is to throw data at the problem, resulting in a wide variation of complex results that are classified into "high-medium-low" scenarios focused on a single indicator such as NPV or LCOE, which do not quantify risk.

The best way to analyze risk, and therefore evaluate options for minimizing it, is to develop scenarios to quantify the risks. One operator developed scenarios that each projected the cumulative effect over 30 years of standardization benefits, technology benefits, economies of scale, competition benefits, learning curve benefits, and project management risks. The company compiled eight scenarios involving the aforementioned parameters that varied according to the number of suppliers, the type of technology, and the use (or not) of an EPC firm. It then valued each scenario and identified the lowest- cost option for the medium term, and the lowest-cost option for the long term (the lowest-cost option in the medium term was different from the lowest-cost option in the long term).

Sometimes statistical and probabilistic tools, such as Monte Carlo simulation, are used to assess risks. However, there are several problems that limit the effectiveness of such analyses. First, they rely on critical assumptions that are often arbitrary, for example the shape of the probability distribution (normal, binomial, geometric, etc.) and the coefficients of likelihood of certain events (such as the probability of supplier bankruptcy or technological failure).

Once risks are analyzed and quantified, project owners need to make a decision on how to best mitigate the risks. Offloading risk is becoming an insufficient strategy, as EPC firms are no longer accepting the risk (or are charging high premiums to accept it).

Sequential decision-making is a relatively simple and still robust way to structure and manage risk. A sequential decision-making framework involves decisions along five dimensions: time horizon for supplier commitments ("Term"), degree of bundling and solution buying ("Bundling"), degree of supplier intimacy ("Intimacy"), mode of relationship ("Rapport"), and rules of competition ("Tendering"). For each dimension there is a matrix of options that highlights the degree of risk and risk-sharing that accompanies each option. For example, a flowchart helps to select the optimal form of project management, and a matrix helps to assess how much risk is involved in engaging one, two, and three suppliers for the major categories of expenditure.

A related, risk management approach is to use "real options" by building physical assets that can be expanded. One way to keep as many options open as possible is to overbuild just enough to provide a platform for future expansion at minimal incremental cost. One example would be building a multipurpose installation vessel that costs more than a single-purpose one in order to have the capacity for any type of installation, even though only one use is currently anticipated (for example, cable laying). An interesting but infrequently used strategy for acquiring options is to purchase equity in a supplier in exchange for a formal option to supply in the future. UK utility SSE took a 15% stake in BiFab in exchange for capacity up to 130 jackets per year plus an option on 50 more foundations per year for 10–12 years.[533]

Capital Project Risks and Mitigation: Using Bid Bonds, Performance Bonds, and Payment Bonds in Construction Contracts

From a supply chain management point of view, savvy construction contract management is essential to mitigating financial and operational exposure.

Terms and conditions of the contract can align (or drive apart) the interests of the owner and the construction contractor. Liquidated damages clauses for delayed completion are often subject to negotiation; a cap of 20% of contract value may serve as a benchmark. The need for extra work and delays are both subject to negotiation, depending on the situation and the possible causes of project extension, expansion, or delay. Certain other clauses tend to be more standard, including for example indemnification, liens for obligations not yet fulfilled, and termination for cause or convenience.[534]

Bid bonds, performance bonds, and payment bonds assure contractor fidelity and integrity. The bid bond ensures that the contractor performs the work as bid if awarded the contract. Performance bonds ensure that the contractor fulfills obligations once engaged. Payment bonds make sure the contractor pays employees and subcontractors. Insurance coverage and insurance pass-throughs such as naming additional insured parties on the contractor's policy deserve careful attention to prevent ambiguity or lack of coverage in the event of incidents or accidents.

Sustainability Trade-offs: Evaluating Fuel-Switching, Co-Firing, and Hybrid Fuels

The term fuel-switching is usually used to refer to a long-term conversion of the fleet of coal-fired plants to natural gas-fired plants, not by retrofits but by retirement of the old coal-fired plants and the construction of new replacement plants fired by natural gas instead of coal. Retrofits to burn completely different fuels are usually cost-prohibitive, and it is assumed here that combustion efficiency and waste heat recovery options have been maximized.

This being said, even when a power plant is designed for coal or gas, there are sometimes options to add renewable co-production to coal- or natural gas–burning power plants to reduce CO_2 and particulate emissions. While the primary and fundamental analysis would rest with engineering, managers from other departments pertaining to supply chain management may be called upon to evaluate the logistical and economic implications of such options. Examples of this include:

- Adding biowaste to the natural gas–fired boiler process (co-firing), either by directly adding biomass to the boiler operation of a gas-fired plant, which is the most common method, or by installing a biomass gasifier that converts the biomass into a fuel gas, which is burned in the coal boiler furnace, which has been done in the in the Zeltweg plant in Austria, the Lahti plant in Finland and the AMERGAS project in the Netherlands. There is a much less common third approach, which is to install a completely independent biomass boiler and use its steam output to run a steam turbine, which has been done in the Avedøre Unit 2 Project in Denmark.[535]
- Using hybrid power generating technologies. A hybrid coal-solar power plant is already in operation in the United States, at the Xcel Cameo Generating Station in Colorado. It essentially uses CSP to preheat water entering the boiler, to reduce the amount of coal required by the boiler to produce the same amount of electricity.[536]

At the national level, some governments are funding a transition from coal and oil to cleaner derivatives of the same. For example, the US Department of Energy (DOE) is funding research and development of alternate uses for coal that is currently mined in 12 coal basins around the United States. The hope is that over time, utilities and other end users will adapt to the new fuels, and the local economies that had thrived on mining the old coal and oil will produce newer, cleaner, high-tech energy products from the same resource base as before.[537]

Supply Unavailability Risks and Mitigation: Estimating Delivery Time of Turbines and Generators

Turbine generators (gas or steam) typically have the longest lead time and are the most expensive single part of a power plant. However, ancillary systems from material handling to emissions control also affect operations and maintenance on an ongoing basis, for example: coal receipt and preparation, coal combustion and steam generation, environmental protection, the condenser and feedwater system, and heat rejection (including the cooling tower).[538]

Outsourcing Risks and Mitigation: Qualifying Suppliers for Major and Minor Component Contracts

The risks of outsourcing can be significantly mitigated by comprehensive supplier qualification procedures. The usual ratings and criteria apply, as discussed in the CapEx principles and practices chapter. However, to identify suppliers with long-term potential that will ensure that they will be able to support a 20–40 year project (or longer), an evaluation of the success factors for the current major players may be helpful. Top suppliers of electrical equipment such as Siemens, ABB, Schneider, and GE have relied on their long historical track records to secure a preponderance of supply of major equipment for electric utilities.[539]

Supplier Partnering Risks and Mitigation: Communicating with Leading Equipment Suppliers as Partners

Treating vendors as partners is one of the best forms of supply risk mitigation. Speaking of the importance of collaborative procurement, Peter Hessler, author of *Power Plant Construction Management*, explains that "although most company procedures require the typical vendor selection process to follow a 'three quote and select the lowest bidder' scenario, that is exactly what often drives the relationship to be adversarial. An alternate approach is to use the bidding process only for identifying and prequalifying the suppliers. Then, the next step would be geared to maximizing value creation, as opposed to reducing costs by squeezing suppliers' margins and scope. However, entering into a search for mutual value creation requires an understanding of each party's objectives and finding ways to achieve fair resolutions to common issues."[540]

Partnerships can strengthen performance reliability and lower cost at fossil fuel–fired power plants and the supply chain for its main production units. China Light & Power considers its suppliers to be an integral part of its business, so it seeks mutual benefits and shares its vision and goals with them as key business partners. The company adopted a risk-based supplier assessment system, which provides systematic performance feedback, to evaluate its major contractors and suppliers.[541] Mitsubishi Heavy Industries (MHI) formed a joint venture with Hangzhou Steam Turbine & Power Group Company, in which MHI benefited from increased sales opportunities in China and lower labor costs, and Chinese buyers benefited from shorter lead times for high-tech equipment. MHI exported key components from Japan, to ensure quality and protect copyright.[542]

Procurement Bundling Trade-offs: Deciding When to Use an EPCM Contractor

For large or complex projects, owners of developers sometimes organize to specifically oversee Engineering, Procurement, and Construction (EPC) operations. In these cases, the project management firm is called an EPCM (Engineering, Procurement, and Construction Management) company, EPCM consultant, or Program Management Office (PMO). For example, Black & Veatch (B&V), one of the largest power industry EPC firms, served as the technical manager for Glow Energy Public Company Ltd. in the construction of a combined cycle cogeneration plant near Rayong, Thailand. Marubeni Corporation of Japan led a consortium of partners to execute the actual EPC work at B&V's direction.[543]

Unbundling installation from the construction may reduce cost, but this places the burden of commissioning and the liability for faulty integration upon the buyer, and typically the owner is potentially responsible for cost overruns, not the EPC, in contrast to how a lump-sum turnkey EPC project would work. Financial and supply chain managers must weigh the costs and benefits of the alternative ways to structure the EPC relationship on a case-by-case basis depending on the characteristics of the project at hand.

Contract Term Risk and Mitigation: Extending the Length of Long-Term Contracts

Because gas- and coal-fired power plant technologies and contractual norms are so mature, proven, and bankable, plant owners often sign very long-term Operation & Maintenance (O&M) and service agreements (Long-Term Service Agreements, or LTSAs). For example, in 2018, Mexico's Federal Electricity Commission (CFE) signed a 25-year Operation and Maintenance (O&M) agreement and a Contractual Service Agreement (CSA) with GE Power Services. Together, the contracts will provide plant original equipment manufacturer (OEM) services and plant operation and maintenance for the 907 MW Norte III power plant, in the municipality of Ciudad Juárez, in the state of Chihuahua.[544]

Trade-offs and Management Techniques in Operations & Maintenance

Like any other process manufacturing operation, production management and maintenance go largely unnoticed if all goes well, but face intense scrutiny if systems fail or if there are accidents.

Cybersecurity Risk Management: Adhering to and Benefiting from Power Industry Cybersecurity Standards

The North American Electric Reliability Company (NERC), an industry oversight group connected to the Federal Energy Regulatory Commission (FERC), has developed security standards that can mitigate cyber risks, on topics such as provenance, cloud computing, open-source software, software development risk management, vendor risk management, and secure equipment delivery, and ensuring patch integrity. The standards as well as the materials produced by the working committees are excellent resources that all operators should take advantage of.[545]

Internet of Things (IoT) and Artificial Intelligence (AI) Technology Choices: Reliability Engineering and Predictive Maintenance

Reliability Engineering

Reliability engineering is essential to keep uptime high. Involvement of personnel keeps a focus on root cause problem solving. A time study of maintenance workers can identify the percentage of active work being performed, in order to see if the right amount of analysis is being done relative to hands-on work. Such studies break observations of work into categories such as: Not working, Operating, Repairing, Maintaining (preventive), and Maintaining (predictive).

Good root cause analysis will inevitably result in a pareto chart of the root causes of failure. In one case, as an example, the principal causes were identified, in order of occurrence, as boiler repair, turbine repair, generator, feedwater, controls, ash water, baghouse, blowing air, and circulating water.[546]

Life-cycle cost management is critical to optimizing the cost of rotating equipment. Unfortunately, life-cycle cost analysis is complex when dealing with turbine systems, and many utilities still struggle with data analysis. One turbine supplier has developed an economic valuation model that calculates the operating costs for two comparable units to make apparent how much one would cost versus the other under a multitude of specific operating conditions, applications, and parameters. It takes into account, for example, hours at base load, number of starts, and operating temperature. It computes an output that shows the amount

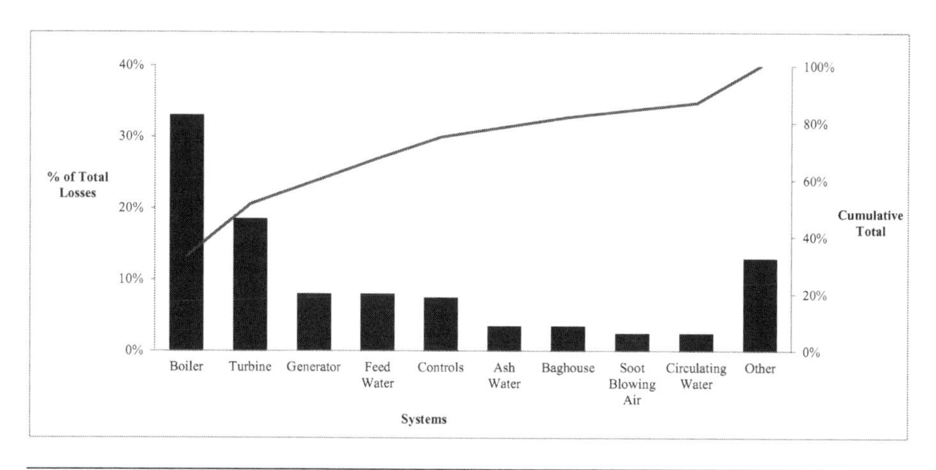

Figure 77. Pareto Chart of Causes of Boiler Failure
(*Source:* Jim August, *Applied Reliability-Centered Maintenance,* Tulsa: PennWell, 2000, p. 117.)

of power produced, the efficiency rate, the emissions rate, and the uptime of each system.

Predictive maintenance

Applied information technology is increasing the productivity of plant operations everywhere and is especially useful in predicting impending failures. Sensors and actuators play a big part in providing valuable information on impending performance degradations, and therefore they play a big role in increasing uptime and reducing cost.

Steam turbine sensors and actuators provide information on over-heat and over-pressure situations, and sometimes provide automated responses. When integrated with logical algorithms that can interpret patterns of information, they can provide better diagnostic information than manual operators can.

Steam turbine blade sensors can help avoid unnecessary preventive maintenance by detecting blade cracks without opening the case. For example, Monitran Technology developed a blade monitoring system that collects condition data and transmits it via a "traffic light" system. The information obviates the need for detailed turbine blade testing during maintenance.[547] GE is also applying "intelligent condition monitoring" to motors through a Turkish acquisition called Artesis, which makes intelligent condition monitoring products designed to detect anomalies on various motor types, sizes, and loads.[548]

Peak Capacity Strategies: Variable and Digital Power Generation Scheduling and Inventory Planning

Production management in power plants is analogous to master scheduling and load planning in a factory.

Long-range planning is handled by "block loading" total energy or load over a period of seasons or years. Fuel budgeting and operations planning is managed by "incremental loading" over a period of weeks or months. Staffing and power pooling plans are generated on a weekly, daily or hourly basis. Simulations using probability can test what would happen under various loads in order to estimate long-run production costs. They can also be used to determine how to get the most power out of the least amount of fuel, giving rise to a multitude of optimization techniques.

Short-run planning has traditionally been based on one of three production scheduling methods: 1) operating at peak utilization and exporting power when it's not needed; 2) operating at low utilization rates and increasing output when necessary; and 3) operating at a constant rate.[549] Technologies are increasingly allowing variable rates of production that can meet demand or supply, with a dramatic impact on cost efficiency by leveling demand patterns and allowing variable production rates, both of which further alleviate the peaking problem.

Smart grid technology including virtual power plants (see the Utility Scale Energy Storage chapter) has, to some degree, mitigated the peak demand challenge. The improved ability to efficiently distribute electricity over the transmission grid has reduced the criticality of the network configuration issue (i.e., site location and the decision about how much to centralize power production). Whereas ten years ago physically decentralized power units may have been needed to optimize electricity distribution (serious consideration was given to widespread implementation of a fuel cell in every house), today the smart grid offers a major step toward pooling power.

Variable speed drives have made a difference in power plant operation, as in many other applications. ABB improved the Energy Intensity Index (EII) of an FCC unit by 10.5% within one year.[550]

Behind the meter, demand response programs price energy to encourage off-peak consumption, and capacitors and compact fluorescent lighting technologies can smooth consumption

MRO Inventory Management

At the simplest level, utilities should pool MRO inventory across multiple operating companies. Boston Strategies International optimized inventory for six operating companies of an electric and gas utility, leading training in replenishment philosophies and mechanics, demand triggers, and reorder points. Through the effort, the utility identified and wrote off obsolete inventory, set inventory targets at the SKU level, and managed a reduction program.

Vendor-managed inventory frequently results in lower inventory management and carrying costs. ABB worked with its own suppliers to establish updated delivery and stocking procedures, whereby the suppliers would manage the inventory instead of ABB doing it. Benefits included shorter lead times and stock reduction.[551]

Parts kitting can reduce field service costs by assembling frequently used kits in an "assembly line" fashion by lower-cost employees in a standardized setting, as compared to having kits assembled by higher-cost field labor on an ad hoc basis.

Rapid inventory replenishment, even if expensive, can sometimes reduce costs by getting equipment up and running sooner. Weir implemented a rapid response program for hydraulic valve controls and experienced a 25% increase in orders in 12 months—proof that customers found the concept worthwhile.[552]

Performance-based logistics contracts can provide suppliers a strong incentive to ensure that uptime and output targets are met. Some utilities are implementing bonus compensation whereby suppliers get paid more if capacity exceeds the nameplate amount, or less if it falls below. In a representative case of below-expectations performance, the Indian government purchased turbines that failed repeatedly at a high-visibility power project, and the ordeal devolved into a blame game. Careful negotiation of accountability for system uptime can improve overall system performance. There are examples of performance-based logistics contracts in the US military. For example, the US Army held GE to engine availability targets for its GE's T700 Engine Program, and GE used lean management to cut engine turnaround times from 265 to 70 days.

Asset Management

Asset management information systems, combined with tags and sensors, can help improve maintenance productivity and uptime in power plants. Benefits include the ability to:[553]

- Locate aging fixed assets (pipelines, between walls, underground, etc.)
- Distinguish one valve or piece of equipment from another similar one nearby, to avoid repairing or replacing the wrong equipment (RFID is used in medical equipment for a similar reason—there have been unfortunate cases of serious medical surgeries being performed on the wrong patients)
- Access repair status and history on an asset, to facilitate better diagnostics
- Manage assets, so expensive tools don't disappear (one turbine maintenance facility "lost" millions of dollars of equipment shortly after a layoff)
- Track and manage inventories of spare parts
- Track employees
- Track vehicles

- Schedule maintenance
- Access management and infrastructure security through access/entry controls

Sourcing Trade-offs: Integrating Purchase and Operating Agreements

A popular form of bundling for power utility capital projects is combined purchase and operating/maintenance agreements related to gas and steam turbines. Many combined agreements now include ancillary services such as commissioning, testing and calibration, user training and engineering support, operation, maintenance, and management of spares and repair parts inventories. The aircraft industry refers to outsourced lifecycle agreements like this such as "power by the hour" agreements.

Examples of bundling purchasing and operating agreements are instructive for those wishing to benchmark their supply configuration and performance. For example, Rolls Royce signed an eight-year contract with BP to provide maintenance for its 28 RB211 turbines in BP's Azerbaijan plant,[554] and Capstone Turbine developed a fixed-price maintenance program for scheduled and unscheduled maintenance.[555]

The decision whether to engage suppliers on this type of basis is partly based on cost-effectiveness, and partly based on the strategic importance of repair and maintenance operations to the operator's business mission. More often than not, operators feel that operations and maintenance is a core competency. Moreover, some want the capability not only for their own operations, but to extend to other power companies on a third-party basis.

TCO Trade-offs: Standardizing Equipment

Although facilities and equipment are highly engineered, a certain degree of standardization is possible and economical.

Rolls Royce pioneered the concept of modular design in the 1970s to reduce the cost of customized parts and to simplify maintenance. The modular design concept helped Rolls Royce achieve economies of scale in R&D, production, and inventory management.[556] Based on the modular design concept, the Rolls Royce RB211 gas generator is made of five modules that can each be replaced individually for maintenance. The units can also be retrofitted effectively by replacing only the modules that have been updated.[557] Similarly, Rolls Royce's RCB barrel compressor is based on a modular design wherein both stationary and rotating components can be removed simultaneously.[558]

More recently, motivated in part by long lead times for turbines, GE and Siemens have been standardizing their offerings by reducing engineering and design customization to reduce construction times. Siemens has also modularized steam turbine design. Siemens has twelve basic power plant combinations including four combinations for simple cycle gas turbines, six for combined cycle plants and remaining two for coal-fired steam power plants. The modularization and design changes have the potential to reduce construction times and lower costs of construction and operation.[559]

The concept applies to pipes and valves as well. A US electric and gas utility lowered the cost of its pipes, valves, and fittings through standardization. After reviewing 1,116 items across six operating companies and 47 stocking locations, a cross-functional team of engineers, consultants, and procurement professionals identified 24% savings by standardizing on MDPE pipes. Many LDPE pipes were upgraded to MDPE in order to achieve the scale and volume purchasing leverage that suppliers would offer due to standardization. The team also standardized corrosion coatings according to a three-part matrix based on the application, pipe diameter, and operating company. The effort led to a consolidation from 46 to 13 suppliers.

HSE Considerations: Taking Advantage of External Resources and Specialists

There are plenty of institutional safeguards for power plant safety, such as risk manuals, hazard identification studies, emergency response plans, fire drills, job-specific training, municipal permits, and hardhat and footwear requirements. However, a safety culture is the most impactful aspect of any HSE program. Safety starts with driving safely in the parking lot, allowing adequate time for important tasks, and looking out for others' health and safety. Organization development consultants have audits that can gauge the solidity of the safety culture. However, regardless of whether it occurs internally or externally, a periodic safety audit that includes safety culture can help prevent accidents, disruptions, and injuries.

Uniper Energy, one of the largest power generation and trading companies in Europe, offers a full package of hazard identification, risk assessment and Safety Integrity Level (SIL) methods including Hazard Identification (HAZID), Hazard and Operability (HAZOP), Layer of Protection Analysis (LOPA), Fault Tree and bowtie analysis, Occupied Building Risk Assessments (OBRA), Control of Major Accident Hazards (COMAH) assessments, and safety and reliability assessments following major incidents or as part of a plant modification process.[560]

Supply Chain Roadmap for Gas- and Coal-Fired Power

High-potential improvements can make gas- and coal-fired power supply chains more synchronized, customized, rationalized, and innovative. As displayed in Figure 72, Predictive Maintenance and reliability engineering can better align production with demand, and cybersecurity can assure stability of production. Co-production can better tailor combustion processes to the availability of local renewable resources while reducing CO_2 emissions. Performance-Based Agreements and Lifecycle Cost Management can help to minimize capital and operating costs. And innovations in the uses for coal may sustain jobs that are in jeopardy as a consequence of the shift toward clean power.

Predictive Maintenance can reduce the cost of maintenance as well as the disruptive impact on operations that unplanned repair can have. While maintenance has gone through an extensive evolutionary process, from preventive to predictive to prescriptive maintenance, the injection of artificial intelligence and machine learning is accelerating this curve, and many plants and vendors have a long way to go to achieve the state of the art. This is good news, as it heralds a large and ongoing opportunity for improvement at most companies.

Performance-Based Agreements can assure reliability and reduce cost, or at least offload it to equipment OEMs and third-party repair and maintenance vendors. Traditional performance-based agreements that guarantee uptime of specific equipment are good, and enhanced performance-based agreements that guarantee uptime of the system as a whole are better.

Reliability Engineering / Life-Cycle Cost Management may seem like a core competency of the Engineering department, but with the advent of smart equipment previous understandings need to be revisited. Assumptions about equipment standardization, maintenance intervals, and spares inventories, to cite just a few examples, will need to be revisited.

Cybersecurity has become a mission-critical competence and budget item and needs to become embedded in all aspects of increasingly smart plants and grid infrastructure.

Co-production, while not an option for many plants, should be evaluated as a possibility for refurbishments and greenfield plants, especially co-firing processes involving renewable fuels in addition to gas or coal. This could be required to meet corporate or regional mandates, to enhance profitability, or both.

Alternative Uses for Coal. As coal plants are increasingly replaced by gas and renewables, the coal supply chain can either be dismantled or repurposed. Several initiatives are underway to repurpose it through innovation and technology.

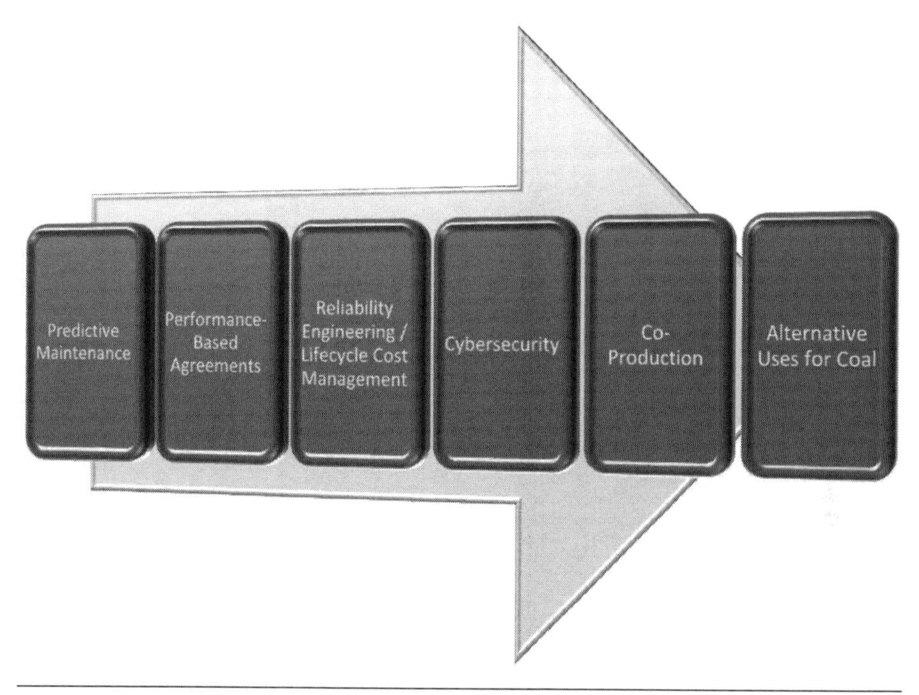

Figure 78. Emerging Supply Chain Opportunities for Gas- and Coal-Fired Power

Hydropower

Chapter Highlights

1. Hydroelectric is a major power generating and energy storage technology, with thousands of plants throughout the world. The presence varies widely by region and country due largely to natural conditions.
2. Climate change is interfering with water levels and flow rates, and some hydroelectric plants are experiencing power shortages due to lack of water flow. The problem has become common enough that an insurance company designed an insurance product for the situation.
3. Due to the size of capital project required and the long lifespan of the plants (30 years is normal), many governments turn to well-respected international EPC firms for construction, and global turbine and generator manufacturers for long-term operating and maintenance contracts.
4. To ensure employment of local labor, outsourcing contracts often involve a local engineering and construction or operations and maintenance partner.
5. Long-term operating and maintenance contracts should include one or more options to terminate for convenience due to unforeseeable circumstances, such as political and social conditions that have been known to impact operations.
6. Due to the cost of unplanned downtime and power outages, hydroelectric plants should use Predictive Maintenance technologies including the use of artificial intelligence and sensors to detect noise and vibration in mechanical parts long before they fail.
7. As electromechanical control systems become a more important part of the plant, firewalls, and unidirectional gateways should be used, among other cybersecurity measures, to prevent unauthorized access to operational and maintenance information systems.
8. Significant opportunities for supply chain optimization include: 1) Predictive Maintenance with Vibration Sensing; 2) Energy Storage; 3) Financial Hedging for Decreased Flow; 4) Hybrid Technologies including Green Hydrogen; and 5) Wave & Current Energy Capture.

Introduction

Hydropower plants make use of the kinetic energy of water flowing in rivers to turn turbines to generate electricity. The large part of the value chain is concentrated around civil works involving construction of various structures such as forebays, reservoirs and penstock. Forebays or reservoirs (sometimes a single structure works as both) are constructed in such a way so that the stored water generates enough mechanical force to turn turbine blades when released into penstock. The hydraulic turbines attached to the generator produce electricity, which is transmitted to the grid via transformer through transmission cables. Figure 79 gives a glimpse of the geothermal power generation value chain from construction and generation to transmission and distribution, notably: digging and earthworking, installation of equipment at field site, channeling of water pressure, production of mechanical energy, and conversion of mechanical to electrical force via a generator, power transmission, and power distribution.

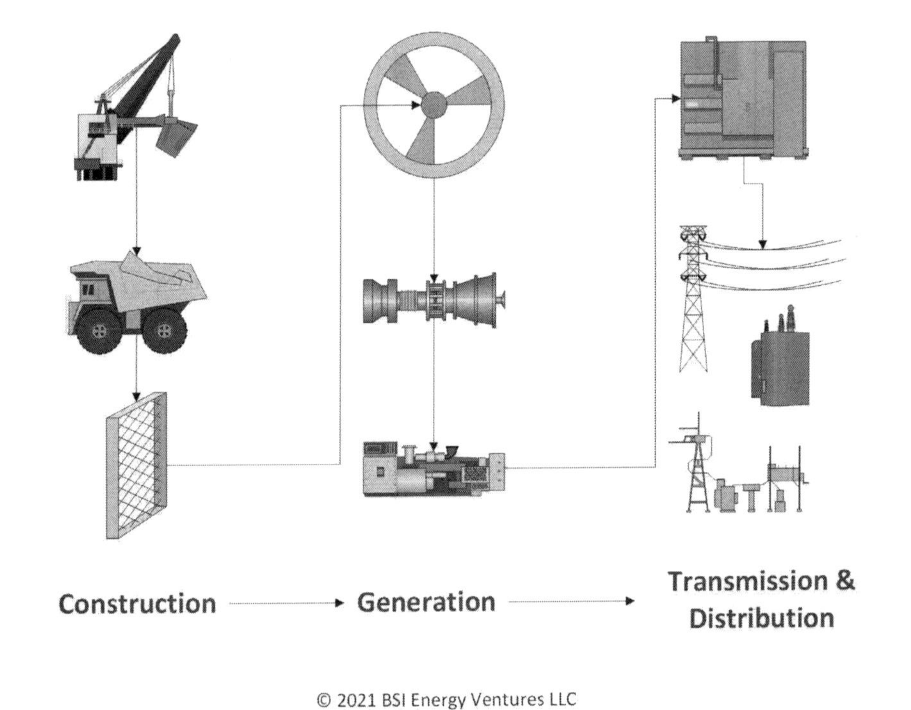

Figure 79. Hydropower Value Chain (Illustrative)
(*Source:* BSI Energy Ventures)

Hydropower constituted approximately 16% of worldwide power generation in 2020, but the share varies widely by region and country, and even by state within countries. It makes up only 6% of electricity generation in large Asian economies like Australia, Japan, and South Korea, while it makes up 46% of Latin America's power. Other continents like North America have wide variations, with Canada producing 60% of its power[561] and 83% of its renewable power[562] with hydroelectric generating stations, while the United States produces only 6% of its power this way.[563] Even within the US, almost half of the hydropower is concentrated in Washington, California, and Oregon, while most of the other 47 states have almost none.[XI] Even in countries with a small percentage of electricity generated by hydropower, there are often many hydropower plants in operation, including 2,200 in the United States today.[564]

Part of the reason for the variety in presence of hydropower is due to physical landscape, which varies by region and locality. An Impoundment-type hydropower station requires a sizable body of water and an elevation drop, around which a dam can be constructed, and turbines and generators can convert the force of the falling water to electricity. A diversion-type hydropower station requires a sort of natural dam, sometimes called "run-of-river," which can support the construction of a canal to channel the water, and a similar installation of turbines and generators. Pumped hydropower storage, which is discussed in the Energy Storage chapter, is located between two bodies of water sitting at different elevations. The ideal conditions for a hydroelectric power station involve strong natural water force and minimal engineering and construction required to capture it. Some natural landscapes are more conducive to this than others.[565]

There are many other types of hydropower—scientists and entrepreneurs have invented countless ways to squeeze energy from water—including transferring thermal energy from stratified layers of ocean currents, capturing the kinetic energy from ocean waves, and capturing the energy from the force of volumetric expansion and contraction within water at different depths and pressures, just to name a few. As of this writing, most are not nearly efficient enough to gain commercial traction, so this chapter focuses on traditional impoundment- and diversion-type hydropower stations.

At least one operator is combining hydropower with hydrogen generation for fuel cells, creating "green hydrogen"—hydrogen production generated through renewable energy forms—and significantly greening the net carbon footprint of hydropower. Hydro-Quebec, the largest energy producer in Canada, is embarking on a $160 million electrolysis plant near Montreal that will use hydroelectricity to power a utility-scale electrolyzer that will convert water into more than 11,000 tonnes of green hydrogen and 88,000 tonnes of oxygen per year.[566]

Supply chain cost is a relatively small component of hydropower—$0.023 per kWh—assuming that supply chain management does not typically have responsibility or influence over the civil engineering or civil works, which make up 70% of supply chain cost, as defined by the methodology outlined in the Preface. Most of the rest of the supply chain cost lies in midstream transportation and distribution.

[XI] For further detail, reference the author's analysis of data from DNV-GL.

Note that while this chapter cites wave energy and ocean thermal energy management, both of which are rising in popular interest and have commercial pilot programs underway, its primary focus is on hydroelectric dams, which make up the bulk of the installed capacity today.

Supply chain management exerts influence over decisions and contracts related to the procurement of electrical and mechanical equipment, which could be subject to strategic sourcing, as well as transmission and distribution assets and their operation and maintenance. Line losses in T&D apply similarly to hydropower as to other central power plant generation models, and supply chain managers are responsible for the procurement and installation of T&D infrastructure and systems that vary in cost and efficiency. Accordingly, this chapter places heavy emphasis on the strategic sourcing of electric and mechanical components of the dam or river-run plant.

Trade-offs and Management Techniques in Capital Project Management

Technology Choice Risk Management: Engineering Hydropower for Declining Water Availability

At the feasibility study stage, the possibility of declining reservoir water availability should be considered. In addition to the fact that smaller plants can be susceptible to low water flow rates based on the topography and engineering of the systems, the incidence of droughts affecting water availability at large hydropower plants is increasing due to climate change. Both topographic and climate-change-induced water shortages have led to instances of underperformance among large hydro projects.

The potential for low water flow can impact the choice of turbines, design and size of reservoir, and decision on upgrading of many ageing plants across the world. Below are some examples of drought conditions affecting hydro power generation.

- California. Hydroelectric power from dams usually provides about 15% of California's electricity needs. But in 2015, at the zenith of the worst drought in California's recorded history, it supplied only 6%.[567]
- Ghana. In 2016, Ghana's main source of energy, the Akosombo Dam, operated at minimum capacity due to drought. In the same year, drought hit Brazil, affecting hydroelectric power producers such as the Itaipu Dam. This forced the country to turn to more expensive and more polluting thermoelectric plants.[568]
- Malawi. For much of 2017, Malawi in southeast Africa experienced intermittent blackouts as a result of low water levels at the country's largest hydropower plant. In December, large areas of Malawi turned

to complete darkness as the country was left with just 150 megawatts (MWs) of power available out of the 300 MW it normally generates.[569]

- Venezuela. The fourth largest power plant in the world, the Guri dam in Venezuela, is so short of water that power outages may last up to eight hours a day, forcing factories to close early, driving down production, and creating darkened havens for muggers and other criminal activity.[570]
- Indonesia. Relatively small hydro plants are increasingly experiencing operational problems due to shortages of water flow. In Lampung in September 2012, for example, towards the end of the dry season, two small hydropower stations operated by the state-owned utility Perusahaan Listrik Negara, or PLN, (with a total capacity around 120 MW) stopped generating power due to shortage of water flow.

Capital Project Risks and Mitigation: Mitigating Ecological Risks and Impacts

Hydropower is widely viewed as an environmentally friendly power generation technology, and in order to live up to that expectation, extra efforts are needed to manage supply chain–related ecological risks and minimize the environmental footprint caused during construction and operations, especially if the primary goal of the project is clean power.

The air- and water-related footprint of the plant should be measured and managed so as to minimize CO_2, and noise emissions caused by fossil fuel–powered earth-moving equipment. Also, construction should be overseen so as to avoid causing methane eruptions; water intake should be designed so as not to disrupt ecosystems of birds and fish; and effluent should not negatively impact dissolved oxygen levels which marine life depend upon.

A group called IHA Sustainability Limited, affiliated with the International Hydropower Association (IHA), offers a comprehensive sustainability assessment tool for evaluating ESG gaps versus best practice in aspects such as project management structures and processes, procurement practices, social issues such as labor conditions, indigenous peoples and resettlement, biodiversity, erosion and sedimentation, and reservoir preparation and landfilling. A related group called the Hydropower Sustainability Assessment Council is preparing a Hydropower Sustainability Standard that will rate performance along criteria similar to those in the self-assessment tool.

Additionally, the potential for drought-related low water levels should be evaluated at the planning stage. After sustained drought and low rainfall in Uruguay and Brazil decreased hydroelectric production and forced national governments to purchase power on the spot markets at high prices, a financial services company developed an insurance product that safeguards developers and operators against the hazards of low rainfall for hydroelectric projects. The insurance acts as a financial hedge that offers compensation in the event of low resource availability.

The company, GCube, mines historical weather data to create an index and an index trigger point, beyond which the policy begins to pay out, compensating operators for the difference between the value of the typical and the actual output.[571]

Behind the scenes, some insurers are hedging one plant's output with complementary and inversely correlated energy output. For example, holding positions in one region where there may be a scarcity of production and an opposite position in a different region where there is likely to be production at that time. This may apply not only within power types but across them. For instance, in Colombia wind resources increase when rainfall drops, according to an industry expert there.

Supply Unavailability Risks and Mitigation: Waiting for the Dam Cement

Civil works and equipment make up 75–90% of the total investment cost (TIC) of hydroelectric power plants, and cement is one of the largest externally purchased materials, so ensuring adequate supply, economical logistics, and transport to the job site, as well as timely delivery, is an important logistics requirement. Ideally, long-term agreements with contractors and cement suppliers would assure the products and services needed to complete construction of a hydroelectric dam and power plant without delays.

For example, India's National Hydroelectric Power Corporation (NHPC) has signed a memorandum of understanding with the Cement Corporation of India (CCI) to meet the cement requirements for its 2880 MW Dibang Multipurpose Project. The project, once installed, would be the country's largest dam. CCI has announced plans to install a clinker grinding unit near the project area, which would fulfil the cement requirements of the project and also generate employment opportunities.[572]

Outsourcing Risks and Mitigation: Securing Performance Guarantees from O&M Subcontractors

Hydroelectric power operators around the world frequently outsource operations and maintenance to third parties that have advanced and specialized capabilities in preventive maintenance for the turbine and generator systems. In addition to ensuring equipment-specific skills and guaranteeing uptime or output targets, outsourcing often reduces or eliminates the potential for budget variations, which makes the PPA more bankable.

Third parties often commit to ensuring prescribed levels of uptime or performance, which are sometimes mandated by law in the interest of providing power to the local population of residential, commercial, and industrial customers. Performance could otherwise be affected by physical equipment failures or by other factors such as labor strikes or management errors. Budget variations could otherwise be caused by labor or materials cost increases, which are often eliminated by a fixed price or predetermined contract escalation rates. The firm and legal commitment that results from an outsourcing contract allows an operator to set reliable rates and dividends years or even decades into the future.

Supplier Partnering Risks and Mitigation: Working with International EPC Firms

Since building and operating a hydro plant entails large capital commitments, long time frames, and financial commitments that are often tied to sovereign loans, most projects rely on trusted international firms with well-established reputations for engineering, procurement, and construction. In addition to the benefit of reliability, contracts with these firms typically reduce the possibility of graft and corruption, which have been known to be associated with large-scale government contracts signed off on by high-level government officials.

In a $159 million contract, SEB Power Sdn. Bhd commissioned GE Power India, head of an alliance consisting of GE Hydro France, GE Renewable Malaysia Sdn. Bhd., and Sinohydro Corporation (M) Sdn. Bhd., for the Main Electrical and Mechanical Works for the 1285 MW Baleh Hydroelectric Project in Sarawak, Malaysia.[573]

In a $471 million contract, the Gambia River Basin Development Organisation hired a consortium headed by VINCI Construction Grands Projets and consisting of VINCI Construction Terrassement, subsidiaries of VINCI Construction (75%), and Andritz, the Austrian turbine manufacturer (25%), to build the Sambangalou dam in southeast Senegal, close to the Guinean border.[574]

Procurement Bundling Trade-offs: Assembling a Consortium of OEMs, Engineering Firms, and Local Construction Capability

Operating and maintenance contracts in developing economies are frequently awarded to consortia involving both major global engineering and equipment providers and their local partners. This assures local employment and knowledge transfer, in addition to ensuring reliability and bankability.

For example, in 2020 the Pakistani government awarded DESCON Power Solutions of Pakistan the construction contract for the Mohmand Dam hydropower project, along with partners China Gezhouba Group Co. Ltd., a subsidiary of Energy China, and Voith Hydro Shanghai Limited. The consortium combines EPC services (China Gezhouba), equipment (Voith Hydro), and local manpower

(DESCON). The Pakistani government views the project as a social development initiative that will create 6,000 jobs at the peak of construction.[575]

Also, Indonesian state-owned utility Perusahaan Listrik Negara (PLN) contracted for multiple hydroelectric dams. In one case, for the Bakaru I and II hydropower projects, PLN contracted with a joint venture co-led by Tractebel Engineering and Japan's NEWJEC, alongside local project management consultant PT Connusa Energindo.[576] In another case, PLN contracted with Mitsubishi Corporation and Voith Hydro to build the 174 MW Asahan 3 hydropower project in Indonesia. Mitsubishi is the lead contractor and EPC firm, and again Voith will be responsible for the design, manufacture, and supply of the electromechanical package. The package includes two 87 MW vertical Francis turbine and generator units and the related control system, the electrical and mechanical "balance of plant" (BOP), and a 150-kV switchyard.[577]

Contract Term Risk and Mitigation: Using Contract Renewal Options

The duration of operations and maintenance contracts is influenced by the length of PPAs. Hydropower PPAs are often 20 to 40 years in duration. For example, in Canada, BC Hydro's PPA is 20 to 40 years long. In Ethiopia, the Ethiopian Electric Power Enterprise (EEP) signs 25-year PPAs. In Tanzania, the Energy and Water Utilities Regulatory Authority (EWURA) proposes a 30-year term in their model PPA for hydro power more than 10 MW.[578]

Operations and maintenance contracts related to hydropower are also often 20 to 40 years in length, unsurprisingly mirroring the PPAs. For example, in Brazil, Statkraft Energias Renovaveis took over an operations and maintenance service agreement with specific milestones and performance targets over a 25-year period. In Nigeria, the Nigerian government contracted with Mainstream Energy Solutions Ltd. (MESL) for operations and maintenance of the Kainji-Jebba hydropower complex for a 30-year term.

Due to the long term of the agreement, the O&M contracts sometimes have break points. In Pakistan, Laraib Energy Limited, a subsidiary of the first hydropower Independent Power Producer (IPP) in Pakistan, and owner and developer of the 84 MW New Bong Escape (NBE) Hydroelectric Power Complex on the Jhelum River in Azad Jammu and Kashmir, terminated its O&M contract after five years with TNB REMACO Pakistan in accordance with the terms of its agreement.

Trade-offs and Management Techniques in Operations & Maintenance

Cybersecurity Risk Management: Using Unidirectional Gateways

Hydropower plants have started deploying cybersecurity solutions, given how large infrastructures supply significant power to the grid, especially in countries like China, US, Brazil, and Canada with a large number of utility-scale projects. For example, a major Canadian hydropower project operator contracted Waterfall Securities, an Australia-based cybersecurity service provider, to secure IT/OT interconnections of the hydropower operations and infrastructure, thus complying with standards such as NERC CIP (North American Electric Reliability Organization—Critical infrastructure Protection).

Accordingly, one of the key accomplishments was to install "Unidirectional Gateways", which is recognized by the NERC CIP standards. This security solution, which carries out unidirectional information exchange, sits between two networks operating like a check valve. The main function is to allow all data to pass in the forward direction, while blocking all data in the reverse direction. The fiber optical connection prevents data from travelling in the opposite direction. Since it is not software, it cannot be directly attacked by any malware, which results in high security.[579]

Internet of Things (IoT) and Artificial Intelligence (AI) Technology Choices: Using Sensors to Detect Turbine Noise

Landsvirkjun, the national power company of Iceland, launched a pilot project on intelligent noise analysis in hydropower plants focused on developing predictive maintenance capabilities and thereby reducing the probability of equipment failure and consequently long-run operating and maintenance costs.[580]

To prevent unplanned shutdowns, Voith is installing an acoustic monitoring system in the Budarhals hydropower plant in Iceland to detect turbine noise that deviates from normal conditions. In addition, continuous analysis of machine data is designed to optimize maintenance work order scheduling. Microphones mounted in the plant will record ambient noise and data for preprocessing, and then selected additional data interpretation will be done offsite with a more intensive analytic engine. For calibration purposes, the system records all acoustic signals in an initial learning phase. The data collected is also benchmarked against data from other hydropower plants. In the event of detected abnormalities, the system flags a warning and notifies one of the power plant operator's service technicians.

Peak Capacity Strategies: Storing Energy with Pumped Hydropower

Hydro plant production varies as a function of water level and pressure, which does not generally impact the supply chain. Also, independent hydro plants must release water when they reach spill levels, which are water levels and pressures at which safety considerations suggest that the water could cause damage to the plant or the surroundings. This is analogous to curtailment on a wind or solar farm.[581] This is primarily an engineering problem.

However, when hydropower is used as a form of energy storage, this affects the production control and dispatch of electricity. When Pumped Hydropower Storage is used (see the Utility Scale Energy Storage chapter), it has an enormous impact on the capacity of whatever primary power generating technology is used. It offers the primary power system the ability to meet peak demands by using off-peak power. The "duck curve" shows that California's load averaged 19,000 MW throughout the day and jumped from 12,000 to 26,000 MW between noon and 9 p.m. Without energy storage, the system capacity would have to be rated at the peak (26,000) instead of the average (19,000), a difference of 7,000 MW or 37% of the average capacity, so in a way one could say that without PHS the primary energy generation plants would have to be built to sustain 37% more production on average.

Also, when hydropower feeds the same grid as wind and solar, the optimal dispatch modeling for combined hydroelectric, wind, and solar power should consider all of the potential sources and their marginal costs and revenues.[582]

Sourcing Trade-offs: Hiring Local Equipment and Services

Local content can be a political lightning rod, and in some cases hydropower plant management is expected to practice socially responsible sourcing practices. In India, hydropower stations and transmission lines will have to purchase equipment locally, as per guidelines issued by the government. The norms apply to power plants and transmission lines set up by government-owned firms as well as private projects being funded by state-run Power Finance Corp. and Rural Electrification Corp.[583]

Because hydropower plants are built on national land, and often by federal authorities or agencies, they are often considered nationalized enterprises, and outsourcing contracts are sometimes expected to be awarded to domestic providers. In 2014, the Finnish power operator Kemijoki Oy Hydro outsourced operations and maintenance to Caverion, another Finnish company, without fanfare. And Swedish operator Fortum outsourced its operations and maintenance to the Swedish company Bilfinger. But in 2020, when the president of the Tennessee Valley Authority argued for outsourcing some aspects of operations and maintenance to non-US companies, he was fired by then-President Donald Trump.[584] In conclusion, managers should consider the political as well as the economic aspects of outsourcing decisions related to hydropower operations, maintenance, and also construction.

TCO Trade-offs: Refurbishing Turbines and Generators

The average hydropower plant lasts 30 years before requiring a major refurbishment. Still, planning for the refurbishment should be done periodically and iteratively, as it can not only extend the lifespan of a hydropower plant, but also increase output, with appropriate upgrades, and increase power generating efficiency.[585]

The main objects of refurbishment are turbines and generators, so initial purchase decisions including the manufacturer and the turbine type (mostly of the Francis or Kaplan types) should consider the potential for later upgrade. Two-thirds of upgrades and refurbishments involve turbines, and three quarters involve generators. Other equipment upgrades involve spillway gates, transformers, and control systems.

Supply Chain Roadmap for Hydropower

As shown in Figure 80, significant opportunities for supply chain optimization through improved reliability of electricity production include: 1) Predictive Maintenance with Vibration Sensing; 2) Energy Storage; and 3) Financial Hedging for Decreased Flow. In addition, substantial Innovation is being underwritten in hybrid energy technologies including Green Hydrogen, and eventually potentially Wave & Current Energy Capture; these may change the footprint and landscape of hydropower as a power generation technology, but most have not yet been proven economical at scale.

Predictive Maintenance with Vibration Sensing has the potential to detect repair and maintenance requirements earlier, and thereby mitigate their direct and indirect costs. Knowing repairs in advance can reduce damage to major components such as turbines and generators, and allow them to be scheduled conveniently. It can also provide sufficient lead time to arrange for replacement of major components during routinely scheduled heavy maintenance operations.

Energy Storage, by buffering against both demand fluctuation and supply variability, can improve supply resilience and flexibility, and can help to lower the average cost and sometimes level the prices of electricity for customers on the grid.

Financial Hedging for Decreased Flow can help to assure electricity availability even if water resources run short. As such it can substitute for operational reliability.

Hybrid technologies including Green Hydrogen may have high potential for hydropower plants due to their renewable characteristics and high operating availability rates. For example, in Canada, Evolugen, a business unit of Brookfield Renewable, and Gazifère, an Enbridge company, launched a project to build a 20 MW hydropower-driven electrolysis ("green hydrogen") plant in Quebec.[586]

Wave & Current Energy Capture, although not directly supply chain–related, may have the potential to open new revenue sources for hydropower operators, thereby leveraging what may have been an operational technology into one or more new revenue streams, and distribute power generation capacity across the grid, thereby reducing transmission networks and line losses.

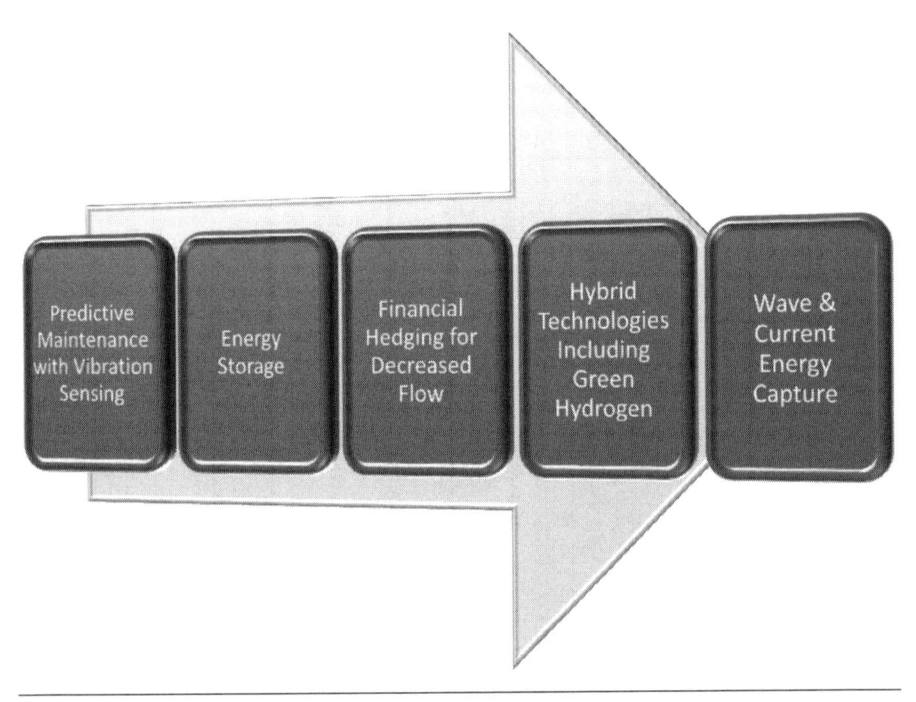

Figure 80. Emerging Supply Chain Opportunities in Hydropower

Nuclear

Chapter Highlights

1. Most of the 440 nuclear power plants across the globe were built between 1962 and 1987. Between 1987 and 2018 net additions were small or negative, as capacity removals offset capacity additions in most years. However, 50 reactor newbuild or upgrade projects are currently underway, for nuclear is seen as a green form of energy that could compete with solar and wind.
2. We are on the fourth generation of fission reactors, which are distinguished from earlier generations primarily by safety features, coolant types, and the chemical composition of their waste materials. The most common type is the Pressurized Water type.
3. Seventy to 600 MW floating nuclear reactors, developed by Rosatom, have received strong interest from the market.
4. Characteristic financial overruns increase the capital project risk profile of nuclear compared to other power generating technologies.
5. Although it is "clean" in that it does not emit CO_2, nuclear is not a renewable power technology. Almost all nuclear power plants run on U-235 uranium fuel, which must be mined and processed.
6. Forty percent of the world's uranium production comes from Kazakhstan. In early 2020 the uranium market was facing a 30 million–pound shortage.
7. Development partnerships are common, especially for countries new to nuclear power. However, outsourcing is problematic due to overwhelming safety concerns.
8. The nuclear power industry is rapidly creating digital twins to improve predictive maintenance, and these movements toward digitalization and AI increase the need for cybersecurity. There have already been major cybersecurity incidents at a nuclear power plant in Iran.
9. There is a major movement to develop accident-tolerant fuel, which would melt without disastrous consequences in the event of a

breach of the containment vessel. This may impact the technological evolution and procurement practices requiring specialized logistics and storage expertise.

10. Fueling and maintenance, performed at intervals of 12–24 months, interrupt power generation and potentially require complete shutdown and turnaround (depending on the plant design).
11. High-impact supply chain imperatives include: 1) Safer Fuels, Waste & Disposal; 2) AI, Predictive Maintenance, & Remote Digital Twins; 3) Modularization & Mini-Nuclear Design; and 4) Floating Nuclear.

Introduction

Nuclear energy constitutes 10.3% of world electricity production today, a figure that is projected to decline to 9% by 2030.[XII] As of 2020, Europe has the most installed nuclear generating capacity—23.3% of total electricity generation—while Latin America, India, the Middle East, Southeast Asia, and Africa have the least—less than 3% of electricity generated is nuclear in each of these places.[587] In addition to energy economics, safety concerns and geopolitics of access to uranium and nuclear technology play a role in which countries proceed with nuclear power plants.

Almost all nuclear plant production occurred between 1962 and 1987, with only a few being built, and a few being retired each year between 1987 and 2018.[588] A recent resurgence of interest in nuclear power as a potential source of clean energy has led to a spate of new projects. In 2018, nuclear power plant production was in a net decline, with three of the four freshly commissioned nuclear power plant construction projects occurring in China.[589] However, as of 2020, about 50 power reactors are currently being built in 16 countries, notably China, India, Russia, and the United Arab Emirates (some of these were upgrades rather than newbuilds), and despite the slight decline in the share of world electricity production mentioned earlier (from 10.3% in 2020 to 9% in 2030), the IEA forecasts a 15% growth in nuclear-powered GW between 2020 and 2040.[590]

Nuclear plants in operation today operate on the principle of fission, and there are three principal types of nuclear fission reactors in operation today:

- Over two-thirds of existing reactors (301 of 441) are of the Pressurized Water type.[591] Pressurized water flows over enriched (about 3.2% U235) uranium dioxide in zirconium alloy cans, and into a steam generator. The whole unit is encased in a concrete shield and a containment vessel.[592] These make up 72% of nuclear power production as measured by global gigawatts of electrical output. They are located in the US, France, Japan, Russia, China, and South Korea.

[XII] Based on author's analysis of data from DNV-GL.

- Boiling water–type reactors make up the next-largest proportion of reactors operating today (15% of the total by number). These have no steam generator; instead, the steam rising from the boiling water in the reactor circuit is used directly to turn a turbine generator. These can be found in the US, Japan, and Sweden.
- The third-largest nuclear technology is the Pressurised Heavy Water Reactor (PHWR, or CANDU). Making up 11% of existing reactors by number but only 6% of electrical power output, these reactors hold unenriched uranium dioxide in zirconium alloy cans loaded into alloy tubes. Pumped heavy water cools the tubes under pressure towards a steam generator, which generates power.

Other less common types include Advanced Gas-Cooled Reactors (AGR), Light Water Graphite Reactors (LWGR), and Fast Neutron Reactors (FBR).

The main components of a nuclear fission power plant are the fuel rods, the moderator that controls the rate of fission; control rods made of material such as cadmium, hafnium or boron; a coolant (which is usually water); a pressure vessel that contains the core and the moderator; the steam generator; and the containment vessel.

At the construction phase, procurement needs to manage the supply contracts to build the reactor, including assuring supply of the core elements such as tubes, cooling towers, piping for water ingress and egress, and electrical contracting services for process control equipment and systems. During plant operation, logistics managers need to assure fuel supply, beginning at the uranium mining and continuing through uranium enrichment and processing, including transportation to the plant. Maintenance and operations are particular to nuclear plants, including shutdowns and turnaround at refuelings. Waste must be safely transported to storage locations designated by governmental authorities in each jurisdiction. Figure 81 shows the fuel delivery and power generation schematically: the ore, transport of the raw ore to the processor, uranium enrichment, transport of the enriched uranium to the plant, radioactive waste generation, transport to a secure site, and disposal.

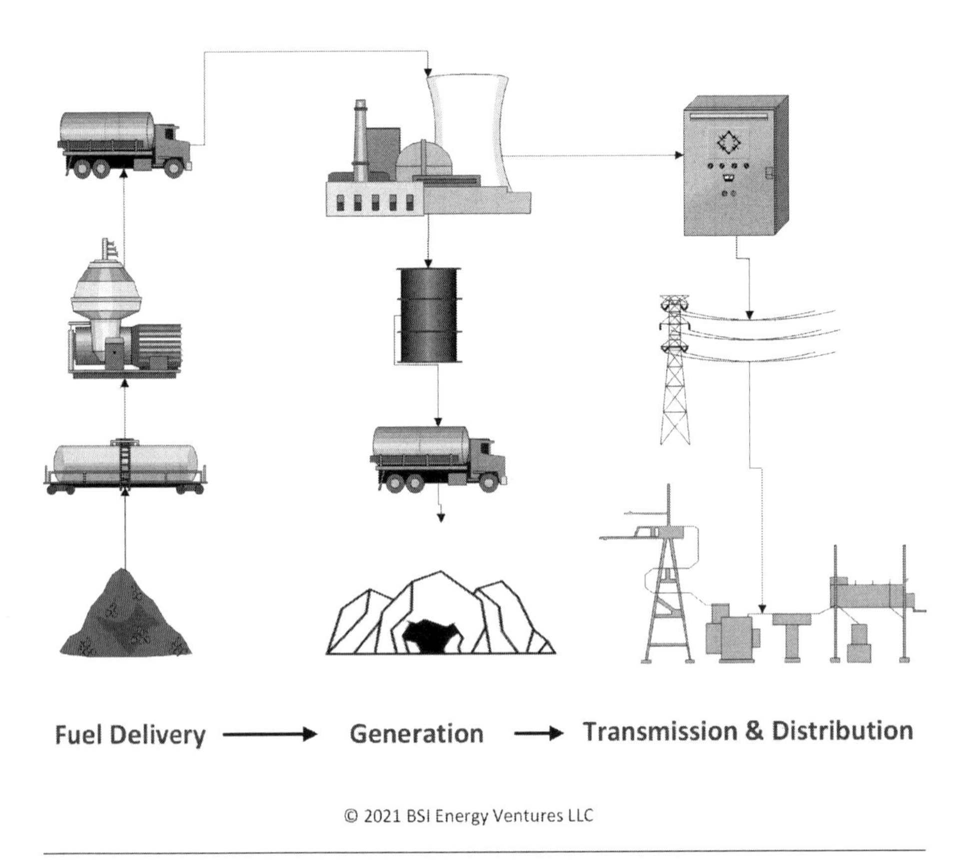

Figure 81. Nuclear Power Value Chain (Illustrative)
(*Source:* BSI Energy Ventures)

Trade-offs and Management Techniques in Capital Project Management

Technology Choice Risk Management: Designing Uniquely Different Value Chains for Small, Large, or Floating Plants

A number of advanced nuclear fission designs are underway, leading to categorizations as Generation I, II, III, and soon IV. The differences between them chiefly relate to safety features, coolant types, and the chemical composition of their waste materials.

There is also growing interest in developing small, modular-base-load nuclear reactors that are more affordable than nuclear projects have ever been before. For example, US-based NuScale Power, along with Utah Associated Municipal Power Systems (UAMPS), will deploy the NuScale Small Modular Reactor at Idaho National

Laboratory (INL). The unit can house four, six, or 12 individual power modules and is rated at 77 MW of electricity using pressurized water reactor technology. NuScale's was the first small modular reactor to receive design approval from the US Nuclear Regulatory Commission, and the Department of Energy has put $1.3 billion behind the project. Actual construction is expected to start in 2025.[593]

Floating nuclear power plants (FNPP) are an intriguing form of technology, due to their mobility and to their offshore nature, which could allow them to serve countries and localities where nuclear power is not practical or allowed onshore. Rosatom (Russia) has set up a subsidiary to produce FCNPPs in the 70 to 600 MWe range.[594] They are built on an icebreaker, and the primary application is in the Arctic, but early in the development program 15 countries such as China, Indonesia, Malaysia, Algeria, Sudan, Namibia, Cape Verde, and Argentina indicated interest in the technology.[595]

One such plant being commissioned by Russia's Rosatom State Corporation, Akademik Lomonosov, and consisting of a 144-meter barge and its two nuclear 35-MW reactors, posed high technological risk.

The ambitious project lasted 12 years and cost more than five times its original budget: $740 million versus the projected $140 million, according to official government records.[596] Construction of the floating reactor began in 2007 and had to overcome a messy financial situation including the threat of bankruptcy in 2011; the plant was finally commissioned in December 2019. The project is based on the small modular reactor (SMR) design: this is a type of nuclear fission reactor that is smaller than conventional reactors.[597]

Safety is also a major concern with the design. As Russia continues to develop plans for "fleets" of FNPPs, many analysts have raised concerns about safety, security, and risks associated with operating fleets of ships in far-flung locations. Since this concept of operations is entirely new for civilian power infrastructure, Russia will have to ensure safeguards in conformance with IAEA standards.[598]

Meanwhile, extensive research is being conducted into nuclear fusion, including the International Thermonuclear Experimental Reactor (ITER) and Chinese fusion research projects. However, these are generally not expected to be commercializable until at least 2050.

Capital Project Risks and Mitigation: Managing the "Capital Project Supply Chain"

The capital cost of construction accounts for almost all supply chain costs. Capital costs for the plant itself (equipment and installation) account for about two-thirds of upfront capital expenditure. Supply chain costs for nuclear fusion projects amount to $0.16 per kwh of output, and due to the high proportion of capital in this total, the supply chain cost is almost entirely upstream. Civil works are next, at 20% of the total investment. EPC adders are about 12%.

Cost overruns in the construction phase can be fatal to the profitability of the nuclear plant project, since "construction lead times and costs, together with the cost of capital, determine a plant's economic performance. Once a plant is built its operational costs are low and predictable."[599]

Despite the fact that they can turn return on investment negative, cost overruns and delays are common in nuclear power projects, usually due to engineering change requests that are so frequently occurring as to regrettably be considered common among large capital projects in the energy industry.[600 601]

For example, EDF's new Flamanville nuclear reactor capital costs have significantly gone up, recording a delay for an eighth year. EDF confirmed that target construction costs had already risen by 14% to $13.6 billion and would not be able to load nuclear fuel before the end of 2022 on account of needing to repair 66 welds. The almost-completed plant, which is already seven years behind schedule, won't be able to load nuclear fuel before the end of 2022.[602]

Sustainability Trade-offs: Transporting Small Amounts of Potent Fuel in Accordance with Defined Protocols

In stark contrast to the unit trains of coal that are transported thousands of miles to coal-burning thermal power plants, transportation of uranium to nuclear plants is environmentally pristine. Although regulations are strict, and the raw material comes from around the world, the quantities are tiny—a typical plant only requires a few canisters of fuel per year. In addition, the thermal power plants must maintain very large stockpiles of coal, which incurs substantial inventory carrying cost and blights the landscape. In this way, choosing to build a nuclear power plant instead of a coal-fired thermal plant is a major sustainability decision with cost and safety trade-offs.

Given the health and safety risks radioactive feedstock entails, strict handling, packaging and transportation protocols are defined. The latest edition of Regulations for the Safe Transport of Radioactive Material was released by the International Atomic Energy Agency (IAEA) in 2012, which has been widely adopted the world over.

Accordingly, most requirements of packaging and transportation relevant to nuclear power plants fall under Type A and Type B protocols. Type A packages are used for the transport of relatively small quantities of radioactive material and are designed to withstand accidents. Type B packages are used for MOX fuel (mixed oxide fuel containing more than one oxide of fissile material), nuclear HLW (high-level waste), and used fuel. They range from drum size to truck size and maintain shielding from gamma and neutron radiation, even under extreme accident conditions. The logistical cost of carrying these radioactive fuels is exorbitant. For example, a single type B shipment can cost up to $1.6 million.[603]

Supply Unavailability Risks and Mitigation: Planning Access to Processed Uranium

Fission reactors are dependent on uranium fuel, which is located disproportionately in Kazakhstan, home of 40% of the world's uranium production. The COVID-19 pandemic exacerbated the pre-existing supply shortage of uranium, and Kazakh mining company Kazatomprom slashed its 2020 production forecast by up to 10.4m pounds, equivalent to 8% of global supply, because of government-imposed measures to mitigate the spread of COVID-19. This caused the price of uranium to rise more than 20% in just about a month since pandemic hit the world.[604] In addition, due to the closure of some of the other large uranium mines for almost a month (e.g. Cigar Lake mine, located in northern Saskatchewan, Canada), 30–35% of global uranium production has now been curtailed by virus-related shutdowns, prompting miners to draw further on their inventories. Overall, the pandemic increased the magnitude of the existing shortage by about a third.

Production cuts at Kazatomprom ripple through to other producers such as Canadian producer Cameco, a uranium refiner and processor, which had expected to purchase almost 5 million pounds of uranium from Inkai, a mine it jointly owns with the Kazakh company in southern Kazakhstan. If supplies from Inkai are hit, that could force Cameco to source replacement metal directly from the cash market or from Kazatomprom's stockpile.

In general, operators procure nuclear fuel through a mix of long-term and short-term contracts, with about 85% of the demand being met through long-term contracts. They also maintain at least two years' stockpile for security.

Outsourcing Risks and Mitigation: Containing Risk by Minimizing Outsourcing

Nuclear power operations are among the most critical of all energy value chains because of high intrinsic safety risk involving radioactive feedstock.

A study released by a French parliament committee acknowledged the growing safety risk from conventional outsourcing practices. In a statement on outsourcing, it concluded that France's nuclear plants are a safety threat because of their excessive reliance on outsourcing. France is the world's most nuclear-reliant country, with state-owned EDF generating 75 percent of its power from 58 aging nuclear reactors spread around the country, many of which have suffered safety scares in recent years. Up to 80 percent of maintenance at the plants is handled by contractors, leading to a loss of internal competence at the utility.[605]

The issue has been brought to the fore by activist groups and quality failures. French nuclear plants have been the target of numerous break-ins by Greenpeace militants, which have highlighted their vulnerability, and EDF has had to close several reactors and delay the opening of a new one because of quality issues with some suppliers.

Supplier Partnering Risks and Mitigation: Forming Strong Alliances with Experienced Nuclear Plant Engineering and Construction Firms

In apparent contrast to the outsourcing dilemma, partnerships are essential mechanisms for knowledge transfer and risk-sharing in capital projects, especially in the developing countries that are new to nuclear power.

India's state owned nuclear power company, NPCIL (Nuclear Power Corporation of India Ltd.), chose EDF, the French power company, for design and EPC work on six EPR nuclear reactors at the Jaitapur site, which is set to be the biggest nuclear project in the world, with a total power capacity of around 10 GW. In order to support this mega project, EDF signed engineering and capacity-building agreements.[606] The engineering agreement involves four additional firms—Assystem, Egis, Reliance, and Bouygues. Together, the five companies will define the contribution of each and set up a joint venture, with EDF holding the majority stake and leading the engineering integration. The second agreement, with Larsen & Toubro, AFCEN, and Bureau Veritas, is for training and capacity building of local companies to achieve technical standards (specific to nuclear power plants) required to manufacture equipment for the Jaitapur project.

Procurement Bundling Trade-offs: Bundling O&M Services to Minimize Interfaces and Handoffs

The technical complexity and safety considerations inherent in nuclear plants lead to increased bundling of operations and maintenance (O&M) services, compared to other conventional and less safety-critical power platforms. Framatome signed a long-term service contract with the Taishan Nuclear Power Joint Venture Company Limited (TNPJVC) to support operations of two EPRs at the Taishan Nuclear Power Plant in China. The contract covers nuclear plant outage and maintenance work, including spare parts supply and engineering services for eight years.[607]

Contract Term Risk and Mitigation: Contracting for Fuel in Increments of Five Years

Due to the aforementioned supply interruptions involving uranium, some operators have opted for long-term supply contracts. For example, India decided to renew its pact with Kazakhstan for supplies of uranium from 2020 to 2024. By 2019, Kazakhstan had already supplied a total of 10,000 tonnes of uranium under two five-year contracts.[608]

Trade-offs and Management Techniques in Operations & Maintenance

Cybersecurity Risk Management: Protecting the Reactor from Determined Hackers

Nuclear plants are even more critically sensitive to cyberattacks than conventional and renewable power plants. Accordingly, cybersecurity should be considered from the start.

A major cybersecurity incident occurred in 2010, when Iran's Nuclear Power Plant (NPP) at Natanz was hit with the Stuxnet cyber worm designed to exploit SCADA systems. The worm damaged 984 uranium enriching centrifuges, which led to an estimated 30% decrease in enrichment efficiency. Stuxnet targeted the Siemens control systems and had the ability to reprogram the PLC (programmable logic controllers), which allows automation that controls machinery and processes including gas centrifuges for separating nuclear material.[609]

Stuxnet demonstrated the effects of maintaining unsecured systems that can let vulnerabilities in and inflict damages on facilities. The worm was very potent, which is evident from the fact that it led to five zero-day vulnerabilities, having kicked off from infected USB flash drives. This malware stole hard-coded password from the Siemens database (CVE-2010- 2772) and even gained control of the SCADA system files, besides tampering with the frequency of the frequency drivers affecting the centrifuges. The intelligent bug also managed to stay concealed by means of driver signing keys, which were stolen from RealTek and JMicron.[610]

Internet of Things (IoT) and Artificial Intelligence (AI) Technology Choices: Developing Digital Twins and Mining Operational Data

The application of artificial intelligence, in the form of expert systems and neural networks, to the control room activities in a nuclear power plant has the potential to reduce operator error and increase plant safety, reliability, and efficiency. Furthermore, there are a large number of non-operating activities (testing, routine maintenance, outage planning, equipment diagnostics, and fuel management) in which artificial intelligence can increase the efficiency and effectiveness of overall plant and corporate operations. Keeping the vitality of AI in mind, especially for the next-generation nuclear projects, governments in the US and France are investing in developing digital twin infrastructure.

A US Department of Energy (DOE)–funded program called Generating Electricity Managed by Intelligent Nuclear Assets (GEMINA) was announced in 2019 to fund R&D efforts to develop artificial intelligence and advanced modeling controls that introduce greater flexibility in nuclear reactor systems, increased autonomy in operations, and faster design iteration.

Nine projects have been awarded $27 million to develop digital twin technology for the Kairos, Xe-100, BWRX-300, and the SSR-W designs, with the aim of achieving a tenfold decrease in their operations and maintenance (O&M) costs through predictive maintenance and model-based fault detection.[611]

One example of nine projects developed under the GEMINA umbrella is "AI-Enabled Predictive Maintenance Digital Twins for Advanced Nuclear Reactors", which received $5.4 million of funding. The project will help nuclear plants move from a time- to a condition-based predictive maintenance framework, using GE Hitachi's BWRX300 boiling water reactor as the reference design. GE will develop digital twins that enable continuous monitoring, early warning, diagnostics, and prognostics for the reactors. The algorithms are supposed to send the reactors into a default known safe operation mode in the event of anomalies.[612]

Digital twins of nuclear plants are also progressing in France. The Digital Reactor Structuring project (le Projet Structurant Pour la Compétitivité, PSPC) launched in September 2020, and consisting of EDF, Framatome, French Alternative Energies, and Atomic Energy Commission (CEA), and six additional organizations from academia and the French nuclear sector, aim to digitally clone all of France's nuclear plants within four years. The digital twins will evolve in line with the design and modifications of each plant.[613]

Peak Capacity Strategies: Shutting Down and Restarting for Fueling and Maintenance

Since nuclear power operations are often supported by large capital costs and long-term supply contracts, and they maintain a buffer stock of uranium for two years on average, they are essentially all "peaker plants."

Fueling and maintenance operations, however, cause major operational disruptions. Most reactors need to be shut down for refueling in order to open up the reactor vessel. About a quarter to a third of the fuel assemblies are replaced with fresh ones every 12, 18, or 24 months. Some plant designs (the CANDU and RBMK types) allow fueling under load, by allowing individual pressure tubes to be disconnected and reconnected.[614]

One potential benefit of the aforementioned floating nuclear barges is that they can be towed out of service and replaced by other floating barges during their maintenance cycles. The "Lomonosov" is designed to operate in three 12-year operational cycles, at the end of which the vessel is required to be towed back to the RosAtomFlot shipyard in Murmansk for almost a year for repairs, defueling, refueling, and radioactive waste removal. If replacement units can be towed to the location to serve the power needs during this year of maintenance, then the avoidance of the need to buy alternative power during the year of the maintenance turnaround cycle may possibly lower the operator's long-term operating costs.

Sourcing Trade-offs: Being Willing to Pay Extra for Safety

Operators will face increasing pressure to adopt safe fuels, even if they cost more. There is a significant industry initiative underway "to develop accident-tolerant fuels which are more resistant to melting under conditions such as those in the Fukushima accident, and with the cladding being more resistant to oxidation with hydrogen formation at very high temperatures under such conditions."[615] Predictably, fuel buyers will need to liaise with executive management on the trade-offs between cheap fuel and potentially more costly accident-resistant fuel. Insurance products and financial instruments may emerge to equalize the cost.

TCO Trade-offs: Extending MTBR with Smart Maintenance

While high capital cost at the construction phase is often relatively unquestioned due to safety priorities, cost savings can often be reaped through efficient use of technology during maintenance and repair.

In a bid to lower the total cost of ownership of Electrabel's seven Belgian nuclear power plants, with a combined capacity of 5.9 GW, Engie launched a two-year program to implement AP 913 performance standards in order to improve equipment reliability and reduce maintenance costs. This program was called the Equipment Reliability Implementation Project (ERIP), under which Engie worked towards reducing Operations and Maintenance (O&M) costs by 5%.[616]

Under the ERIP, the classification of some 400,000 components will drive maintenance strategies, and synergies will be maximized when working on similar components to ensure benefits are gained as early as possible. The ERIP will first focus on the non-nuclear section of the plant, where issues account for around 30% of power availability losses. Classification will use a four-tier system based on criticality. A key driver of performance improvements will be reducing failures of Single Point Vulnerability (SPV) components. Engie will also perform an equipment qualification program to identify ageing equipment, and obsolete components and data resources will be improved by identifying and retrieving missing data. The operator has also performed a risk-based analysis of long-term component costs which has increased the Mean Time Between Repair (MTBR) and reduced the number of regular outages over the long term.

Supply Chain Roadmap for Nuclear Power

Nuclear supply chains can become more customizable through mini-nuclear and floating nuclear, more rationalized by optimizing economies of scale, more innovative via pioneering fuels and waste technologies, and more synchronized by reducing the bullwhip effect of the capital investment cycles inherent in megaprojects and gigaprojects (see Figure 82).

Safer Fuels and Waste & Disposal can be game-changers for nuclear power by lessening the safety concerns that led Germany to effectively ban nuclear power and decommission its existing plants and prompted many other countries to oppose initiating or continuing nuclear power development. Safer fuel composition would reduce inbound logistics complexity and cost, as well as potential dependency on uranium-rich countries. Safer waste disposal would reduce the complexity and cost of outbound logistics. The strategic and operational implications of safer fuels and waste disposal make supply chain an integral theme in the ongoing development of nuclear power.

AI, Predictive Maintenance, & Remote Digital Twins are increasing reliability as well as public and Operations & Maintenance staff safety. The ability to visualize components and operate the plant from a dashboard means less intervention and earlier detection and correction. The nuclear industry can leverage off AI initiatives in the oil & gas industry to accelerate these advances.

Modularization & Mini-Nuclear Design offer the potential to make nuclear cost-effective at a smaller scale, thereby mitigating the effect of infamous capital construction cost overruns. This could spur nuclear's revival, just as mini-mills offered the US steel industry a second wind in the late 1980s and early 1990s. As every aspect of capital and operations would change with modularization and miniaturization, supply chain professionals need to redesign operations, maintenance, procurement, and logistics costs and refresh best practices as inputs to the optimal scale and configuration of mini-nuclear plants.

Floating Nuclear not only benefits from the mini-nuclear concept described above, but also provides the ability to reposition power where it is needed and to swap power sources during refueling operations, thereby eliminating shutdowns and in some cases the need for backup conventional power options. The combined benefits are exceedingly financially attractive. Supply chain is integral to the emergence of floating nuclear, including the dispatch and movement of vessels, the optimization of refueling and plant maintenance schedules, and the logistics of turnarounds and "black starts."

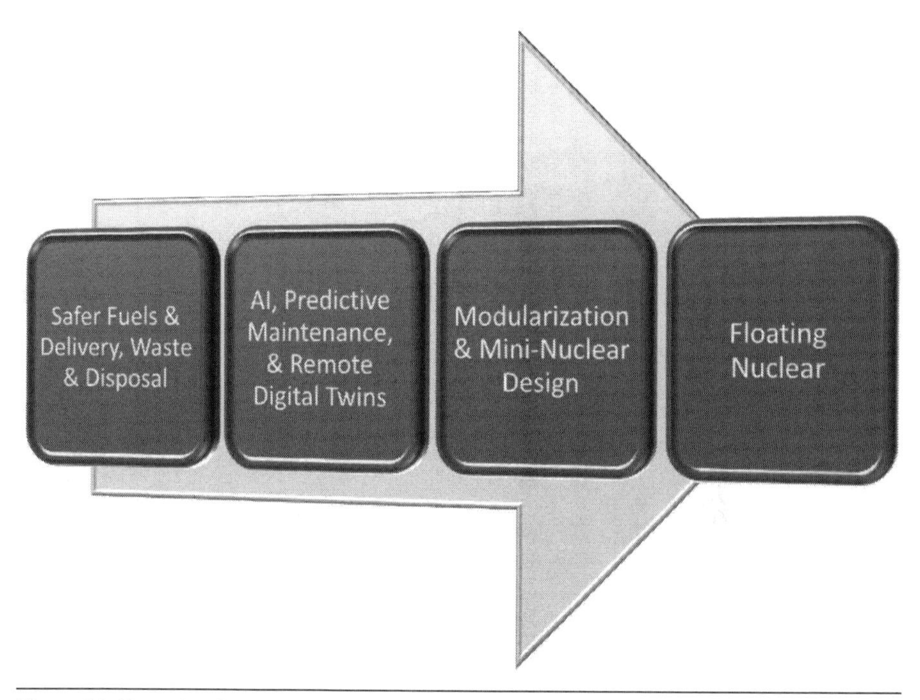

Figure 82. Emerging Supply Chain Opportunities for Nuclear Power

Conclusion

The rapid emergence of information technology and the push for clean energy have disrupted the energy business over the last decade, creating new value chain imperatives.

Supply chain strategies and management are enabling new energy product and market opportunities. The modularity and small-scale production potential of some renewable power technologies is creating new opportunities to customize and innovate, not just rationalize and improve asset productivity of centralized plants. In this way, supply chains are the "tipping point" for the new energy future. Eventually, discretely manufactured power cells—from fuel cells to batteries to solar pico solar systems—may be embedded in everyday devices from automobiles to rooftops to dishwashers, phones, and office buildings. In the long run, energy will consist of many widely varying technologies, production modes and distribution channels.

Every executive and manager who takes or influences decisions is challenged to make skillful and appropriate use of the unique supply chain types, profit margins, economies of scale, growth rates, and asset intensity characteristics of each technology. While there are similarities between asset-intensive, continuous production energy operations such as oil & gas, thermal power, nuclear, and hydropower, several of the emerging technologies present very different supply chain types that are either hybrids of conventional continuous production operations (for example, hydrogen production coupled with fuel cells) or discrete mass production operations like solar PV. Below are four unifying principles to remember.

1. The supply chain strategy should be consistent with the economic leverage points of the business. Oil & Gas (O&G), Thermal, Nuclear, Hydropower, Geothermal, and certain Biomass plants focus heavily on CapEx and asset reliability, drilling, earth-working, and excavation. In contrast, the economics of distributed technologies such as energy storage and fuel cells will require a supply chain characterized by discrete manufacturing and weighted toward activities close to the point of power consumption, with capabilities for distribution, installation, commissioning, permitting, and charging networks more than on centralized generation or transmission. Wind and solar firms can adopt any of the four supply chain strategies, and the right one depends on their business strategy.

2. The value chain should be clean, green, and smart. As energy policy is becoming an area of increasing international political interest, managers will need to become more flexible and adaptable to a variety of regulatory directives and regimes. Many governments, non-governmental organizations, and companies have established carbon-neutrality target dates, and these continue to creep forward with a direct impact on plant technology and design as well as construction and retirement schedules. In addition, presidential and regional directives can influence imports and exports of fuel and power, and tariffs can change the supply chain practically overnight. Project managers need to be able to reconfigure their plant technology, network, and supply base more rapidly than before in response to this new geopolitical reality.
3. Information technology should make operations and maintenance automatic, digital, smart, and secure. Automation, the Internet of Things, artificial intelligence, and machine learning have in many cases already supported the digital replication of mechanical operations and are on their way to becoming the primary interface, rather than a shadow of the actual operation. To keep up, engineering, operations, and maintenance staff need to be digitally trained and resourced. Furthermore, due to the breadth and depth of the digital footprint they especially need to maintain cybersecurity across their enterprises and with supply chain partners.
4. An end-to-end value chain vision (managing your suppliers and those suppliers' suppliers) should include economic targets, and more importantly, ensure safety and environmental stewardship. Demand planning, multi-echelon network modeling, and unbundling analyses require visibility up and down the value chain. Moreover, safety with every power technology but especially oil & gas, nuclear, hydrogen fuel cells, and CSP requires coordination with suppliers and customers, or else "faultlines"—small misalignments and omissions—will become exposed and could have severe consequences.

These first principles provide a foundation for more specific guidance. Although the actions with the greatest potential to increase supply chain value will vary by company and over time, this book has pointed toward some areas where producers, suppliers and investors may find particular value.

For **Hydrogen/Fuel cells**, supply chain reliability, efficiency and safety may drive the ability of the technology to compete with other technologies. High-potential areas include:

1. Reducing the potential for leaks and flammable conditions;
2. Realizing economies of scale in the gdl and the fuel stack;
3. Sourcing platinum to ensure adequate supply;

4. Protecting the FCCU from cyberattack;
5. Developing optimal chemistries for various power ranges and environments; and
6. Creating a market for replacement membrane elements.

For **Energy Storage**, game-changing supply chain improvements are likely to stem from eight axes of innovation:

1. Working with suppliers to minimize or eliminate the risk of overheating and fire;
2. Integrating energy storage with vehicle charging networks;
3. Developing economical battery recycling approaches;
4. Ensuring availability of critical minerals such as cobalt;
5. Reducing the cybersecurity attack surface;
6. Proactively managing battery costs throughout the supply chain ("should-cost");
7. Realizing multiple benefits simultaneously to increase the ROI of energy storage investments ("value stacking"); and
8. Deploying AI for battery operating system and load management ("Smart Load Management").

For **Wind power**, supply chain techniques can reduce cost and enhance the reliability of power. Supply chain imperatives include:

1. Adopting construction standards like those from IEC;
2. Bundling turbine acquisition and service agreements;
3. Structuring ownership with Design-Build-Operate in mind;
4. Using options in contracts;
5. Deploying machine learning algorithms that intelligently adjust energy storage, turbine and blade directionality, and electricity trading actions based on real-time physical and market conditions;
6. Developing and scaling Wind+Storage solutions;
7. Ensuring cybersecurity (through standards such as those from NERC for wind farms over 75 MW); and
8. Disposing of end-of-life blades in an environmentally friendly way.

For **Solar power**, many emerging innovations will open up new horizons that require supply chain analysis and support, including:

1. Maximizing benefits of IoT/AI, especially smart tracking, to optimally align the modules towards the sun;
2. Engaging a multi-stakeholder effort to secure the polysilicon supply chain;
3. Optimizing energy storage cost-efficiently by adjusting equipment parameters and specifications to minimize lifecycle cost of the integrated system;

4. Steering clear of low-priced equipment that entails lower operating performance and potential component failure;
5. Integrating rooftop solar and home vehicle charging systems;
6. Embedding solar receptors in building materials, ensuring safety and connectivity;
7. Improving the safety, scale, and reliability of floating solar installations;
8. Aligning regulatory frameworks and installer capacity to make distributed solar more viable;
9. Building supply chains that ensure human rights in silicon mining and component manufacture;
10. Avoiding environmental waste and emissions throughout the extended supply chain; and
11. Managing end of life disposal of toxic PV waste.

For **Biomass**, our recommended roadmap includes three steps:

1. "Value stacking" by adopting hybrid technologies, especially CHP and pyrolysis, to increase yield;
2. Evaluating multi-stage process routes such as co-firing and anaerobic digestion pre-treatments to increase efficiency; and
3. Regrowing natural resources to minimize CO_2 footprint.

For **Upstream—Oil & Gas**, technological solutions can significantly improve efficiency, safety, and environmental footprint. Supply chains can be strengthened by the following measures:

1. Evaluating the viability of Carbon Capture and Storage (CCS) solutions widely wherever it is technically and economically viable;
2. Digitalizing operations with the help of smart IoT devices including optical readers, sensors, virtual reality, artificial intelligence, pattern recognition, and full digital twins;
3. Using renewable energy + storage to power drilling operations and production platforms;
4. Converting product purchases to product-as-a-service contracts to incentivize suppliers to create value, not just reduce prices;
5. Deploying robust cybersecurity protections and safeguards, including penetration tests; and
6. Using remote equipment such as drones and robots to conduct operations that may be safety risks for people.

For **Midstream—Oil & Gas**, value chains can be made more efficient, customized, and secure by:

1. Continuing to increase LNG terminal and infrastructure capacity and optimize physical flows to increase scale economies in natural gas distribution;

2. Deploying modular gas processing solutions to offer customized natural gas process routes and delivery options;
3. Using drones to surveil and inspect pipelines worldwide to reduce labor and improve safety; and
4. Making SCADA systems smarter and more resistant to cyberattacks.

For **Downstream—Oil & Gas**, most roadmap actions relate to digitalization, and especially AI, IoT, and cybersecurity, due to the susceptibility of downstream facilities to intrusion. We recommend four vectors of investment to protect and enhance supply chain value:

1. Monitoring and governing multi-tier supply chain risk more rigorously;
2. Ensuring cybersecurity for IoT devices and systems;
3. Building smarter predictive maintenance algorithms; and
4. Producing low sulfur fuel for midstream transport.

Geothermal power supply chains may yield cost efficiency and asset productivity by:

1. Retrofitting binary plants instead of building new flash plants;
2. Integrating hybrid energy technology formats such as Solar PV and district heating; and
3. Adding energy storage where needed to make uptime even more reliable.

Gas and coal-fired power operators can continue to take action to ensure reliability, cost-efficiency, and security, for example, by:

1. Improving predictive maintenance and reliability engineering to better align production with demand;
2. Strengthening cybersecurity to assure stability of production;
3. Co-producing with local renewable resources where possible to reduce CO_2 emissions;
4. Establishing performance-based agreements with vendors to assure reliability and reduce cost;
5. Using lifecycle cost management principles and tools to minimize capital and operating costs; and
6. Researching innovations in the uses for coal to sustain jobs that are in jeopardy as a consequence of the shift toward clean power.

Hydropower supply chains can be improved to lower cost and unavailability risks, sometimes in combination with other power generating technologies. Major axes of progress should include:

1. Investing in predictive maintenance with vibration sensing, which detects repair and maintenance requirements earlier.
2. Buffering against both demand fluctuation and supply variability with energy storage.

3. Hedging against decreased water flow with novel financial instruments.
4. Evaluating the return on investment from hybrid power generation especially green hydrogen.
5. Researching wave & current energy capture as potential product and revenue extensions.

For **Nuclear power**, high-impact supply chain imperatives include:

1. Standardizing on the use of safer fuels and fuel waste & disposal;
2. Using AI, predictive maintenance, & remote digital twins more extensively to increase safety and reliability.
3. Modularizing design, including mini-nuclear plants, which can widen adoption; and
4. Extending pilot programs involving floating designs to reduce downtime from scheduled heavy maintenance.

Hopefully this book has inspired you with new ways of thinking of supply chain value creation, and has been informative regarding how to manage organizations, procedures, and processes to maximize the potential of the extended energy value chain. If you have any feedback or questions, or suggestions for the next edition, please email them to david@bostonstrategies.com.

Appendix 1

Bullwhip in the Oil and Gas Supply Chain: The Cost of Volatility

Filling an Important Research Gap

Since Jay Forrester described the bullwhip effect in 1958, a vast body of literature has examined how to avoid bullwhip in an industry-agnostic way. This literature has explored, among other topics, the effect of gaming and adjusting order placement mechanisms. For example, Kimbrough, Wu, and Zhong explored how artificial agents can dampen the bullwhip effect (Kimbrough, et al., 2002), and Ouyang tested four different ordering methods (order up-to, kanban, generalized kanban, and order-based (Ouyang, et al., 2006).

Many authors have explored evidence of this effect in consumer products and electronics. A consulting study assessed the potential savings from implementation of efficient consumer response (ECR) in that industry at $30 billion, or 12.5–25% of cost (Poirier, et al., 1996). Procter and Gamble discovered wide swings in the production and inventory of diapers despite relatively smooth demand (Lee et al., 2006). Italian pasta maker Barilla's volume discounts and promotional pricing induced volatility in a system with inherently low end-user demand (Lee, Whang, 2006). HP measured the bullwhip effect by measuring the standard deviation of orders at the stores to the standard deviation of production at the upstream suppliers (Lee et al., 2006). Network systems manufacturer Cisco studied the impact of the right to cancel orders on the magnitude of the bullwhip effect (Lee, Whang, 2006).

Few published works have applied system dynamics to the oil and gas industry, and none has quantified the cost of volatility on the upstream oil and gas supply chain. The existing literature in oil and gas has focused on predicting the price of oil rather than analyzing the effect of volatility on the cost of the oil and gas supply chain. For example, Mashayekhi built a model of oil price drivers, showing that as demand increases, oil price rises, which causes producers to expand capacity, forcing prices down and depressing demand, in a feedback loop (Mashayekhi, 2001). John Sterman came the closest to studying the bullwhip effect in the oil and gas industry (exploration and production, or "E&P") when he noted that "oil and gas drilling activity fluctuates about three times as much as production." (Sterman, 2006).

This study shows evidence of the bullwhip effect in the upstream oil and gas supply chain and demonstrates that over time the cost of gasoline is 10% higher as oil producers, oil refiners, heavy equipment suppliers, and their component suppliers pass on the costs of inventory overages and shortages, poorly timed capacity investments, and inflationary prices.

Evidence of the Bullwhip Effect

The pattern of drilling activity and capital investment on the other hand is characterized by oscillation and amplitude magnification, and it appears that oil price shocks are the root cause. From 1949 until 1973, the average annual price of oil fluctuated within a 7% band, but from 1981 through 2008 the variation leapt to almost 10 times that amount. The 1973 and 1979 oil crises and the sharp escalation and crash of oil prices between 1998 and 2009 introduced a new and seemingly systemic unpredictability to oil prices.

Since the recent period of oil price volatility that began in 1998, oil drilling investment and activity has tracked the price of oil and has in some cases exaggerated the pattern set by the oil price.[XIII] The price of oil rose and fell by 52%[XIV] from its peak of $97 per barrel in 1980 to a low of $17 in 1998 and rose again to reach a new high of $97 in 2008. As the price of oil rose and fell between 1998 and 2008, capital expenditure for major oil companies rose and fell by 25–63%[XV] and the total US rig count (oil and gas) rose and fell by 36%.

Regardless of the reason behind the *initial* shocks—some think the pattern is cyclical, and there is evidence that it could be chaotic[XVI XVII]—the variation from a steady state historical demand clearly induced oscillating and increasing reverberations in production, capacity, and inventory from 1995 to 2009 in markets for oil & gas field machinery and equipment, including turbines & turbines generator sets, motors, electrical equipment, and iron castings.

Bullwhip effect is evident in six oil and gas supply markets between 1995 and 2009, as indicated by the measurements in Figure 83 below, where:

- Amplification Ratio = Variance [production] / Variance [demand]
- Amplification Difference = Variance [production] - Variance [demand]

and demand is measured by orders received and production (or supplier's production capacity) by units manufactured. Bullwhip effect exists if the Amplification Ratio is greater than 1 and the Amplification Difference is positive.

[XIII] Analysis of data from Baker Hughes and EIA.

[XIV] Based on annual data; using monthly data from 1998–2009, the swing was 66%.

[XV] Chevron: 63%; Royal Dutch Shell: 41%; Exxon Mobil: 26%; BP: 25%.

[XVI] Many factors have contributed to the recent volatility, including political crises, financial speculation, and a sharp increase in demand from developing economies.

[XVII] Two-thirds of respondents to a 2009 Boston Strategies International survey felt that oil prices are caused by speculation by commodity traders and distortions in financial markets.

Type of Equipment	Metric	Variance	Amplification Ratio	Amplification Difference
Oil & Gas Field Machinery & Equipment	Demand	0.002	5.763	0.0094
	Production	0.0114		
Turbines & Turbines Generator Sets	Demand	0.0057	3.2063	0.0127
	Production	0.0184		
Motors and Generators	Demand	0.0076	2.2201	0.0093
	Production	0.017		
Engine Electrical Equipment	Demand	0.0057	2.275	0.0073
	Production	0.0131		
Standard Malleable Iron Castings	Demand	0.0057	4.6989	0.0212
	Production	0.027		
Steel Investment Foundries	Demand	0.0036	3.5495	0.0093
	Production	0.0129		

Figure 83. Evidence of Bullwhip Effect in the Oil & Gas Equipment Industry (*Source:* Boston Strategies International)

The swings in capital investment by oil companies caused even bigger swings in the equipment supply chain, causing oscillation in production, inventory, and backlog. While production of turbines and engines declined by 7% between 1998 and 2008, inventories rose by 24%, as shown in Figure 84. In an analogous period when new orders spiked three times in 12 years, the backlog of turbine generators tripled and then plummeted to nearly zero twice.[XVIII]

This bullwhip effect causes four types of economic inefficiency at oil companies and their heavy equipment suppliers:

1. Oil companies pay higher prices that are set when markets are overheated and never rolled back when recession hits.
2. Equipment manufacturers hold excess inventory during the boom and take a long time to draw it down whena recession hits.
3. Equipment manufacturers make excessive capacity investments near the peak and then suffer a low or negative return on investment on it.
4. Component and parts suppliers lose orders that they are not able to fulfill at the peak due to inadequate capacity and long lead times caused by large backlogs.

[XVIII] Based on sales of the three largest turbine generator manufacturers between 1948 and 1962. Source: Exhibit material, Ohio Valley Electric vs. General Electric, Civil Action 62 Civ. 695, Second U.S. District Court of New York, 1965.

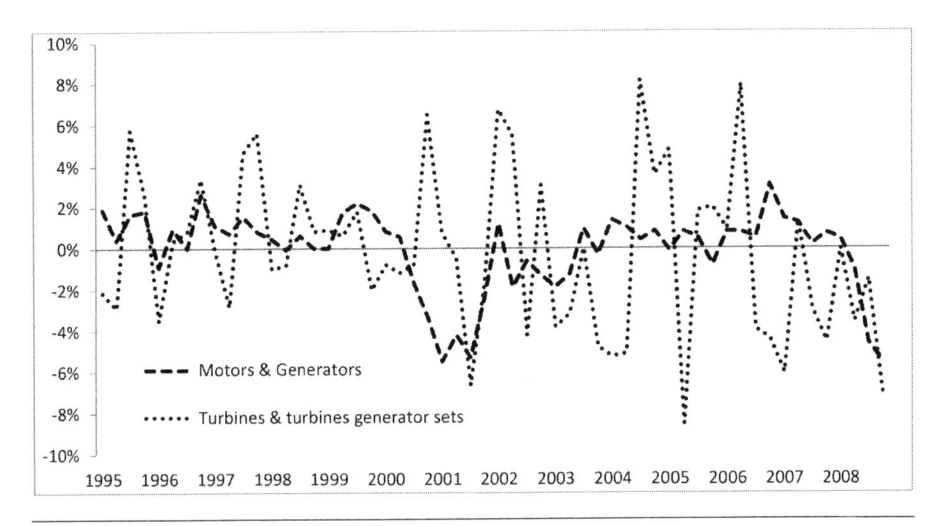

Figure 84. Demand Volatility of Motor, Generator, and Turbine Sales

(*Source:* Boston Strategies International Inc. after IHS Global Insight. David Steven Jacoby, The oil price 'bullwhip': problem, cost, response, *Oil & Gas Journal,* March 22, 2010.)

The Cost to E&P Companies, Refiners, OEMs, and Component Suppliers

The authors constructed a system dynamics simulation of a four-tier supply chain in order to test for the presence of, and quantify the impact of, bullwhip in the upstream oil and gas supply chain. The four tiers were oil producers, oil refiners, heavy equipment suppliers, and their component suppliers, as illustrated in Figure 85.

We used the model to run two scenarios: a flat oil price scenario and a volatile oil price scenario. In the flat oil price scenario, we simulated an initial shock and traced the after-effects on the supply chain. In this case the initial shock was an increase in the price of oil from \$30 to \$60/barrel. The price of oil rises to a peak of \$90/barrel, drops to \$30/barrel, and then rises back to \$60/barrel to complete a sine wave, with a cycle of 20 years (the whole simulation lasted 43 years). In the volatile oil price scenario (see Figure 86), after the initial shock, oil price fluctuates in a sine wave with the same overall amplitude as under the Smooth Price Cycle scenario but with random oscillation.

Over an extended time frame, the initial increase in demand for oil raises the production levels of refined oil and drilled crude oil, which translates into increased demand for oil field equipment such as oil and gas compressors and turbines. Excess production is higher at the refiner than at the driller, higher at the OEM than at the refiner, and higher at the component supplier than at the OEM during most periods.

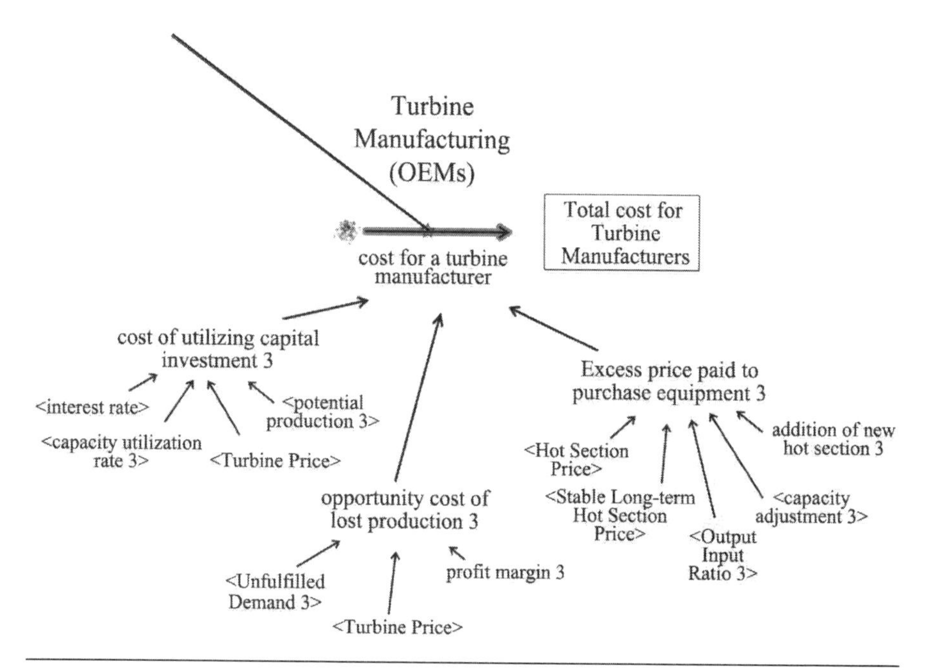

Figure 85. Supply Chain Simulation Architecture
(*Source:* Boston Strategies International)

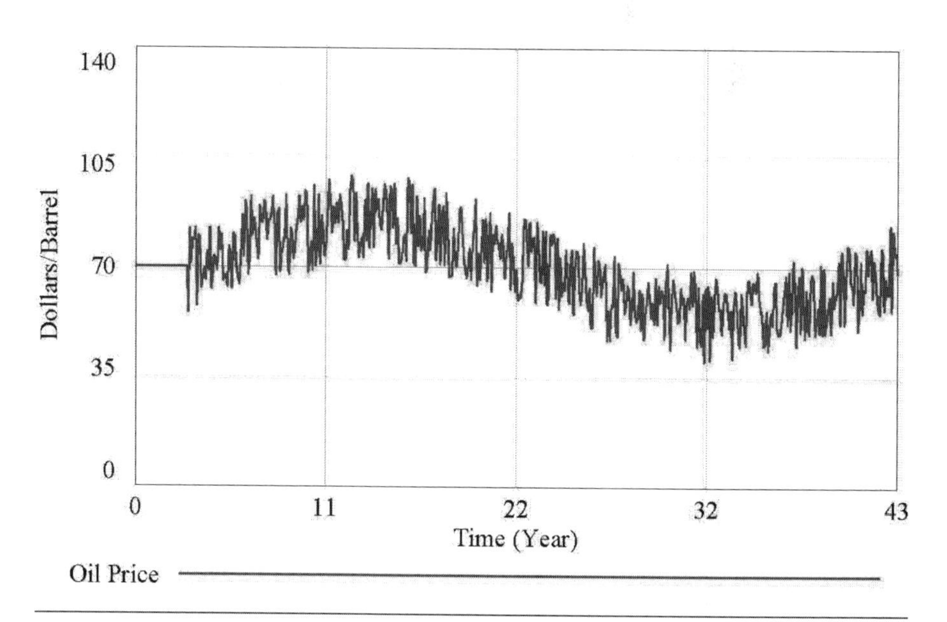

Figure 86. Oil Price in the Volatile Oil Price Scenario
(*Source:* Boston Strategies International)

Refiners and upstream oil producers pay higher prices that are set when markets are overheated and not de-escalated when recession hits. These price hikes add 5% per year to the cost of the equipment, materials, and services that producers buy, after adjusting for inflation caused by metals prices and pure commodity inputs. Moreover, equipment and service prices keep rising even as the price of oil falls, equipment orders drop, capacity utilization drops 15%, and lead times decline to manufacturing throughput time. Capacity adjusts, with a lag, as orders and production fluctuate. The annual cost for refiners is the highest under rough price scenario from years 8 to 17.

Equipment OEMs incur high costs in Years 11 through 22 as orders grow due to rising oil prices, which causes equipment OEMs to make excess capacity investments and pay high prices for components as those costs inflate as well. In fact, prices of turbine hot sections double over a 22-year period in the simulation. The capacity additions weigh heavily on the OEMs finances as orders and backlog decline and bottom out, and the OEMs carry that excess capacity for four years too long (although to a lesser degree each year). The equipment manufacturers also hold excess inventory, which adds 8% to the cost of the equipment, similar to the way in which OEM manufacturers doubled their inventory between 2004 and 2008, which then became redundant when orders dropped off, and took 12 months to draw it down when the recession hit.

Component suppliers lose orders on the upswing and hold excess inventory on the downswing. Component suppliers are the last ones to see backlogs decline due to their upstream role in the supply chain, and the approximate halving of their backlog amounts to a depletion of inventory. So, for most of the time during Years 11 and 22, they are depleting inventory. This inventory carrying cost is their prime supply chain cost. Component and parts suppliers also lose orders that they were not able to fulfill at the peak due to inadequate capacity and long lead times caused by their large backlogs.

Average annual supply chain costs over a 43-year period in a flat oil price scenario total $8.3 billion, while in the volatile oil price scenario they are $10.3 billion (see Figure 87). The difference, $2 billion, spread across 85 billion barrels of oil consumed per day, equates to roughly 6.4 cents per barrel. Considering that turbines and compressors represent only 5.8% of oil companies' total external expenditure on equipment, materials, and services, the impact extrapolated to all equipment and services in the oil and gas supply chain is $1.09 ($0.064/0.058). This was approximately 10% of the weighted average cost of producing a barrel of oil in 2008.[XIX] The chart below shows the cumulative cost of supply in the volatile oil price scenario for the first ten years of the cycle.

XIX Based on Boston Strategies International 2010 calculations of the all-in cost of purchased materials and services.

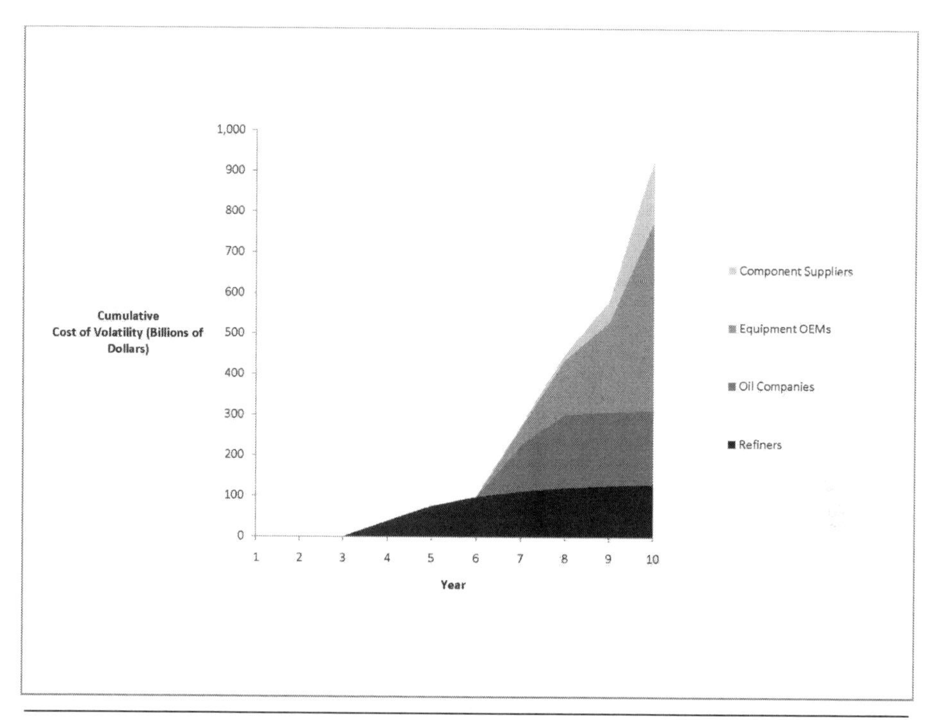

Figure 87. Cumulative Cost of Supply in the Volatile Oil Price Scenario
(*Source:* Boston Strategies International)

Suggestions for Further Research: How to Mitigate the Costs of Bullwhip

A stable environment would help to establish steadier prices and operating profits across the upstream oil and gas supply chain. Stable oil prices and input costs would minimize layoffs during downturns and rehiring during upturns, thereby reducing long-term operating costs. It would also encourage more stable R&D investments, which would result in higher exploration, refining, and distribution productivity due to faster and more consistent advances in oil & gas equipment technology. This would translate to a higher Return on Assets (ROA).

Therefore, further research on stabilizing strategies would be helpful in determining the course(s) of action that would most effectively mitigate the cost of bullwhip. The research should evaluate the effectiveness of strategies such as: 1) long-term supply agreements that are long enough to bridge demand-price-capacity cycles, which may last 20 years or more; 2) sharing of production plans between exploration and production companies, refiners, equipment OEMs, and component manufacturers; 3) sharing of capital investment; and 4) sharing of supply risk through price indexing and the use of options and futures contracts.

Appendix 2

Common Categories of Externally Purchased Equipment and Services

		O&G Upstream	O&G Midstream	O&G Downstream	Energy Storage	Solar	Geothermal	Hydropower	Wind	Gas- and Coal-Fired	Hydrogen/Fuel Cells	Biomass	Nuclear
1	**Exploration**												
1.1	Seismology	X					X						
1.2	Reservoir / flow modeling	X					X	X					
2	**Construction**												
2.1	EPC Services	X	X	X	X	X	X	X	X	X		X	X
2.2	Offshore Construction & Installation	X	X	X		X			X				
2.3	Onshore Construction & installation	X	X	X	X	X	X	X	X	X	X	X	X
2.4	Civil Works					X	X	X	X			X	
2.4.1	Iron & Steel	X	X	X	X	X	X	X	X	X	X	X	X
2.4.2	Non-Ferrous Metals	X	X	X	X	X	X	X	X	X	X	X	X
2.4.3	Rebar	X						X	X				
2.4.4	Pipefitting	X	X	X			X	X	X	X	X	X	X
2.4.5	Steel Fabrication/Erection	X	X	X	X	X	X	X	X	X	X	X	X
3	**Drilling**												
3.1	Tubulars excluding line & pipe	X					X						
3.2	Fluids	X					X						
3.3	Downhole Tools	X					X						
3.4	Bits	X					X						
3.5	Surface Equipment	X					X						

3.6	BOPs	X					X						
3.7	Muds / special chemicals	X											
3.8	Pressure Pumping and Cementing	X											
3.8.1	Centrifugal Pumps	X	X	X		X	X	X				X	X
3.8.2	Reciprocating Pumps	X											
3.9	Well Fracturing & Stimulation	X											
3.9.1	Flowlines & tiebacks	X											
3.1	Fluid Services	X											
3.10.1	Solids control	X											
3.10.2	Waste management	X		X						X	X	X	X
3.11	Formation Evaluation	X	X				X	X	X				
3.11.1	Wireline and mud logging	X											
3.11.2	MWD/LWD	X											
3.11.3	Coring	X											
3.12	Well Completions services	X											
3.13	Service Tools	X											
3.13.1	Coiled tubing	X											
3.13.2	Stimulation tools	X											
3.14	Fishing	X											
3.15	Rig operation	X											
4	**Production**												
4.1	Wellheads	X					X						
4.2	Muds and Completion Fluids	X											
4.3	Oil & Gas Production Chemicals	X											
4.4	Electrical Submersible Pumps	X					X	X					
4.5	Water Control/Profile Control	X											
4.6	Enhanced Oil Recovery (EOR)	X											
4.7	Artificial Lift	X											
4.8	Multiphase Flow Modeling	X											
4.9	Well Servicing	X											
4.9.1	Patch	X											

4.9.2	Casing	X											
4.1	Plug and Abandon, Decommission	X											
4.11	Contract Compression Services	X	X										
4.11.1	Centrifugal Compressors	X	X										
4.11.2	Air Compressors/Air Separation Units	X		X							X		
4.11.3	Reciprocating Compressors	X											
4.12	Laboratory Services	X											X
4.13	Production Chemical Services	X											
4.13.1	Corrosion	X	X	X	X	X	X	X	X	X	X	X	X
4.13.2	Sweetening	X											
4.13.3	Hydrate Inhibition	X											
4.13.4	Water Treatment	X											
4.13.5.1	Field Separation Systems	X						X					
4.13.5.2	Macrofiltration Equipment	X											
4.13.5.3	Media	X											
4.13.5.4	Membranes and Membrane Systems	X											
4.13.5.5	Flowmeters	X	X	X			X	X		X			
5	**Marine Services**												
5.1	Mooring	X							X				
5.2	Tug	X							X				
5.3	Anchoring	X							X				
5.4	Crew	X							X				
5.5	Tankering (VLCC, ULCC, etc.)	X											
6	**Offshore**												
6.1	Foundations								X				
6.1.1	Jacket								X				
6.1.2	Tripod								X				
6.1.3	Tripile								X				
6.1.4	Gravity Base Foundation								X				
6.1.5	Monopile								X				
6.1.6	Floating					X			X				

7	**Rotating Equipment**												
7.1	Compressors	X	X	X			X						
7.1	Pumps	X	X	X			X	X					X
7.1	Turbines	X	X	X			X	X	X	X		X	X
8	**Electrical Systems Installation and Management**												
8.1	Electrical Distribution & Control	X	X	X	X	X	X	X	X	X	X	X	X
8.1	Generators	X	X	X	X	X	X	X	X	X	X	X	X
8.1	Large Motors	X	X	X			X	X		X			
8.1	Lighting and Fixtures	X	X	X			X	X		X			
8.1	Wire and Cable	X	X	X	X	X	X	X	X	X	X	X	X
8.1	Uninterruptible Power Supply (UPS) and Batteries	X	X	X			X	X					
9	**Safety**												
9.1	Spill Prevention and Control	X	X	X									
9.1	Intrusion Security Systems	X	X	X	X	X	X	X	X	X	X	X	X
9.1	Fire Suppression Equipment and Systems	X	X	X		X	X	X		X	X	X	X
9.1	Respirators	X	X	X			X			X	X	X	X
9.1	Gas Detection	X	X	X						X	X	X	X
10	**Inspection Services**												
10.1	Inspection	X	X	X	X	X	X	X	X	X	X	X	X
10.2	Non-destructive evaluation	X	X	X	X	X	X	X	X	X	X	X	X
10.3	Coating	X	X	X	X	X	X	X	X	X	X	X	X
11	**Breakbulk and Project Logistics**												
11.1	Trucking/transport	X	X	X	X	X	X	X	X	X	X	X	X
12	**Processing/Refining**												
12.1	Pressure vessels			X			X						
12.2	Storage tanks			X			X						
12.3	Heat exchangers			X			X						
12.3.1	Fired Heaters for General Refinery Services			X			X						
12.3.2	Shell-and-Tube Heat Exchangers			X									

12.3.3	Air-Cooled Heat Exchangers			X									
12.4	Fired heaters			X			X						
12.5	Piping system			X			X						
12.5.1	Pipes and Valves			X			X						
12.5.2	Seals and Gaskets			X			X						
12.6	Pumps, Compressors, and Turbines			X			X						
12.6.1	Pumps	X	X	X		X	X	X		X		X	X
12.6.2	Compressors		X	X			X						
12.6.3	Steam Turbines			X		X	X	X	X	X		X	
12.6.4	Gas Turbines			X		X	X	X	X	X			
12.6.5	Steam System			X			X						
12.6.6	Cooling Water System			X			X						X
12.6.7	Other Equipment			X			X						
12.6.8	Vibration Monitoring Systems			X			X						
12.9.2	Squirrel-cage Induction Motors			X									
12.1	**Refinery Consumables**												
12.10.1	Chemicals and Catalysts			X									
12.10.2	Fuel / Electricity			X									
12.10.3	Claus Sulfur			X									
12.11	Desalter			X									
12.12	Atmosphere Crude Still			X									
12.13	Vacuum Pipe Still			X									
12.14	Coker			X									
12.15	Middle Distillate HT			X									
12.16	FCC Unit			X									
12.17	FCC HT			X									
12.18	HC Unit			X									
12.2	Naphtha HT			X									
12.21	Reformer			X									
12.22	Isomerization			X									
12.23	Alkylation			X									
12.24	Polymerization			X									
12.25	H2 Unit			X									

12.26	Satd, gas plant			X									
12.27	Amine Treater			X									
12.28	SCOT Unit			X									
13	**Power Generation—Capital**												
13.1	Boilers			X		X	X			X		X	
13.2	Turbines	X		X						X		X	
13.2.1	Steam Turbines			X		X	X			X		X	
13.2.2	Gas Turbines	X		X		X	X			X			
13.2.3	Microturbine	X											
13.3	Engines and generators	X											
13.3	Valves	X	X	X	X	X	X	X		X		X	X
13.3	Process Piping	X	X	X	X	X	X	X		X		X	X
13.3	Electronic control	X	X	X	X	X	X	X	X	X	X	X	X
13.3	Transmission	X	X	X	X	X	X	X	X	X	X	X	X
13.3.1	Converter	X	X	X	X	X	X	X	X	X	X	X	X
13.3.2	HV Transformers	X	X	X	X	X	X	X	X	X	X	X	X
13.3.3	Switchgear	X	X	X	X	X	X	X	X	X	X	X	X
13.3.4	Circuit Breakers	X	X	X	X	X	X	X	X	X	X	X	X
13.3.5	Substations	X	X	X	X	X	X	X	X	X	X	X	X
13.3.6	Array cables/Infield Cables	X	X	X	X	X	X	X	X	X	X	X	X
13.3.7	Export Cables HVAC & HVDC	X	X	X	X	X	X	X	X	X	X	X	X
13.4	Distribution	X	X	X	X	X	X	X	X	X	X	X	X
13.4.1	Vegetation management / Brush removal	X	X	X	X	X	X	X	X	X	X	X	X
14	**Power Consumables & Indirects**												
14.1	Fossil Fuels	X	X	X			X			X			
14.2	Coal									X			
14.4	Uranium												X
14.5	Biomass and Waste											X	
14.6	Maintenance Services	X	X	X	X	X	X	X	X	X	X	X	X
15	**Utility-Scale Energy Storage**												
15.1	Battery (Li-ion cells/ stacks)	X	X	X	X	X	X	X	X	X	X	X	X
15.2	Power Conversion System	X	X	X	X	X	X	X	X	X	X	X	X

15.3	Battery Management System	X	X	X	X	X	X	X	X	X	X	X	X
15.4	Monitoring & Controls	X	X	X	X	X	X	X	X	X	X	X	X
15.5	AC Transformer	X	X	X	X	X	X	X	X	X	X	X	X
15.6	AC Breaker	X	X	X	X	X	X	X	X	X	X	X	X
16	**Solar**												
16.1	Solar PV Cells/ Array/ Modules					X							
16.1.1	Monocrystalline Solar Panels					X							
16.1.2	Polycrystalline Solar Panels					X							
16.1.3	Amorphous Thin-Film Solar Panels					X							
16.1.4	Bi-facial Panels					X							
16.2	Power Inverter					X							
16.3	Switches	X	X	X	X	X	X	X	X	X	X	X	X
16.4	Cables	X	X	X	X	X	X	X	X	X	X	X	X
16.5	Battery	X	X	X	X	X	X	X	X	X	X	X	X
16.6	Charge Controller	X	X	X	X	X	X	X	X	X	X	X	X
16.7	Transformer	X	X	X	X	X	X	X	X	X	X	X	X
16.8	Mounting structures					X							
16.9	Solar tracker					X							
16.1	Solar PV meters					X							
17	**Geothermal**												
17.1	Drilling	X					X						
17.1.1	Drilling bits	X					X						
17.1.2	Protective tubes	X					X						
17.1.3	Inner drill pipes	X					X						
17.1.4	Ring drilling crowns	X					X						
17.1.5	Drive flanges						X						
17.1.6	Discharge sockets						X						
17.1.7	Changeovers and Kelly stubs						X						
17.2	Steam Production Unit						X						
17.3	Steam Water Separator						X						
17.4	Control Valve	X	X	X			X	X		X		X	X
17.5	Steam Turbine/ Generator	X		X			X	X	X	X		X	X
17.6	Condenser						X						

17.7	Cooling Tower			X			X						
17.8	Transformer					X	X	X	X	X	X	X	X
17.9	Transmission Power Lines					X	X	X	X	X	X	X	X
18	**Hydropower**												
18.1	Concrete for construction of Dam, Forebay, intake structure, reservoirs etc.							X					
18.2	Penstock pipes (made of steel or concrete)							X					
18.3	Valves	X	X	X			X	X		X		X	X
18.4	Gates	X	X	X			X	X		X		X	X
18.5	Hydraulic Turbines							X					
18.5.1	Impulse Hydraulic Turbines							X					
18.5.2	Reaction Hydraulic Turbines							X					
18.6	Generator	X	X	X	X	X	X	X	X	X	X	X	X
18.7	Draft Tube							X					
18.8	Power House							X					
19	**Wind**												
19.1	Wind Turbine								X				
19.1.1	Tower								X				
19.1.2	Hub								X				
19.1.3	Rotor								X				
19.1.4	Blades								X				
19.1.5	Shafts								X				
19.1.6	Wind vane								X				
19.1.7	Nacelle								X				
19.1.8	Generator	X	X	X	X	X	X	X	X	X	X	X	X
19.1.9	Shafts								X				
19.1.9.1	Low-speed shaft								X				
19.1.9.2	High-speed shaft								X				
19.1.10	Brake								X				
19.1.11	Gearbox	X	X	X		X	X		X				
19.1.12	Bearings	X	X	X	X	X	X	X	X	X	X	X	X
19.1.13	Brushes & Slip Rings								X				
19.1.14	Yaw Motor & Drive								X				
19.2.15	Transformers	X	X	X	X	X	X	X	X	X	X	X	X

19.3.16	Electronic (AC/DC) Converters	X	X	X	X	X	X	X	X	X	X	X	X
19.4.17	Cables	X	X	X	X	X	X	X	X	X	X	X	X
20	**Fuel Cells**												
20.1	Electrolytes				X						X		
20.1.1	Polymer Electrolyte Membrane										X		
20.1.2	Aqueous Potassium Hydroxide										X		
20.1.3	Phosphoric Acid										X		
20.1.4	Molten Carbonates										X		
20.1.5	Solid Oxide (Zerconia)										X		
20.2	Catalysts			X							X		
20.2.1	Anode catalyst, usually platinum powder				X						X		
20.2.2	Cathode catalyst, often nickel				X						X		
20.3	Bipolar Plates										X		
20.4	Electrodes (Anode & Cathode)				X						X		
20.5	Hydrogen (fuel)										X		
20.6	End Plates										X		
20.7	Gaskets										X		
21	**Biomass**											X	
21.1	Fuel											X	
21.1.1	Feedstock (organic waste)											X	
21.1.2	Fuel Storage Tankers											X	
21.1.3	Fuel Handling Equipment											X	
21.2	Combustor/ Furnace											X	
21.3	Boiler			X		X	X			X		X	
21.4	Pumps	X	X	X		X	X	X		X		X	
21.5	Fans											X	
21.6	Steam Turbine	X	X	X		X	X	X		X		X	X
21.7	Generator					X	X	X	X	X		X	X
21.8	Condenser											X	
21.9	Cooling Tower			X								X	X
21.1	Exhaust/ Emission outlet tower											X	
21.11	System Controls	X	X	X	X	X	X	X	X	X	X	X	X

22	**Nuclear**												X
22.1	Fuel: Uranium												X
22.2	Moderator: Grahite or heavy water												X
22.3	Control rods or blades												X
22.4	Pump	X	X	X		X	X	X		X		X	X
22.5	Generator					X	X	X	X	X		X	X
22.6	Turbine	X	X	X		X	X	X	X	X		X	X
22.7	Condenser						X						X
22.8	Cooling tower			X									X
22.9	Reactor												X
22.1	Heat exchanger			X			X			X			X
22.11	Pressuriser												X
22.12	Alternator				X	X	X	X	X	X	X	X	X
22.13	Transformer				X	X	X	X	X	X	X	X	X

Appendix 3

Registration, Evaluation, Authorisation, and Restriction of Chemicals (REACH)

REACH is a European Community Regulation on chemicals and their safe use (EC 1907/2006). It deals with the Registration, Evaluation, Authorisation, and Restriction of Chemical substances. The new law entered into force on 1 June 2007. REACH addresses the production and use of chemical substances, and their potential impacts on both human health and the environment.

The REACH Regulation gives greater responsibility to industry to manage the risks from chemicals and to provide safety information on the substances. Manufacturers and importers are required to gather information on the properties of their chemical substances, which will allow their safe handling, and to register the information in a central database run by the European Chemicals Agency (ECHA) in Helsinki. The Agency acts as the central point in the REACH system: it manages the databases necessary to operate the system, coordinates the in-depth evaluation of suspicious chemicals, and runs a public database in which consumers and professionals can find hazard information. REACH requires all companies manufacturing or importing chemical substances into the European Union in quantities of one tonne or more per year to register these substances with the European Chemicals Agency (ECHA).

REACH places a substantial reporting burden on any company that is shipping substances to or from or within the EU. Supply of substances to the European market which have not been pre-registered or registered is illegal. Companies must notify the European Chemicals Agency of the presence of SVHCs in articles or ingredients of articles if the total quantity used is more than one tonne per year and the SVHC is present at more than 0.1% of the mass of the object. Some uses of SVHCs may be subject to prior authorization from the European Chemicals Agency, and applicants for authorization will have to include plans to replace the use of the SVHC with a safer alternative (or, if no safer alternative exists, the applicant must work to find one)—a process known "substitution." To complicate matters, the filing companies must file through an "Only representative." Only Representatives are independent EU-based entities that must comply with REACH (Article 8) and should operate standard, transparent working practices. Often companies choose one party to be the "lead registrant" in order to simplify reporting and reduce the chance of overlap and redundancy. Huntsman was the lead registrant for over 40% of the chemicals that it buys or sells, running up to the November 30, 2010, deadline of pre-registration of high volume, dangerous, and intermediate products.[617]

The related legislation "Classification, Labeling, and Packaging of chemical substances" (CLP) (European regulation (EC) No 1272/2008) implemented the Globally Harmonised System (GHS) in Europe. The CLP will replace the current Dangerous Preparations Directive (1999/45/EC) and the related Dangerous Substances Directive (Directive 67/548/EEC) of 2015. Another recent change affects US shipments: the US Department of Transportation's Final Rule 76 FR 3308 attempts to harmonize the US laws with international standards, in addition to enacting numerous detailed changes to the previous body of law. Although this one piece of legislation was more substantial than many, this law is just the latest in a series of near-monthly updates, all of which are important to shippers. Linde, the industrial gases supplier, launched a program to help its customers adopt the new Globally Harmonized System of Classification and Labeling of Chemicals. Since most countries have independent systems for the classification of hazardous chemicals and gases to regulate the safety of transportation, storage, use and disposal of such chemicals, there is a need to align these with the GHS.

Recommended action steps for shippers shipping to or from Europe are:[618]

- Verify conformance with previous registration deadlines (2008 and 2010).
- Register new substances that you buy, sell, or ship with the European Chemicals Agency (ECHA) within one month.
- Update the pictograms used for shipping labels.
- Check the substance-naming conventions, as some of them have changed.
- Update the format of Safety Data Sheets (SDS) or Material Safety Data Sheets (MSDS) sheets.
- Apply for approval of low-risk and lower-volume (from 100 to 1000 metric tons per year) substances.

Recommended action steps for shippers within the US would include:

- Determine if you have a placardable shipment, and if so, register it. A placardable shipment is a bulk shipment of a substance that falls into one of the Pipeline and Hazardous Materials Safety (PHMSA)'s nine hazard classes, or a shipment greater than 1,000 pounds, or any shipment of agents or toxins identified in 49 CFR § 172.101.
- Check to see if you are affected by any recent changes in PHMSA rulings, especially Final Rule 76 FR 3308. That rule made several important changes and reclassifications. Here are a few examples:

  - Sour crude oil is now addressed by the Hazardous Materials Regulations.
  - Corrosive substances (such as UN 3485, just to name one example) have been added to the list of regulated chemicals.
  - Other substances (such as Bromates like UN3213) need to have the letter "G" added to their shipping labels.
  - Quantity thresholds were aligned with international standards.

Summary management insights of a technical paper recently presented at OTC[619] offer a basic response to what needs to be a fully developed strategy for compliance to REACH:

- Designate a lead person in charge of compliance if you don't already have one.
- Scan the lists of substances in various risk categories to determine your exposure.
- See if one of your supply chain partners could serve as the lead registrant.
- Register the necessary chemicals by the stipulated deadlines.
- Update warehousing and materials management paperwork such as safety data sheets (SDS or MSDS) sheets and shipping labels so they conform to the new labeling standards.

Glossary of Terms, Acronyms, and Abbreviations

3PL: third-party logistics service provider. Service provider that manages transport carriers on behalf of a shipper.

4PL: fourth-party logistics service provider. Service provider that manages transport carriers and all other aspects of the logistics function—such as demand and shipment planning, information systems, and inventory management—on behalf of a shipper.

AFE: authority for expenditure.

ANP: National Agency of Petroleum, Natural Gas, and Biofuels (Brazil).

ANSI: (American National Standards International): an international body that prescribes product and process standards to ensure design quality and consistency and increase interoperability. Also, the parent organization of the electronic interchange that serves as the clearinghouse for US electronic data interchange (EDI) standards.

API: American Petroleum Institute

ATO: assemble to order. A subset of made to order (see MTO) in which standard components are pulled from stock to produce the product once an order is received, and customer- or order- specific elements are manufactured and assembled at the last possible point in time (postponement helps to reduce inventory and increase flexibility in production if implemented correctly).

bbl: barrel.

Bid Slate Development: the process of developing a “short-list” of qualified suppliers that will be invited to bid on a project.

BOM: bill of materials.

BOO: build, own, operate.

BOOT: build, own, operate, transfer.

bpd: barrels per day.

BSI: British Standards Institute.

BTO: build to order. See MTO.

Btu: British thermal unit. Unit of energy, approximately equal to 1,055 joules.

Bullwhip Effect: the amplification of changes in inventory or cost upstream in a supply chain generated by an initial disturbance (often a small change in demand) downstream in the supply chain. Bullwhip is caused by delays in the transmission of information throughout an extended supply chain, and exacerbated by overcorrection, promotions, batching, and tweaking of demand forecasts. Theoretically it can be eliminated by synchronizing the supply chain, but the degree to which it can be dampened is limited by the length of the supply chain (physical and over time).

Capacity reservation: a proactive approach whereby the scheduling system allocates a certain amount of the resources' capacity to attend to specific service types.

Capacity utilization: output as a percentage of possible production at current investment levels.

CAPEX: capital expenditure.

Category management: a process that defines the desired end state of procurement activities for each family of externally purchased goods or services.

CCGT: combined-cycle gas turbine.

CCS: carbon capture and sequestration.

CDP: Carbon Disclosure Project. A not-for-profit organization that enlists companies and municipalities to share their carbon footprint data, in an effort to reduce greenhouse gas emissions and encourage sustainable water use.

CFM: cubic feet per minute.

CHP: combined heat and power.

CNG: compressed natural gas.

COGS: cost of goods sold.

Concession agreements: a right granted by a government or parastatal organization to a private operator, by which the operator commits to output and reliability levels and agrees to adhere to certain rules.

Constraints management (also Debottlenecking): the process of successively identifying the binding constraint, eliminating that constraint, aligning other processes to the new throughput levels, and pursuing the next constraint.

Critical path: the longest combination of consecutive activities in a project plan. The project duration is constrained by the critical path.

DBO: design, build, operate.

DBOM: design, build, operate, maintain.

DBOO: design, build, own, operate. debottlenecking. See constraints management.

Debottlenecking (also Constraints Management): the process of successively identifying the binding constraint, eliminating that constraint, aligning other processes to the new throughput levels, and pursuing the next constraint.

Design Simplification: reduction in the complexity that reduces cost, increases performance reliability, or both.

Digital Oilfields: the application of information technology to obtain real-time intelligence about downhole conditions help operators minimize fluid loss, manage flows within the reservoir, reduce the need for interventions, thereby maximizing production.

DOE: Department of Energy.

Downstream: refining (production of gasoline, kerosene, diesel, and other distillate fuel oils, residual fuel oils, and lubricants) and marketing (operation of a retail distribution chain including points of sale).

EMEA: Europe, Middle East, and Africa.

E&P: exploration and production.

E-procurement: catalogue-based electronic purchasing that may include one or more of the following processes: requisitioning, authorizing, ordering, receiving, and payment.

EOQ: Economic Order Quantity: algorithm that determines the number of units of an item to be replenished at one time based on minimizing the total costs of acquiring and carrying inventory, which are defined by the unit cost, usage, ordering costs, and inventory carrying cost.

EOR: Enhanced Oil Recovery. Any process for increasing the rate of recovery of hydrocarbon resources from a reservoir after the primary production phase.

EPA: Environmental Protection Agency. Government agency that establishes regulations pertaining to natural resources (air, water, etc.).

EPC: Engineering, Procurement, and Construction. Contracting arrangement that involves engineering, procurement, and construction, also used to classify a type of company that routinely executes contracts in this mode. EPC firms provide the engineering design, purchase all the equipment and materials needed for the project, and build or oversee the construction.

EPCI: Engineering, Procurement, Construction, and Installation. EPC contracting arrangement that also includes installation, which is particularly important in offshore construction, where installation of rigs, platforms, and foundations can be a highly technical and expensive project involving specialized vessels and equipment.

EPFIC: Engineering, Procurement, Fabrication, Installation, and Commissioning.

ERM: Enterprise Risk Management. Process of identifying, assessing, and managing risks across all the processes and organizational units, to increase sustainability and long-term financial health.

ERP: Enterprise Resource Planning. Companywide information system that centralizes data for increased visibility, accuracy, elimination of redundancy, and rapid access to information that supports decision-making at all levels.

ETO: Engineer to Order.

EVA: Economic Value Added. A measurement of financial strength that accounts for both cash flows (which are income statement factors) and efficiency of asset utilization (which is balance sheet factors). EVA is defined as net profit after tax less the cost of capital employed. The calculation is as follows: EVA= NOPAT - (capital × cost of capital), where NOPAT is the net operating profit after taxes.

FCC: fluid catalytic cracking.

FEED: front-end engineering and design.

FGSO: floating gas storage and off-loading unit.

FID: Final Investment Decision. Decision on the part of an organization to definitively move forward with a capital project, assuming all the legal and financial risks attached thereto, including commitments to suppliers and supply chain partners. Organizations may budget for a project internally before they reach an FID.

FPSO: Floating Production, Storage, and Offloading vessel framework agreement. Stipulation of the terms and conditions that will govern a relationship that both parties expect will consist of multiple contract awards over an extended period of time. While non-Europeans may use the term generically to refer to agreements that cover some but not necessarily all of the terms of subsequent contracts, the European Commission's Directive 2004/18/EC and Directive 2004/17/CE for utilities lay out specific legal requirements for framework agreements as they should be applied in public procurement.

Framework Agreement: an agreement that stipulates the terms and conditions that will govern a relationship that both parties expect will consist of multiple contract awards over an extended period of time.

GCC: Gulf Cooperation Council. Group founded in 1981, comprising Saudi Arabia, Kuwait, the United Arab Emirates, Qatar, Bahrain, and Oman. (Also called Cooperation Council for the Arab States of the Gulf [CCASG].)

Global sourcing. Procurement process involving identifying, pre-screening, and possibly qualifying potential sources in foreign countries, either to lower the cost of externally purchased materials and services by switching suppliers or applying competitive pressure on existing suppliers, or to increase quality or technological advantages by introducing new capabilities.

GOSP: gas/oil separation plant; or gas/oil separation process.

GTC: gas to commodity.

Green Hydrogen: a process whereby renewable energy is used to break water into hydrogen and oxygen, with the resulting hydrogen being used to power a fuel cell, all without any use of fossil fuels.

GTL: gas to liquids.

GTP: gas to power.

GTS: gas to solids.

GW: gigawatt.

HHI: Herfindahl-Hirschman index. Measure of market concentration that is calculated by adding the squares of the market shares of each firm competing in the market. If the HHI is between 1,000 and 1,800, the market is moderately concentrated; if it exceeds 1,800, the market is concentrated.

hp: horsepower.

HSE: Health, Safety, and Environment. Standards, regulations, policies, and organizations intended to protect workers and the environment.

IEA: International Energy Agency.

IEC: International Electrotechnical Commission

IMO: International Maritime Organization.

Integrated supply: when a vendor provides resources at the customer site to handle materials and inventory management services for its own and possibly other vendors' products, often in conjunction with a vendor-managed inventory program.

Inventory turns: measure of the velocity of inventory, calculated as the cost of goods sold (COGS) divided by the average inventory level.

IOC: international oil company.

IPP: independent power project.

ISM: Institute for Supply Management.

ISO: International Standards Organization. Group of not-for- profit institutions from 155 countries that develops and publishes standards with the goals of defining international best practices and transcending differences between national standards. Higher-level international standards facilitate and accelerate implementation of best practices worldwide and often eliminate redundancies and wasted resources.

JIT: just-in-time: a manufacturing philosophy that aims to reduce inventory to a one-piece flow involving no buffer stock. JIT is closely related to concepts that enable its execution, such as total quality, total productive maintenance, root-cause problem-solving, single-minute exchange of die (fast changeovers), level loading, Kanban, and Kaizen.

Joint process improvement: a formal and concerted effort to achieve business benefits by working collaboratively with supply chain partners. Typical benefits are lower overhead or direct costs, shorter cycle times, higher quality levels, and increased reliability and uptime.

Kanban: a tool used in just-in-time (see JIT) production whereby cards are attached to replenishment bins or containers to signal the consumption of one unit, thereby triggering a replacement unit. Kanban is often deployed within a batch production environment to trigger production of replenishment parts or components.

kg: kilogram.

KPI: key performance indicator.

KSF: key success factor.

KTPA: one thousand tons per annum.

kW: kilowatt.

LCCS: low-cost country sourcing. Procurement strategy to lower costs by switching from high-cost suppliers in mature economies to lower-cost suppliers in emerging economies.

Lean Management: a set of principles based on delivering what the customer wants without extra steps or waste of any kind. Lean programs such as JIT, Kaizen, and quality management share the same goal, but are distinguished by their focus on values and attitudes, including worker involvement and empowerment, that are needed to sustain continuous improvement in the long run.

Lean Six Sigma: deployment of lean methods, to eliminate waste and increase process efficiency, followed with Six Sigma methods, to reduce process variation.

LEC: levelized cost of electricity. Constant level of revenue necessary to recover all the expenses over the life of a power plant.

Life-cycle cost: total cost of a product, service, or solution from design through disposal, most frequently applied to equipment, in which initial purchase cost needs to be balanced against operating and maintenance costs over the duration of the useful life of the unit(s). Life-cycle cost includes indirect costs, such as the cost of downtime resulting from equipment unreliability, which may result from inferior design or insufficient or improper maintenance. (Also called total cost of ownership [TCO].)

LNG: liquefied natural gas.

LPG: liquefied petroleum gas. (Also called liquefied propane gas.)

LSTK: lump sum/turnkey.

LTA: long-term agreement. Contract with a term that is longer than most of the agreements to which an organization has historically committed with suppliers of similar products or services. Many buyers enter into LTAs to achieve lower cost, to secure guarantees, and to obtain better service from their suppliers. Many suppliers enter into LTAs to reduce their long-term cost of sales, to lock in stable revenue streams, and to solidify relationships with strategic customers. The length of contract itself varies widely among organizations since each has a different baseline or historical norm.

LTSA: long-term service agreement. Agreement between a customer and a manufacturer under which the original equipment manufacturer (see OEM) is responsible for maintenance, repair, and/or operation of the equipment for a fixed fee. (Also called contractual services agreement.)

M&A: mergers and acquisitions.

Macondo: the Macondo crisis refers to the largest marine oil spill in the history of the petroleum industry, which took place April 20–July 15, 2010. It involved an explosion and subsequent fire on the Deepwater Horizon, a semisubmersible mobile offshore drilling unit, which was owned and operated by Transocean and was drilling for BP in the Macondo Prospect oil field in the Gulf of Mexico. The explosion resulted in the deaths of 11 operators and subsequent spillage of 4.9 million barrels of crude oil.

MENA: Middle East and North Africa.

Midstream: transportation and storage of hydrocarbons via pipelines, oil tankers, LNG tankers, and tank farms, including related handling activities.

MMBtu: million Btus.

Monte Carlo simulation: computer-based repetition of theoretical events, using specific probability distributions, that is designed to provide decision-makers with data on likely outcomes, based on a set of possible choices and stochastic or random influencing variables.

MRO: maintenance, repair, and operating supplies. Parts or components used to support operations and maintenance activities in a manufacturing organization, including consumables and spare parts.

MRP: materials requirements planning.

MSDS: Material Safety Data Sheet. A document that must by law accompany products that contain certain hazardous materials. The MSDS provides information on how to respond if the product spills, if there is a fire, or if people are exposed to it unintentionally. (Also called PSDS [Product Safety Data Sheet].)

MT: metric ton. (Also called tonnes.)

MTBR: Mean Time Between Repairs. Interval between planned or unplanned repairs to a piece of equipment. MTBR is thus a measure of operating reliability.

MTO: Made To Order. Production paradigm whereby production of goods or service does not occur until the organization receives an order from a customer. (Also called BTO [build to order].)

MTPA: metric tonnes per annum; or, million tonnes per annum. MW: megawatts.

NACE International: the organization formerly known as National Association of Corrosion Engineers.

Network Optimization: a process of modifying the configuration of a set of nodes (e.g., plants or warehouses) and flows between the nodes (e.g., inbound or outbound shipments) to minimize the total cost of operating the system (see life-cycle cost).

Nm: newton-meter.

NOC: national oil company.

NPV: net present value.

OCTG: oil country tubular goods.

OECD: Organisation for Economic Co-operation and Development.

OEM: Original Equipment Manufacturer. Maker of equipment that sells directly to end customers or through a distributor or a value-added reseller. Distributor and value-added resellers are not OEMs.

OM: Operations Management. Process of adjusting demand and capacity at work centers throughout a constrained system, to generate output that satisfies customer goals such as cost, quality, and speed. OM is broader than supply chain management (see SCM); it applies to both product and service flows, whereas the latter applies primarily to physical flows.

OPEC: Organization of Petroleum Exporting Companies. operator. Producer of energy.

Operator: a producer of energy (in the context of this book)

OPEX: operating expenditure.

Opportunity cost: cost of an alternative that must be forgone in order to pursue a certain action.

OSHA: Occupational Safety and Health Administration. owner. Equity shareholder of an energy producer (in the context of this book).

Owner: an equity shareholder of an energy producer (in the context of this book)

PAO: polyalphaolefin.

Parcel tanker: tanker ship with multiple holds that are designed to carry many bulk substances, such as chemicals or different grades of petroleum, at one time.

PBL: Performance-Based Logistics. An arrangement under which a logistics provider commits to an agreed level of process performance for a fixed price, possibly including incentives and penalties, rather than conventional billing modes such as time and materials or hourly rates.

Power: any of the following industries: coal-fired thermal power plants (simple cycle: boiler + turbine + generator design); natural gas–fired thermal power plants; cogeneration plants (producing steam for an industrial process, e.g., refining); combined heat and power (CHP); combined cycle (gas turbine generates steam, which is recovered and fed to a steam generator); integrated power and desalination plants; nuclear; hydroelectric; wind; solar; biomass-fueled power plants; municipal solid waste; methane; and electrical distribution & control equipment (including monitoring devices, sensing switches, controls, connectors, motor controls, relays, timers, and temperature controls, programmable logic controllers, and circuit and load protection devices, especially transformers and switchgear).

PPM: parts per million.

PPP: Public-Private Partnership. Agreement between a government entity and one or more private companies to collaboratively build, develop, and/or operate a large-scale facility or network, usually related to power, water, or transportation infrastructure.

Predictive Maintenance: a type of preventive maintenance that uses statistical analysis and sometimes sensors to predict when required maintenance will need to occur and ensure that maintenance occurs before equipment failure is likely.

Price Indexation: the practice of linking price changes for a product or service to either raw materials costs or an agreed basket of prices for a given period (typically a month, quarter, or year).

psi: pounds per square inch.

RCRA: Resource Conservation and Recovery Act of 1976.

REACH: Registration, Evaluation and Authorisation of Chemical Substances. A European Commission regulation on safe use of chemicals, creating a central database run by the European Chemicals Agency in Helsinki.

Real options: acquisition in the current or a future time frame of tangible assets (e.g., land, property, or equipment) and/or Value-added physical transformations (e.g., construction) that offer or increase the flexibility to expand, contract, or change the nature of an investment commitment at a future date. Real options can increase the value of an investment by reducing downside risk. (See also real-options analysis.)

Real-options analysis: a process of evaluating the benefits and costs of acquiring certain real options.

Reliability Centered Maintenance: a body of knowledge used to shape the development and optimization of repair and maintenance programs, which involves total productive maintenance, as well as analytical tools common to lean, in order to reduce mean time between repairs (see MTBR) and increase uptime.

Reverse auction: an electronic auction in which suppliers bid in descending prices and the lowest price wins.

RFID: radiofrequency identification.

RFx: request for information, proposal, or quotation (RFI, RFP, or RFQ, respectively).

ROA: return on assets.

ROHS: Restriction of Hazardous Substances. European Union Directive, implemented in 2006, that prevents companies from marketing new electrical and electronic equipment containing lead, mercury, cadmium, hexavalent chromium, polybrominated biphenyls (PBBs), or polybrominated diphenyl ethers (PBDE) in quantities exceeding specified concentrations.

ROI: return on investment.

RONA: return on net assets.

ROP: rate of penetration.

rpm: revolutions per minute.

SCADA: supervisory control and data acquisition.

SCOR: supply chain operations reference model. SEAFOM: Subsea Fiber Optic Monitoring Group. SECA: Sulphur Emission Control Areas.

Should cost: a calculated figure that reflects the price that a supplier could offer if it took full advantage of opportunities for standardization and product simplification, economies of scale from volume purchasing, economies of scope from purchasing multiple types of products or services from the same supplier, and potential savings from better coordination of production and inventory planning with its customer(s).

Single Sourcing: intentionally purchasing from only one supplier in order to benefit from close relationships that yield lower cost, higher quality and reliability, and/or shorter lead times through joint problem-solving, than could be obtained with more suppliers.

SKU: stock-keeping unit. An inventoried item with a specific part number.

Smart Grid: an electricity network that gathers information on usage and production from the behavior of users connected to it and uses that information to better balance supply and demand of electricity to lower cost across the whole network by distributing production to the sources and times when it is most economical and incentivizing consumption from sources and times when it is less costly to produce. Smart grid consists of measurement (meters), networking hardware (transmitters, buses, modems, etc.), data management (SCADA system), and distribution management (relays, controls, demand response software, etc.).

Spares: parts, components, or subassemblies kept in reserve for use during the repair of equipment. Spare parts must support an irregular demand based on unplanned repairs, as opposed to service parts, which are used to support planned maintenance and thus have a more consistent replenishment requirement.

STOIIP: stock tank oil initially in place.

Strategic Sourcing: a process for improving the effectiveness of the procurement function. Involves improving spend visibility, understanding supplier capabilities beyond the typical transactional product or service purchase, and developing long-term relationships with strategic suppliers that can help the buyer meet profitability and customer satisfaction goals.

Supplier: contracted provider of services or products.

Supply Chain Management (SCM): the co-ordination of the set of activities involved in moving a product (such as a machine tool) and its ancillary services (such as installation, maintenance or repair) from the ultimate supplier to the ultimate customer so as to maximize economic value added (EVA). SCM includes manufacturing value added as it accrues along the chain, but excludes the initial manufacturing or conversion activity, so an initial basic activity such as extraction or farming would be considered a manufacturing activity (a node), not an SCM activity. (Source: The Economist Guide to Supply Chain Management)

Supply Market Concentration Index: the sum of the squares of the market shares of the suppliers in each category, based on the Herfindahl-Hirschman index (see HHI) methodology.

Supply Risk Mitigation: a set of coordinated actions that act upon factors that might create disruption or unavailability of materials or services to reduce their likelihood or severity.

System Dynamics: a branch of decision science that relies on modeling of feedback loops and delays in complex systems in order to identify and reduce the incidence of uneconomical resource allocation and to improve system performance as measured by target output levels such as cost, flexibility, and service level.

TCO: Total Cost of Ownership. See life-cycle cost.

T&D: Transmission and Distribution.

Third-Party Maintenance: maintenance provided by a vendor other than the original equipment manufacturer (see OEM) or the user. See 3PL.

T&M: time and materials.

TMS: Transportation Management System. A computerized system to manage fleet activities, especially by assigning loads to the lowest-cost qualified carrier. TMS also may include functionality to decide on modes of transport, manage export- import paperwork, and optimize load planning.

Tonnes: short tonnes. tonnes: metric tonnes.

TPA: tonnes per annum. (Also called tonnes per year [tpy].)

TPM: total productive maintenance.

TPY: tonnes per year. (Also called tonnes per annum [tpa].)

TQM: Total Quality Management. Engagement of the workforce in ensuring processes that continuously eradicate quality problems through the use of process analysis tools.

TSCA: Toxic Substances Control Act of 1976. UAV: unmanned aerial vehicle.

ULCC: Ultralarge Crude Carrier.

Upstream: exploration and production of crude oil and natural gas; drilling, completing, and equipping wells; and processing (e.g., separation, emulsion breaking, desilting, and field petroleum gathering, including shale plays and oil sands).

Vendor-Managed Inventory (VMI): a mode of inventory management in which a supplier monitors and regulates the amount of inventory at a customer's location to keep supply and demand in balance.

Vertical integration: acquiring a supplier in order to assure supply or enter into a new or related line of business.

VLCC: very large crude carrier.

WMS: Warehouse Management System.

WTI: West Texas Intermediate crude oil.

WTO: World Trade Organization.

Bibliography

Ackermann, Thomas, *Wind Power in Power Systems,* Wiley, 2005

Al-Qahtani, Khalid Y., and Ali Elkamel, *Planning and Integration of Refinery and Petrochemical Operations,* Wiley-VCH, 2010

Alistair M. Macnab, *The Fundamentals of BreakBulk Shipping: A Primer and Refresher Study for Global Logistics Students and Professional Logisticians, Ashore and Afloat,* Pearson Custom Publishing, 2010

Anderson, Roger N., and Albert Boulanger, *Computer-Aided Lean Management for the Energy Industry,* PennWell, 2008

Association of Energy Engineers, *Integrated Solutions for Energy and Facility Management,* Fairmont Press, 2002

August, Jim, *Applied Reliability-Centered Maintenance,* PennWell, 2000

Bachmann, Rolf, and Henrik Nielsen, *Combined - Cycle Gas & Steam Turbine Power Plants,* PennWell, 1999

Barnett, Dave, and Kirk Bjornsgaard, *Electric Power Generation: A Nontechnical Guide,* PennWell, 2000

Bartnik, Ryszard, and Zbigniew Buryn, *Conversion of Coal-Fired Power Plants to Cogeneration and Combined-Cycle: Thermal and Economic Effectiveness,* Springer, 2011

Baxter, Richard, *Energy Storage: A Nontechnical Guide,* PennWell, 2005

Ben Ebenhack, *A Nontechnical Guide to Energy Resources: Availability, Use, and Impact,* PennWell, 1995

Bloch, Heinz P. and John J. Hoefner, *Reciprocating Compressors: Operation and Maintenance,* Gulf Professional, 1996

Boothroyd, Ted, *Fire Detection and Suppression Systems,* International Fire Service Training Association, 2011

Burdick, Donald L, *Petrochemicals in Nontechnical Language,* PennWell, 2001

Carson, Richard T., and Michael B. Conaway, *Valuing Oil Spill Prevention: A Case Study of California's Central Coast,* Springer, 2004

Cedrick N. Osphey, *Wind Power: Technology, Economics and Policies, Nova Science Pub,* 2009

Chambers, Ann, and Barry Schnoor, *Distributed Generation: A Nontechnical Guide,* PennWell, 2011

Chambers, Ann, *Natural Gas & Electric Power in Nontechnical Language*, PennWell, 1999

Chambers, Ann, *Power Primer: A Nontechnical Guide from Generation to End-Use*, PennWell, 1999

Christopher, Martin, *Logistics and Supply Chain Management*, FT Press, 2005

Cipollina, Andrea, and Giorgio Micale, *Seawater Desalination: Conventional and Renewable Energy Processes*, Springer, 2009

Cory, Jordan A., *Combined Heat and Power: Analysis of Various Markets*, Nova Science, 2010

Cox, Andrew, *Strategic Procurement in the Construction Industry*, Thomas Telford, 1998

De Renzo, D.J., *Cogeneration Technology and Economics for the Process Industries*, Noyes Pubn, 1983

Degeare, Joe P., and David Haughton, *Gulf Drilling Guides: Oilwell Fishing Operations: Tools, Techniques, and Rules of Thumb*, Gulf Professional, 2003

Delea, Frank and Jack Casazza, *Understanding Electric Power Systems: An Overview of the Technology, the Marketplace, and Government Regulation*, Wiley-IEEE Press, 2010

Devereux, Steve, *Drilling Technology in Nontechnical Language*, PennWell, 1999

Devine, Michael D., *Cogeneration and Decentralized Electricity Production: Technology, Economics, and Policy*, Westview PR, 1987

Dickson, Mary H., and Mario Fanelli, *Geothermal Energy: Utilization and Technology*, EarthScan, 2003

Dipankar Sen and Prosenjit Sen, *RFID For Energy & Utility Industries*, PennWell, 2005

Drinan, Joanne, *Water and Wastewater Treatment: A Guide for the Non-engineering Professional*, CRC Press, 2001

Emadi, Ali, and Abdolhosein Nasiri, *Uninterruptible Power Supplies and Active Filters*, CRC Press, 2005

European Wind Energy Association, *Wind Energy - The Facts: A Guide to the Technology, Economics and Future of Wind Power*, EarthScan, 2009

Fingas, Mervin, *Oil Spill Science and Technology*, Gulf Professional Publishing, 2011

Finn R. Forsund, *Hydropower Economics*, Springer, 2010

Flynn, Damian, *Thermal Power Plant Simulation and Control, The Institution of Engineering and Technology*, 2003

Forsthoffer, W.E., *Forsthoffer's Rotating Equipment Handbooks: Fundamentals of Rotating Equipment*, Elsevier Inc., 2005

Gary, James and Glenn Handweck, *Petroleum Refining: Technology and Economics*, CRC Press, 2001

Gerwick, Ben C., Jr, *Construction of Marine and Offshore Structures*, Taylor & Francis, 2007

Grace, Robert D., *Blowout and Well Control Handbook*, Gulf Professional Publishing, 2003

Hessler, Peter G., *Power Plant Construction Management: A Survival Guide*, PennWell, 2005

Horlock, J.H., *Cogeneration-Combined Heat and Power: Thermodynamics and Economics*, Krieger, 1996

Hughes, John, *Manhole Inspection and Rehabilitation*, American Society of Civil Engineers, 2009

IFP-School, *Oil and Gas Exploration and Production: Reserves, Costs, Contracts*, Editions Technip, 2004

International energy agency, *Fossil Fuel-Fired Power Generation: Case studies of recently constructed coal- and gas-fired power plants (Cleaner Fossil Fuels)*, OECD Publishing, 2007

Jahn, Frank, *Hydrocarbon Exploration and Production Developments in Petroleum Science Series*, Volume 46, Elsevier Science, 1998

Jeffs, E., *Generating Power at High Efficiency: Combined Cycle Technology for Sustainable Energy Production*, Woodhead, 2008

Jorb, W. and B.H. Joergensen, *Decentralised Power Generation in the Liberalised EU Energy Markets*, Springer, 2003

K.V. Pavithran, *Economics of Power Generation, Transmission, and Distribution, Serials Pub*, 2004

Kaimeh, Philip, *Power Generation Handbook: Selection, Applications, Operation, Maintenance, McGraw-Hill Professional*, 2003

Kaltschmitt, Martin, and Wolfgang Streicher, *Renewable Energy: Technology, Economics and Environment*, Springer, 2007

Kehlhofer, Rolf, and Bert Rukes, *Combined-Cycle Gas & Steam Turbine Power Plants*, PennWell, 2009

Kenyon, Rex, *Process Plant Reliability and Maintenance for Pacesetter Performance*, PennWell, 2004

Kimbrough, Steven O., Wu, D. J., and Zhong, Fang, "Computers play the beer game: can artificial agents manage supply chains?" Decision Support Systems. 33. 2002

Lee, H.L., Padmanabhan, V. and Whang, S, "The Bullwhip Effect: Reflections." The Bullwhip Effect in Supply Chains. Ed. Octavio A. Carranzo Torres, and Felipe A. Villegas Moran. New York, NY: Palgrave Macmillan, 2006

Lee, H.L., Padmanabhan, V. and Whang, S. "Information Distortion in a Supply Chain: The Bullwhip Effect," Management Science, 32, 1997

Lee, H.L., Whang, S, "The Bullwhip Effect: A Review of Field Studies." The Bullwhip Effect in Supply Chains. Ed. Octavio A. Carranzo Torres, and Felipe A. Villegas Moran. New York, NY: Palgrave Macmillan, 2006

Leffler, William L., and Gordon Sterling, *Deepwater Petroleum Exploration & Production: A Nontechnical Guide*, PennWell, 2003

Leffler, William, *Petroleum Refining in Nontechnical Language*, PennWell, 2000

Lorange, Peter, *Shipping Company Strategies: Global Management under Turbulent Conditions*, Elsevier Inc., 2005

Lyons, William, Working Guide to Drilling Equipment and Operations, Gulf Professional Publishing, 2010

Mannrique, Vladimir Alvarado Eduardo, *Enhanced Oil Recovery: Field Planning and Development Strategies*, Elsevier Inc., 2010

Mariani, E., and S.S. Murthy, *Control of Modern Integrated Power Systems (Advances in Industrial Control)*, Springer, 1997

Marsh, W.D., *Economics of Electric Utility Power Generation*, Oxford University Press, 1980

Mashayekhi, Ali, "Dynamics of Oil Price in the World Market". Tehran, Iran: School of Management and Economics, Sharif University of Technology. Paper presented to the 19th International Conference of The System Dynamics Society, July 23 - 27, 2001, at the Emory Hotel and Conference Center, Atlanta, Georgia, USA; downloaded at http://www.systemdynamics.org/conferences/2001/papers/ on September 10, 2010.

Mccrae, Hugh, *Marine Riser Systems and Subsea Blowout Preventers*, Unit 5, Lesson 10, The University of Texas at Austin—Petroleum Extension Service, 2003M

Miesner, Thomas, and William Leffler, *Oil & Gas Pipelines in Nontechnical Language*, PennWell, 2006

Mokhatab, Saeid, and William A. Poe, *Handbook of Natural Gas Transmission and Processing, Gulf Professional Publishing*, 2006

Myers, J. Hutcheon, and S. Oue, *Petroleum and Marine Technology Information Guide: A bibliographic sourcebook and directory of services*, E&FN, 1993

Myers, Philip, A*bove Ground Storage Tanks*, McGraw-Hill Professional, 1997

Najafi, Mohammad, *Trenchless Technology Piping: Installation and Inspection*, McGrawj-Hill, 2010

Nardone, Paul J., *Well Testing Project Management: Onshore and Offshore Operations*, Gulf Professional Publishing, 2009

Naujoks, Boris and Theodor J. Stewart, *Multiple Criteria Decision Making for Sustainable Energy and Transportation Systems*, Springer, 2010

Northcote-Green, James, and Robert G. Wilson, *Control and Automation of Electrical Power Distribution Systems*, CRC Press, 2007

Ouyang, Yanfeng, Lago, Alejandro, and Daganzo, Carlos F, "Taming the Bullwhip Effect: From Traffic to Supply Chains." The Bullwhip Effect in Supply Chains. Ed. Octavio A. Torres, Carranzo, and Felipe A. Villegas Moran. New York, NY: Palgrave Macmillan, 2006

Page, John S., *Cost Estimating Manual for Pipelines and Marine Structures*, Gulf Publishing, 1977

Pehnt, Martin, and Martin Cames, *Micro Cogeneration: Towards Decentralized Energy Systems*, Springer, 2010

Petchers, Neil, *Combined Heating, Cooling & Power Handbook: Technologies & Applications: An Integrated Approach to Energy Resource Optimization*, Fairmont Press, 2003.

Poirier, Charles C. and Reiter, Stephen E., Supply Chain Optimization: Building the Strongest Total Business Network, Berrett-Koehler, 1996

Porter, Professor Michael E., General Electric vs. Westinghouse in Large Turbine Generators (A). Boston, MA: Harvard Business School Publishing, 1980

Rajan, Dr. G.G. *Practical Energy Efficiency Optimization*, Penwell, 200

Raymond, Martin S., and William L. Leffler, *Oil & Gas Production in Nontechnical Language*, Pennwell, 2005

Read, Colin, *BP and the Macondo Spill: The Complete Story*, Palgrave Macmillan, 2011.

Rebennack, Steffen, and Panos M. Pardalos, *Handbook of Power Systems I, Springer*, 2010

Rebennack, Steffen, and Panos M. Pardalos, *Handbook of Power Systems II*, Springer, 2010

Rustebakke, Homer M., *Electric Utility Systems and Practice*, Wiley-Interscience, 1983

Santorella, Gary, *Lean Culture for the Construction Industry: Building Responsible and Committed Project Teams*, Taylor & Francis, 2011

Sathyajith, Mathew, *Wind Energy: Fundamentals, Resource Analysis and Economics*, Springer, 2006

Scheer, Hermann, *The Solar Economy: Renewable Energy for a Sustainable Global Future, EarthScan*, 2004

Sears, Glenn, and Richard Clough, *Construction Project Management: A Practical Guide to Field Construction Management*, John Wiley & Sons, 2008

Seba, Richard D., *Economics of Worldwide Petroleum Production*, Pennwell, 2003

Seddon, Duncan, *Gas Usage & Value*, Pennwell, 2006

Seger, Karl A., *Utility Security: The New Paradigm*, Pennwell, 2003

Shively, Bob, and John Ferrare, *Understanding Today's Electricity Business*, Pennwell, 2010

Shively, Bob, and John Ferrare, *Understanding Today's Natural Gas Business*, Enerdynamics, 2009

Sirchis, J., *Combined Production of Heat and Power*, Routledge, 1990

Slater, Philip, *Smart Inventory Solutions: Improving the Management of Engineering Materials and Spare Parts*, Industrial Press, 2010R

Society of International Gas Tanker and Terminal Operators, *LNG Shipping Suggested Competency Standards: Guidance and Suggested Best Practice for the LNG Industry in the 21st Century*, Seamanship International, 2008

Spellman, Frank R., *Handbook of Water and Wastewater Treatment Plant Operations*, CRC Press, 2009

Spitzer, David W., and Walt Boyes, *The Consumer Guide to Magnetic Flow meters*, Copperhill and Pointer 2003

Spliethoff, Hartmut, *Power Generation from Solid Fuels (Power Systems)*, Springer, 2010

Sterman, John, "Operational and Behavioral Causes of Supply Chain Instability." The Bullwhip Effect in Supply Chains. Ed. Octavio A. Carranzo Torres, and Felipe A. Villegas Moran. New York, NY: Palgrave Macmillan, 2006

Tessmer, Raymond G., and John R. Boyle, *Cogeneration and Wheeling of Electric Power: Opportunities in a Changing Market*, Pennwell, 1995

Thayamballi, Anil Kumar, and Jeom Kee Paik, *Ship-shaped Offshore Installations: Design, Building, and Operation*, Cambridge University Press, 2007

Tiratsoo, John, *Pipeline Pigging and Inspection Technology*, Gulf Professional Publishing, 1999

Tusiani, Michael D., and Gordon Shearer, *LNG: A Nontechnical Guide*, Pennwell, 2007

Underhill, Tim, *Strategic alliances: managing the supply chain*, Pennwell, 1996.

Urboniene, Irena A., *Desalination: Methods, Costs and Technology*, Nova Science, 2010

Venem, Jan-Erik, *Offshore Risk Assessment: Principles, Modelling and Applications of QRA Studies (Springer Series in Reliability Engineering)*, Springer, 2007

Warkentin-Glenn, Denise, *Electric Power Industry in Nontechnical Language*, Pennwell, 2006

Wilson, Jeff, *The Model Railroader's Guide to Industries Along the Tracks*, Kalmbach Publishing Company, 2004

Wood, Allen J., and Bruce F. Wollenberg, *Power Generation, Operation, and Control*, Wiley-Interscience, 1996

Woodruff, Everett, and Herbert Lammers, *Steam Plant Operation*, McGraw-Hill, 2004

Wright, Charlotte J., and Rebecca A. Gallun, *Fundamentals of Oil & Gas Accounting*, Pennwell, 2008

Yunus Çerçi, *The Thermodynamic and Economic Efficiencies of Desalination Plants: Performance Evaluation of Desalination Plants*, Lap Lambert, 2010

"World Energy Investment 2020." International Energy Agency.

"Global Energy Statistical Yearbook 2020." Enerdata.

"Renewable Power Generation Costs in 2019." IRENA.

"2018 U.S. Utility-Scale Photovoltaics-Plus-Energy Storage System Costs Benchmark." By Ran Fu, Timothy Remo, and Robert Margolis. National Renewable Energy Laboratory.

"Q4 2019/Q1 2020 Solar Industry Update, Publication." NREL/PR-6A20-77010. By David Feldman and Robert Margolis. May 28, 2020. National Renewable Energy Laboratory (NREL).

"DNV GL's Energy Transition Outlook 2020," DNV GL (DNV GL AS, 2020).

"Lazard's Levelized Cost of Storage Analysis - Version 6.0" (Lazard)

"Annual Energy Outlook 2020." United States Energy Information Administration.

References

1. N. Sönnichsen, "Number of Nuclear Reactors Worldwide 1954-2019," *Statista* (*Statista*, January 7, 2021), https://www.statista.com/statistics/263945/number-of-nuclear-power-plants-worldwide/.

2. Author's analysis of HPI Market Data 2010, HPI Construction Boxscore; and Economist Intelligence Unit Data Tool.

3. David Steven Jacoby, "Balancing Economic Risks: Tips for a Well-structured Deal," *Middle East Energy*, September 2010, p. 5.

4. "Oil 2020," IEA (International Energy Agency, March 2020), https://www.iea.org/reports/oil-2020.

5. "Oil Information Overview: Statistics Report," IEA (International Energy Agency, July 2020), https://www.iea.org/reports/oil-information-overview.

6. "Greenfield E&P Tenders Are Set to Shrink to USD 60 Billion in 2020, a 2-Decade Low," *Energy Northern Perspective* (*MH Magazine*, May 5, 2020), https://energynorthern.com/2020/05/05/greenfield-ep-tenders-are-set-to-shrink-to-usd-60-billion-in-2020-a-2-decade-low/.

7. "The Oracle of Oil: The man who predicted peak oil." *New Scientist*. By David Strahan. June 1, 2016. https://www.newscientist.com/article/mg23030762-700-the-man-who-predicted-peak-oil/#ixzz6oeWLXb3Y

8. "Energy Prices Re-Shaping the Supply Chain: Charting a New Course? Boston Logistics Group. 2007. http://bostonstrategies.com/images/BSI_-_SS4_Energy_Prices_Reshaping_the_SC.pdf

9. Yergin, Daniel. The New Map: Energy, Climate, and the Clash of Nations." Penguin Random House LLC, 2020. P. 51.

10. Jacoby, David Steven. "The oil price 'bullwhip': problem, cost, response." *Oil and Gas Journal*, March 22,2010. http://bostonstrategies.com/images/Jacoby_Oil_and_Gas_Bullwhip_Article.pdf

11. "Capital Cost and Performance Characteristic Estimates for Utility Scale Electric Power Generating Technologies." Energy Information Administration, February 2020.

12. "World's largest battery storage system now operational." *PV Magazine*. August 20, 2020. https://www.pv-magazine.com/2020/08/20/worlds-largest-battery-storage-system-now-operational/#:~:text=It's%20a%20title%20that%20is,storage%20battery%20in%20the%20world.

13. "Top 10 biggest wind farms." Power Technology, 29 June 2019. https://www.power-technology.com/features/feature-biggest-wind-farms-in-the-world-texas/.

14. "Five of the World's Largest Offshore Wind Energy Farms," SAUR Energy. https://www.saurenergy.com/solar-energy-blog/worlds-largest-offshore-wind-energy-farms

15. "The World's Biggest Solar Plants," Power Technology. https://www.power-technology.com/features/the-worlds-biggest-solar-power-plants/. Last accessed January 25, 2021.

16. Jacoby, David Steven. *The High Cost of Low Prices: A Roadmap to Sustainable Prosperity*. Business Expert Press, 2017.

17. Goldsby, Thomas. Cambridge University Press, 2021. (forthcoming)

18. Jacoby, David Steven. Guide to Supply Chain Management, *The Economist.* 2009.

19. Boston Strategies International, analysis of data from Thomson Financial. "Snapshot" based on 2011 data.

20. "South Korea unveils $43 billion plan for world's largest offshore wind farm." Reuters. By Hyonhee Shin. February 5, 2021. https://www.reuters.com/article/us-southkorea-energy-windfarm/south-korea-unveils-43-billion-plan-for-worlds-largest-offshore-wind-farm-idUSKBN2A512D Accessed February 10, 2021.

21. William L. Leffler and Martin Raymond, *Oil and Gas Production in Nontechnical Language,* Tulsa: PennWell, 2005.

22. David Steven Jacoby, Guide to Supply Chain Management, New York: Bloomberg, 2009, p. 161.

23. Mark Graham, Mark Cook and Frank Jahn, "Project and Contract Management," *Hydrocarbon Exploration and Production 55,* 2001, p. 329.

24. *Ibid.*

25. Anil Kumar Thayamballi and Jeom Kee Paik, Ship Shaped Offshore Installations: Design, Building and Operation, New York: Cambridge University Press, 2007, p. 38.

26. "CAPS Infographic Archive," CAPS Research (Arizona State University), accessed January 22, 2021, https://www.capsresearch.org/infographic-archive/.

27. "Managing the Supply Chain: a Complex but Vital Task Saudi Aramco, accessed January 14, 2012. http://www.saudiaramco.com/en/home/news/latest-news/2010/managing-the-supply-chai.

28. Adapted from William L. Leffler and Martin Raymond, *Oil and Gas Production in Nontechnical Language,* Tulsa: PennWell, 2005.

29. Ali Mashayekhi, "Dynamics of oil price in the world market." In 19th International System Dynamics Conference. 2001.

30. John Sterman, "Learning from Evidence in a Complex World." (*American Journal of Public Health,* 2006).

31. Gordon Shearer and Michael D. Tusiani, *LNG: A Nontechnical Guide,* Tulsa: Pennwell, 2007, p. 125.

32. Baker Hughes Wins Boston Strategies International's 2010 Oil and Gas Award for Excellence in Supply Chain Management," Boston Strategies International, accessed September 22, 2010, http://www.bostonstrategy.com/images/Baker_Hughes_Wins_2010_BSI_Supply_Chain_Award.pdf.

33. "Vattenfall combines wind, solar and batteries in new hybrid energy park." https://group.vattenfall.com/press-and-media/pressreleases/2019/vattenfall-combines-wind-solar-and-batteries-in-new-hybrid-energy-park. Last accessed January 25, 2021.

34. "#NYEW2019 Disrutor's Day winner XL Batteries with the organizing team. Congratulations and much gratitude to all the participants, influencers, and volunteers who made all this possible." https://www.facebook.com/newyorkenergyweek/ Last accessed January 25, 2021.

35. "Announcing the Finalists of Disruptors Day!" New York Energy Week website. http://nyenergyweek.com/announcing-finalists-disruptors-day/ Last accessed January 25, 2021.

36. Jacoby, David Steven. *The Guide to Supply Chain Management* (The Economist, 2009).

37. "Outstanding Supply Chain Performance Wins Qatar Fuel the Boston Strategies Award for Overall Excellence." http://bostonstrategies.com/images/Outstanding_Supply_Chain_Performance_Wins_Qatar_Fuel_the_Boston_Strategies_Award_for_Overall_Excellence_081007a.pdf.

38. "Chevron Wins Boston Strategies International's 2010 Award for Lean Six Sigma Implementation in Oil and Gas Operations." http://bostonstrategies.com/images/Chevron_Wins_2010_BSI_Supply_Chain_Award.pdf.

39. "Bharat Petroleum Corporation Limited Wins Boston Strategies International's 2009 Oil and Gas Industry Award for Excellence in Supply Chain Synchronization." http://bostonstrategies.com/images/Bharat_PR_091006.pdf.

40. "NLC—Q4 2010 Nalco Holding Company Earnings Conference Call," Thomson Reuters, accessed February 9, 2012, http://www.google.co.in/url?sa=t&rct=j&q=nalco+getfit+blue+belt+savings+thomson+reuters+&source=web&cd=1&ved=0CCUQFjAA&url=http%3A%2F%2Fphx.corporateir.net%2FExternal.File%3Fitem%3DUGFyZW50SUQ9MzY2OTM1NHxDaGlsZElEPTQxMTk4MHxUeXBlPTI%3D%26t%3D1&ei=IrgzT5OJIpGnrAed15ibDA&usg=AFQjCNGH7PSmPHYuqvSdxbGsWfke6uEZWg.

41. "Sustainable improvement of cost base," BASF, accessed February 9, 2012, http://www.basf.com/group/corporate/en/investor-relations/strategy/cost-of-capital/cost-reduction.

42. "Boston Strategies International is proud to be working with ... Solar Earth Technologies, which has developed a gem-like photovoltaic cell technology for remote, industrial and off-grid power: it's embedded in roads, walkways, and buildings, and is 30% more efficient than flat solar panels. The technology enables clean power for new construction in less developed areas." LinkedIn post by David Steven Jacoby. November 25, 2019. Last accessed January 25, 2021.

43. Belinda Petty, John Ferrare and Bob Shively, *Understanding Today's Global LNG Business,* New York: Energydynamics, 2010, p. 97.

44. Jacoby, David, and Erik Halbert. "Energy Prices Re-Shaping the Supply Chain: Charting a New Course?" Boston: Boston Strategies International, January 2007.

45. CAPS Research Petroleum Industry 2011 Supply Management Performance Benchmarking Report, Issued August 15, 2011, p. 8.

46. *Ibid.*

47. "Energy—United Nations Sustainable Development," United Nations (United Nations), accessed January 22, 2021, https://www.un.org/sustainabledevelopment/energy/.

48. "Mobilizing for a resource revolution." *McKinsey Quarterly.* Energy, Resources, Material. Page 1 of 11.

49. Teague Egan, "Beating China at the Lithium Game - Can the US Secure Supplies to Meet Its Renewables Targets?," Utility Dive (Industry Dive, February 18, 2020), https://www.utilitydive.com/news/beating-china-at-the-lithium-game-can-the-us-secure-supplies-to-meet-its/572307/#:~:text=China%20controls%2051%25%20of%20the,components%20of%20lithium%2Dion%20batteries.

50. "Executive Order on Addressing the Threat to the Domestic Supply Chain," WhiteHouse.gov (United States of America), accessed 2020, https://www.whitehouse.gov/presidential-actions/executive-order-addressing-threat-domestic-supply-chain-reliance-critical-minerals-foreign-adversaries/.

51. "The Only US-Based Primary Cobalt Mine Is Nearing Production," StreetWise Reports (StreetWise Reports, May 29, 2018), https://www.streetwisereports.com/article/2018/05/29/the-only-us-based-primary-cobalt-mine-is-nearing-production.html.

52. Jake Spring, "Hands off Brazil's Niobium: Bolsonaro Sees China as Threat to Utopian Vision," Reuters (Thomson Reuters, October 25, 2018), https://www.reuters.com/article/us-brazil-election-china-niobium/hands-off-brazils-niobium-bolsonaro-sees-china-as-threat-to-utopian-vision-idUSKCN1MZ1JN?edition-redirect=uk.

53. "Overview of the Elk Creek Project," NioCorp (NioCorp Development Ltd.), accessed January 20, 2021, http://www.niocorp.com/elk-creek-project/#:~:text=Overview%20of%20the%20Elk%20Creek%20Project&text=Located%20near%20Elk%20Creek%2C%20Nebraska,mines%20in%20the%20world%20today.

54. Barite (Barium). Chapter D of Critical Mineral Resources of the United States—Economic and Environmental Geology and Prospects for Future Supply. U.S. Department of the Interior. U.S. Geological Survey. Professional Paper 1802–D. 2017.

55. Yergin, Daniel. *The New Map: Energy, Climate, and the Clash of Nations.* Penguin Press, 2020. p. 56.

56. Ronald H. Coase, *The Nature of the Firm,* Economica 4 (16): p.386 a co 1937. http://www.sonoma.edu/users/e/eyler/426/coase1.pdf.

57. W.R. Grace's 2011 Q1 earnings call.

58. David Steven Jacoby and Bruna Figueiredo, "The Art of High-Cost *Supply Chain Management Review,* May/June 2008, p. 32.

59. "Element 1 Corp licenses methanol-based M-Series hydrogen generator technology to RIX Industries," Bioage Group (Green Car Congress, October 11, 2020), https://www.greencarcongress.com/2020/10/20201011-e1.html.

60. "Licensing agreement and a Li-ion battery facility in Aus," Electronics (Westwick-Farrow Pty Ltd, February 25, 2019), https://www.electronicsonline.net.au/content/business/news/licensing-agreement-and-a-li-ion-battery-facility-in-aus-1036608913#axzz6klcUqb00.

61. "Ballard in China," Ballard in China | About Ballard | Ballard Power (Ballard Power Systems), accessed January 13, 2021, https://www.ballard.com/about-ballard/ballard-in-china.

62. "NOV Announces Joint Venture to Provide High-Specification Drilling Rigs and Advanced Drilling Equipment," NOV Inc (National Oil Varco, June 28, 2018), https://investors.nov.com/news-releases/news-release-details/nov-announces-joint-venture-provide-high-specification-drilling.

63. "MHI Compressor Technology Licensee in China Begins Marketing," Mitsubishi Heavy Industries, accessed February 9, 2012, http://www.mhi.co.jp/en/news/story/1102091407.html.

64. "Fluence Gridstack." Technical Specifications. https://info.fluenceenergy.com/hubfs/Collateral/Gen6/Gridstack%20Tech%20Spec.pdf. Last accessed January 25, 2021.

65. Q1 2011 Flowserve Corp Earnings Conference Call - Final FD (Fair Disclosure) Wire April 28, 2011 Thursday

66. "Overhaul & Maintenance, July 1, 2008, "Innovative Strokes for Readiness," p. 3.

67. *Ibid.*

68. Rabia, SPE paper, A New Approach to Drill Bit Selection, http://www.onepetro.org/mslib/servlet/onepetropreview?id=00015894.

69. David Steven Jacoby, "The oil price 'bullwhip': problem, cost, response," *Oil & Gas Journal,* March 2010, pp. 20–25.

70. David Steven Jacoby, "The oil price 'bullwhip': problem, cost, response *Oil & Gas Journal,* March 2010, pp. 20–25.

71. "Bravo Options Barite Rights to Baker Hughes Bravo Gold Corp., April 16, 2009, http://www.bravogoldcorp.com/en/news/147/bravo-options-barite-rights-to-baker-hughes.php.

72. David Steven Jacoby, *Using Flexible Capacity Techniques to Thrive in a Volatile Economy, Logistics Digest,* February 2010, p. 31.

73. Kuwait Oil Company (Register of Commerce No. 21835), General Conditions of Contract for Lump Sum Turnkey Projects, http://www.google.com/url?sa=t&rct=j&q=register%20of%

20commerce%20no.%2021835&source=web&cd=1&ved=0CCMQFjAA&url=http%3A%2F%2Fmcsetender.kockw.com%2FCommercial%2520Documents%2Fstandards%2FGCC%2520LUMP%2520SUM%2520TURNKEY%2FGCC%2520LSTK%2520%2520March%25202007.pdf&ei=gCY4T5vlEczq0QGyma28Ag&usg=AFQjCNH0CsdnTWzz3BngjwsnWa5OYGyOqQ&cad=rjt. Accessed February 12, 2012.

74. “Goal 9: Industrial Innovation and Infrastructure,” UNDP (United Nations Development Program), accessed January 22, 2021, https://www.undp.org/content/undp/en/home/sustainable-development-goals/goal-9-industry-innovation-and-infrastructure.html.

75. Based on the ratio of human-machine working hours, 2018 vs. 2022. World Economic Forum: The Future of Jobs Report 2018. Centre for the New Economy and Society.

76. “Security Guideline for the Electricity Sector—Supply Chain,” NERC (North American Electric Reliability Corporation, September 17, 2019), https://www.nerc.com/comm/CIPC_Security_Guidelines_DL/Security_Guideline-Provenance.pdf.

77. Nuris Ismail, “Enterprise Asset Management in 2012”. Aberdeen report, December 2011.

78. Adam Hayes, “Understanding Return on Net Assets,” ed. Janet Berry-Johnson, Investopedia (Investopedia, November 27, 2020), https://www.investopedia.com/terms/r/rona.asp.

79. Joseph Juran, *The Complete Guide to Performance Excellence,* McGraw Hill, 2010.

80. Phil Crosby, Philip Crosby's Reflections on Quality: 295 Inspirations from the World's Foremost Quality Guru, McGraw-Hill, 1995.

81. Kaoru Ishikawa, *What Is Total Quality Control?: The Japanese Way,* Prentice Hall Trade, 1991.

82. David Steven Jacoby, “Using Flexible Capacity Techniques to Thrive in a Volatile Economy.” Logistics Digest, February 2010. http://www.logisticsdigest.com/inter-education/inter-opinion/item/4669-using-flexible-capacity-techniques-to-thrive-in-a-volatile-economy.html.

83. 2010 Oil and Gas Award for Excellence in Supply Chain Management awarded to Shell, Boston Strategies International press release, September 22, 2010, p. 1.

84. Profile: Angola, ExxonMobil brochure.

85. “Petrobras' ambitious PROPOÇO a ‘road map’ for optimizing well construction performance, Drilling Contractor,” January/February 2008, p. 138.

86. “Management Committee approves Gazprom's standardization and technical regulation efforts,” Gazprom press release, April 1, 2009, p. 1.

87. Utilization of PETRONAS Technical Standard (PTS), a presentation by By Pau KH, accessed February 13, 2012 at http://www.google.com/url?sa=t&rct=j&q=utilization%20of%20petronas%20technical%20standard%20(pts)&source=web&cd=1&ved=0CCYQFjAA&url=http%3A%2F%2Finfo.ogp.org.uk%2Fstandards%2F09Malaysia%2FPresentations%2F11SharingOnPTSFeb09v10.pdf&ei=K584T56RLYPq0gHi3KyZAg&usg=AFQjCNFIsyiNSSfon00tnhGMf2vfff3RIQ.

88. “International Standardization Workshop,” Petro-Canada presentation, Doha, 3 April 2006.

89. “The Oil and Natural Gas Industry's Most Valuable Resource American Petroleum Institute Standards, http://www.api.org/Standards/faq/upload/valueofstandards.pdf

90. Bruce Murphy, “How Has the Recession Changed the Way we Plan? presentation to APICS, February 2010, p. 12.

91. Phillip Slater, Smart Inventory Solutions: Improving the Management of Engineering Materials and Spare Parts, New York: Industrial Press, 2010.

92. Phillip Slater, Smart Inventory Solutions: Improving the Management of Engineering Materials and Spare Parts, New York: Industrial Press, 2010.

93. "Orders go out the door within 24 business hours," New Pig Corporation, http://www.newpig.com/us/content/why-new-pig?title=Why%2520New%2520Pig%3F.

94. "Orders go out the door within 24 business hours New Pig Corporation, http://www.newpig.com/us/content/why-new-pig?title=Why%2520New%2520Pig%3F.

95. Bill Carreira, *Lean Manufacturing That Works: Powerful Tools for Dramatically Reducing Waste and Maximizing Profits*, New York: AMACOM, 2005, p. 3.

96. Amrik Sohal and Keith Howard, "Trends in Materials Management," *International Journal of Production Distribution and Materials Management* 17, no. 5 (1987), pp. 3–11.

97. "Chevron Wins Boston Strategies International's 2010 Award for Lean Six Sigma Implementation in Oil and Gas Operations," Boston Strategies International press release, September 22, 2010.

98. Tim Underhill, *Strategic Alliances: Managing the Supply Chain,* Tulsa: PennWell, 1996, p. 110.

99. [i] PLS Logistics, "Freight Transportation in the Oil & Gas Industry: Five Mistakes that Cripple Profitability," White Paper, 2012.

100. UNION MEMBERS—2020. United States Bureau of Labor Statistics. January 22, 2021. https://www.bls.gov/news.release/pdf/union2.pdf.

101. Steve Geary, "Performance-Based Outsourcing," University of Tennessee, Presentation to Boston APICS, 2009.

102. Kennaugh, "4PL from a provider's perspective Building and Implementing a 4PL The journey is tough - The reward is mutual," Rob Kennaugh, presentation to TransOman, October 14, 2009.

103. Rice, "4PL from a clients perspective," Presentation to TransOman, October 14, 2009.

104. Warith Kharusi, "Creating Logistics Synergies within Oman," Presentation to TransOman, October 14, 2009.

105. Luke M. Froeb, Lance Brannman, "Mergers, Cartels, Set-Asides, and Bidding Preferences in Asymmetric Oral Auctions," *The Review of Economics and Statistics* 82, 2006, p. 283–90, doi:10.1162/003465300558795.

106. Allan T. Ingraham, "A Test for Collusion between a Bidder and an Auctioneer in Sealed-Bid Auctions," *The B.E. Journal of Economic Analysis & Policy* 4, 2005, http://www.bepress.com/cgi/viewcontent.cgi?article=1448&context=bejeap.

107. Jeffery L. Harrison and Roger D. Blair, *Monopsony in Law and Economics,* Cambridge: Cambridge University Press, 2010, p.60.

108. Boston Logistics Group, *Precision-Guided Sourcing Strategies for Maximum Results,* High-Impact Sourcing, 2005.

109. David Steven Jacoby, *Guide to Supply Chain Management,* New York: Bloomberg Press, 2009, p.64.

110. Luke M. Froeb and Lance Brannman, "Mergers, Cartels, Set-Asides, and Bidding Preferences in Asymmetric Oral Auctions," *The Review of Economics and* Statistics 82, 2006, p.283-90, doi:10.1162/003465300558795.

111. "Ketera Teams With Chevron and Aberdeen Group to Host Live Supplier Enablement Webinar Business Wire, May 24, 2007. http://www.businesswire.com/news/home/20070524005278/en/Ketera-Teams-Chevron-Aberdeen-Group-Host-Live.

112. Craig C. Zawada, Eric V. Roegner and Michael V. Marn, *The Price Advantage,* Hoboken: Wiley, 2004, p. 258, Appendix 2.

113. Craig C. Zawada, Eric V. Roegner and Michael V. Marn, *The Price Advantage,* Hoboken: Wiley, 2004, p. 258, Appendix 2.

114. "US Sparked Global Antitrust Enforcement In 2010 Law 360, January 1, 2011. http://www.law360.com/articles/216998/us-sparked-global-antitrust-enforcement-in-2010.

115. "Justice Department Requires Divestitures in Baker Hughes' Merger With BJ Services," PR Newswire.

116. "Who We Are," Schlumberger Limited, https://www.slb.com/who-we-are.

117. "BHGE Launches Multimodal Facility Expansion in Angola ," Baker Hughes Company (June 07, 2019), https://www.bakerhughes.com/company/news/bhge-launches-multimodal-facility-expansion-angola.

118. Mai El Ghandour—Moslem Ali, "Weatherford inaugurates new facility in egypt with key industry leaders," Weatherford (*Egypt Oil & Gas Newspaper,* May 2019), https://www.weatherford.com/getattachment/fab0b0b1-17f2-4b3a-8ae4-9072b9245de3/WEATHERFORD-INAUGURATES-NEW-FACILITY-IN-EGYPT-WITH.

119. Based on 2008 labor costs in Shanghai, Tianjin, Beijing, and Hangzhou. Source: Boston Strategies International, "China Sourcing—The Long View May 2008.

120. Boston Strategies International, "The Asian Sourcing Boom: How Long Will it Last?" *State of Strategies Sourcing,* 2006.

121. "Manufacturing labor costs per hour for China, Vietnam, Mexico from 2016 to 2020." Statista. https://www.statista.com/statistics/744071/manufacturing-labor-costs-per-hour-china-vietnam-mexico/#:~:text=In%202018%2C%20manufacturing%20labor%20costs,2.73%20U.S.%20dollars%20in%20Vietnam. Last accessed January 26, 2021.

122. "How China's pollution clean-up is driving up prices for shoppers." *South China Morning Post.* November 20, 2017. https://www.scmp.com/news/china/policies-politics/article/2120636/how-chinas-pollution-clean-driving-prices-shoppers Last accessed January 25, 2021.

123. Cathy Hopkins, "DLA Aviation saves money with reverse auctions," DLA Aviation Public Affairs, May 31, 2011. http://www.aviation.dla.mil/externalnews/news/20110531.htm.

124. ONGC Case study by BOB Tech Solutions: A Case Study on Reverse Auction, Purchase of Signal Cables for Oil & Gas projects development.

125. The Cost of Not Acting: The Total Telecom Cost Management Benchmark Report, November 2006, Aberdeen Group, page i.

126. "The Legality of Local Content Measures under WTO Law." *Journal of World Trade.* June 2014. https://www.wto.org/english/news_e/news19_e/trim_06jun19_e.htm#:~:text=China%20%E2%80%93%20local%20content%20provisions%20on%20cybersecurity%20measures&text=A%20requirement%20for%20companies%20to,Tariffs%20and%20Trade%20Article%20III.

127. "Local Content Requirements and the Green Economy."United Nations Conference on Trade and Development (UNCTAD), 2014. https://unctad.org/system/files/official-document/ditcted2013d7_en.pdf.

128. David Steven Jacoby, "Uncovering economic and supply chain success in the new emerging economies," presentation to APICS International Conference, 2011.

129. World Bank LPI data. http://web.worldbank.org/WBSITE/EXTERNAL/TOPICS/EXTTRANSPORT/EXTTLF/0,,contentMDK:21514122~menuPK:3875957~pagePK:210058~piPK:210062~theSitePK:515434,00.html.

130. Nigerian Oil and Gas Industry Content Development Act, Explanatory Memorandum, 2010.

131. "Total Pazflor," FMC Technologies, Inc., accessed February 6, 2012, http://www.fmc technologies.com/en/SubseaSystems/GlobalProjects/Africa/Angola/TotalPazflor.aspx.

132. FMC website, and Willy Olsen, "Maximizing the value of strategic partnerships Introduction—setting the scene INTSOK presentation to National Oil Congress, London, June 2010, p. 63.

133. Rio de Janeiro, "Local Content Framework, The Brazilian Experience," Agência Nacional do Petróleo, February 6, 2011.

134. Maureen Hoch. "New Estimate Puts Gulf Oil Leak at 205 Million Gallons." PBS Newshour, August 2, 2010, http://www.pbs.org/newshour/rundown/2010/08/new-estimate-puts-oil-leak-at-49-million-barrels.html.

135. Bob Canvar, *Disaster on the Horizon,* Chelsea Green Publishing, 2010.

136. "EBARA Group CSR Report 2011 EBARA Group, http://www.ebara.co.jp/en/csr/csr/2011/pdf/ebara_csr_e11-19.pdf.

137. "Form 10-Q for CAMERON INTERNATIONAL CORP," Yahoo Finance, accessed February 10, 2012, http://biz.yahoo.com/e/110801/cam10-q.html.

138. "A Deadly Industry." HSE Today. By Jim Malewitz, March 31, 2015. Based on data from the United States Occupational Safety and Health Administration (OSHA).

139. The Edwards Law Firm, http://www.edwardsfirm.com/Articles/Oil-Refinery-Safety-Violations-Can-Lead-to-Lawsuits.shtml, accessed February 15, 2012.

140. Amy Sinden, Rena Steinzor, Matthew Shudtz, James Goodwin, Yee Huang, and Lena Pons. "Twelve Crucial Health, Safety, and Environmental Regulations: Will the Obama Administration Finish in Time?" Page 10. Washington DC, Center for Progressive Reform.

141. "Huntsman Chooses Dyadem for Operational Risk Management," Dyadem International Ltd., Last modified September 29, 2010, http://www.dyadem.com/press-releases/huntsman-chooses-dyadem-for-operational-risk-management.

142. "Safety and Security of Offshore Oil and Gas Installations," CSCAP Memorandum, No.16, January 2011.

143. David Steven Jacoby, *Guide to Supply Chain Management,* New York: Bloomberg Press, 2009, p. 177.

144. SupplyChainRiskManagementBlog."PredictingandManagingSupplyChainRisks"Accessed February 8, 2012. http://scrmblog.com/review/pred-icting-and-managing-supply-chain-risks.

145. "KNPC Advances Industry Best Practices by Expanding Enterprise-level HSE Information Management," HIS , accessed February 9, 2012, http://www.ihs.com/ar/images/KNPC-2011_Excellence_Award.pdf

146. Government of Alberta, "Alberta to better integrate oil and gas policy and regulatory system." http://alberta.ca/home/NewsFrame.cfm?ReleaseID=/acn/201101/29834CE192CF6-96A7-DCE3-E92C873AAD1CF83F.html.

147. Quality Austria Certification Gulf. "Quality Austria risk Manager certification Training for Qatar Petroleum Senior Managers." http://www.qualityaustriagulf.com/html/newsandevents.html.

148. Edmond Furter."Heads and tails of risk. Edmond Furter interviews TWP Sherq officer Quinton Van Eeden." Miner's Choice Mining Magazine. http://www.minerschoice.co.za/sheq2.html.

149. "A structured approach to Enterprise Risk Management (ERM) and the requirements of ISO 31000," The Public Risk Management Association.

150. Ibid. p. 11.

151. Website of the Committee of Sponsoring Organization of the Treadway Commission. https://www.coso.org/Documents/COSO-ERM-Presentation-September-2017.pdf.

152. "Internal Control—Integrated Framework." Committee of Sponsoring Organization of the Treadway Commission. https://ce.jalisco.gob.mx/sites/ce.jalisco.gob.mx/files/coso_mejoras_al_control_interno.pdf.

153. "Management of Risk in Government: A framework for boards and examples of what has worked in practice." Chief Executive of the Civil Service and Permanent Secretary for the Cabinet Office. January 2017. https://assets.publishing.service.gov.uk/government/uploads/system/uploads/attachment_data/file/584363/170110_Framework_for_Management_of_Risk_in_Govt__final_.pdf.

154. "New Class Notation for Integrated Software Dependent Systems Released DNV Company, May 18, 2010. http://www.dnv.com/press_area/press_releases/2010/newclassnotationforintegratedsoftwaredependentsystemsreleased.asp.

155. Offshore Standards. DNVGL-OS-D203. Integrated software dependent systems (ISDS). July 2017. OFFSHORE STANDARDS DNVGL-OS-D203 Edition July 2017 Integrated software dependent systems (ISDS)

156. "Nissan Green Purchasing guideline Nissan Motor Co., Ltd. http://www.nissan-global.com/EN/DOCUMENT/PDF/SR/Nissan_Green_Purchasing_Guideline_e.pdf.

157. "Nissan Green Purchasing guideline Nissan Motor Co., Ltd. http://www.nissan-global.com/EN/DOCUMENT/PDF/SR/Nissan_Green_Purchasing_Guideline_e.pdf.

158. "Saudi Aramco's Carbon Capture and Sequestration Technology Roadmap: Ali Al-Meshari." Saudi Aramco, January 2008. http://www.cslforum.org/publications/documents/SaudiArabia/T2_2_CSLF_CM_TechRoadMap_Saudi_Aramoc_Jan08.pdf.

159. "Green Production Chemistry," University of Stavanger, June 8, 2010. http://www.uis.no/research/natural_sciences/chemistry_and_environment/oil_field_production_chemicals/.

160. "Gas drillers recycling more water, using fewer chemicals," *Pittsburgh Post-Gazette,* March 1, 2011. http://www.post-gazette.com/pg/11060/1128780-503.stm?cmpid=news.xml.

161. "Environmental Management," Draka, accessed February 9, 2012, http://www.draka.com/draka/lang/en/nav/Sustainable_business/Environmental_management/index.jsp.

162. David Steven Jacoby, *The Guide to Supply Chain Management,* The Economist, 2009.

163. Fred Mayes, "Fuel Cell Power Plants Are Used in Diverse Ways across the United States," EIA (U.S. Energy Information Administration, April 20, 2018), https://www.eia.gov/todayinenergy/detail.php?id=35872.

164. "Scania Fuel Cell Trucks Begin Pilot Operations," OEM Off-Highway (AC Business Media, LLC., January 20, 2020), https://www.oemoffhighway.com/electronics/power-systems/press-release/21111207/scania-ab-scania-fuel-cell-trucks-begin-pilot-operations.

165. "Hydrogen Fuel Cell Electric Bus Pilot Project," MiWay (City of Mississauga, December 2, 2020), https://www.mississauga.ca/miway-transit/about-miway/hydrogen-fuel-cell-electric-bus-pilot-project/.

166. "SK E&C's New Gumi Fuel Cell Plant Opens," FuelCellsWorks (FuelCellsWorks, October 20, 2020), https://fuelcellsworks.com/news/sk-ecs-new-gumi-fuel-cell-plant-opens/.

167. Nicholas Nhede, "Auto Industry Giants to Pilot Vehicle Fuel Cells for Stationary Energy Generation," Smart Energy International (Clarion Energy, December 4, 2019), https://www.smart-energy.com/renewable-energy/auto-industry-giants-to-pilot-vehicle-fuel-cells-for-stationary-energy-generation-mercedes-benz/.

168. "World's Largest Fuel Cell Plant Opens in South Korea." Power Magazine. February 25, 2014. https://www.powermag.com/worlds-largest-fuel-cell-plant-opens-in-south-korea/#:~:text=The%20Gyeonggi%20Green%20Energy%20facility,FuelCell%20Energy%20of%20Danbury%2C%20Conn. Accessed January 31, 2021.

169. "South Korea: Work Begins on World's Largest Hydrogen Fuel Cell Power Plant." FuelCellsWorks. August 5, 2019. https://fuelcellsworks.com/news/south-korea-work-begins-on-worlds-largest-hydrogen-fuel-cell-power-plant-commissioned/ Accessed January 31, 2021.

170. "Hyundai Motor Announces Pilot Project to Generate Electricity from Hydrogen with NEXO Technology," Hyundai (Hyundai Motor Company, April 12, 2019), https://www.hyundai.com/worldwide/en/company/newsroom/hyundai-motor-announces-pilot-project-to-generate-electricity-from-hydrogen-with-nexo-technology-0000016199.

171. Jack Burke, "Power Plant Sets 100% Hydrogen Goal For Turbines," Diesel and Gas Turbine Worldwide (KHL Group Americas LLC. , November 19, 2020), https://www.dieselgasturbine.com/7012513.article.

172. Fred Mayes, "Fuel Cell Power Plants Are Used in Diverse Ways across the United States," EIA (U.S. Energy Information Administration, April 20, 2018), https://www.eia.gov/todayinenergy/detail.php?id=35872.

173. Parts of a Fuel Cell," Hydrogen & Fuel Cell Technologies Office (Office of Energy Efficiency & Renewable Energy, U.S. Department of Energy), https://www.energy.gov/eere/fuelcells/parts-fuel-cell.

174. "Chapter 14—Direct Methanol Fuel Cell." Fereshteh Samimi and Mohammad R. Rahimpour. ScienceDirect. https://www.sciencedirect.com/science/article/pii/B9780444639035000145 Last accessed January 27, 2021.

175. E Gülzow & M Schulze, "Alkaline Fuel Cells," Woodhead Publishing Series in Electronic and Optical Materials (Materials for Fuel Cells, March 27, 2014), https://www.sciencedirect.com/science/article/pii/B9781845693305500038

176. M.Okumura, "Fuel Cells-Phosphoric Acid Fuel Cells | Systems," Encyclopedia of Electrochemical Power Sources (Reference Module in Chemistry, Molecular Sciences and Chemical Engineering, September 09, 2013), https://www.sciencedirect.com/science/article/pii/B978012409547201235X.

177. A.L. Dicks, "Fuel Cells—Molten Carbonate Fuel Cells|Cathode,"The University of Queensland (Encyclopedia of Electrochemical Power Sources, December 08, 2009), https://www.sciencedirect.com/science/article/pii/B9780444527455002653.

178. Earnest Grrison, "Solid Oxide Fuel Cell Cells," Illinois Institute of Technology (Science and Mathematics with Application of Relevant Technology SMART program), https://mypages.iit.edu/~smart/garrear/fuelcells.htm.

179. "Hydrogen Production: Natural Gas Reforming," Energy.gov (U.S. Department of Energy), accessed January 13, 2021, https://www.energy.gov/eere/fuelcells/hydrogen-production-natural-gas-reforming.

180. "Hydrogen—Tracking Report," IEA (International Energy Agency, June 1, 2020), https://www.iea.org/reports/hydrogen.

181. "Types of Fuel Cells," Energy.gov (U.S. Department of Energy), accessed January 20, 2021, https://www.energy.gov/eere/fuelcells/types-fuel-cells.

182. Hydrogen Storage,"Hydrogen & Fuel Cell Technologies Office (Office of Energy Efficiency & Renewable Energy, U.S. Department of Energy), https://www.energy.gov/eere/fuelcells/hydrogen-storage.

183. "Hanwha Energy Celebrates Its Completion of the World's First and Largest Byproduct-Hydrogen-Fuel-Cell Power Plant," Hanwha.com (Hanwha Group, August 7, 2020), https://www.hanwha.com/en/news_and_media/press_release/hanwha-energy-celebrates-its-completion-of-the-worlds-first-and-largest-byproduct-hydrogen-fuel-cell-power-plant.html.

184. Choi Moon-hee, "SK E&C Starts Producing World's Best Eco-Friendly Fuel Cells," Business Korea (BusinessKorea Co., Ltd., October 21, 2020), http://www.businesskorea.co.kr/news/articleView.html?idxno=53569.

185. "Black & Veatch Is EPC Contractor for the First Hydrogen Fueling Station Network in U.S. History," Black & Veatch (Black and Veatch Holding Company), accessed January 13, 2021, https://www.bv.com/projects/black-veatch-epc-contractor-first-hydrogen-fueling-station-network-us-history.

186. Sonal Patel, "How the Energy Transition Is Affecting the EPC Business," POWER Magazine (Access Intelligence, LLC, January 4, 2021), https://www.powermag.com/how-the-energy-transition-is-affecting-the-epc-business/.

187. Cvetelin Vasilev, "Ecolectro: Producing Cost-Efficient Polymers for Fuel Cells and Hydrogen Generation," AZoM.com (AZoNetwork, August 18, 2020), https://www.azom.com/article.aspx?ArticleID=19530.

188. Jackson Chen, "Platinum deficit to reach 1.2 million ounces in 2020 – report," mining.com (Glacier Media group, November 19, 2020), https://www.mining.com/platinum-deficit-to-reach-1-2-million-ounces-in-2020-report/.

189. "Ballard and HDF Energy Sign Development Agreement For Multi-Megawatt Fuel Cell Systems," Cision Canada (CNW Group Ltd., December 10, 2019), https://www.newswire.ca/news-releases/ballard-and-hdf-energy-sign-development-agreement-for-multi-megawatt-fuel-cell-systems-820007754.html.

190. "Toshiba to Accelerate Development of Pure Hydrogen Fuel Cell Module for Vessels and Railroad Vehicles," Toshiba (Toshiba Energy Systems and Solutions Corporation, October 8, 2020), https://www.toshiba-energy.com/en/info/info2020_1008.htm.

191. Scott McMahan, "Bosch Partners on Volume Production of Hydrogen Fuel Cells for Trucks and Cars," EEPower (EETech Media, LLC., April 30, 2019), https://eepower.com/news/bosch-to-partner-on-large-scale-production-of-fuel-cells-for-trucks-and-cars/.

192. "Ballard in China," Ballard in China | About Ballard | Ballard Power (Ballard Power Systems), accessed January 13, 2021, https://www.ballard.com/about-ballard/ballard-in-china.

193. Chelsea Diana, "Plug Power Lands $172M Contract in One of First Steps toward $1 Billion Goal," Albany Business Review (American City Business Journals, January 6, 2020), https://www.bizjournals.com/albany/news/2020/01/06/plug-power-lands-172m-contract-fortune-100.html.

194. "European Shipbuilder Awards Hydrogen Fuel Cell Contract to Proton Motor," Green Car Congress (BioAge Group, LLC., October 31, 2020), https://www.greencarcongress.com/2020/10/20201031-proton.html.

195. "Fuel Cell Control Unit," Bosch Mobility Solution (Robert Bosch GmbH), https://www.bosch-mobility-solutions.com/en/products-and-services/passenger-cars-and-light-commercial-vehicles/powertrain-systems/fuel-cell-electric-vehicle/control-unit-for-powertrain-fuel-cell-systems/.

196. "Doosan Fuel Cell Signs Business Agreement with KT on 'Expanding Fuel Cell Business," FuelCellsWorks (FuelCellsWorks, April 14, 2020), https://fuelcellsworks.com/news/doosan-fuel-cell-signs-business-agreement-with-kt-on-expanding-fuel-cell-business/.

197. Jason Marcinkoski et al., "DOE Hydrogen and Fuel Cells Program Record," Energy.gov (U.S. Department of Energy, September 30, 2015), https://www.hydrogen.energy.gov/pdfs/15015_fuel_cell_system_cost_2015.pdf.

198. "Ballard Expanding MEA Production Capacity 6x to Meet Expected Growth in Fuel Cell Electric Vehicle Demand," Ballard (Ballard Power Systems, September 28, 2020), https://www.ballard.com/about-ballard/newsroom/news-releases/2020/09/28/ballard-expanding-mea-production-capacity-6x-to-meet-expected-growth-in-fuel-cell-electric-vehicle-demand.

199. "Hyundai Motor Company and LS Electric sign hydrogen fuel cell power generation development agreement," Fuel Cells Works (December14, 2020), https://fuelcellsworks.com/news/hyundai-motor-company-and-ls-electric-sign-hydrogen-fuel-cell-power-generation-development-agreement/.

200. "Hydrogen Fuel Cells Will Be a Modest $3 Billion Market in 2030," Industry Week (Endeavor Business Media, LLC., January 8, 2013), https://www.industryweek.com/technology-and-iiot/article/21959256/hydrogen-fuel-cells-will-be-a-modest-3-billion-market-in-2030.

201. "Recycling PEM fuel Cells," Technical Note (Ballard Power System), https://www.ballard.com/docs/default-source/web-pdf's/recycling-technical-note_final.pdf.

202. "How AI Is Transforming The Solar Energy Industry," Utilities Middle East (ITP Media Group, January 2, 2021), https://www.utilities-me.com/news/16391-how-ai-is-transforming-the-solar-energy-industry.

203. "Rocky River Pumped Storage Hydraulic Plant. American Society of Civil Engineers website. https://www.asce.org/project/rocky-river-pumped-storage-hydraulic-plant/

204. "Pumped-Storage Hydropower," Energy.gov (U.S. Department of Energy), accessed January 8, 2021, https://www.energy.gov/eere/water/pumped-storage-hydropower.

205. Energy Storage: key trends and insights." International Energy Agency. Luis Munuera. Presentation to Invest in Net, December 9, 2020.

206. DOE/EPRI Electricity Storage Handbook in Collaboration with NRECA. Abbas A. Akhil et al. Prepared by Sandia National Laboratories. https://www.energy.gov/sites/prod/files/2017/01/f34/Deployment%20of%20Grid-Scale%20Batteries%20in%20the%20United%20States.pdf Accessed January 30, 2021.

207. "Compressed Air Energy Storage," Compressed Air Energy Storage - An Overview (ScienceDirect), accessed January 8, 2021, https://www.sciencedirect.com/topics/engineering/compressed-air-energy-storage#:~:text=Compressed%2Dair%20energy%20storage%20(CAES)%20facilities%20have%20been%20commercially,the%20United%20States%20%5B26%5D.

208. HJ Mai, "Global Redox Flow Battery Market Set to Reach $370M by 2025: QY Research," Utility Dive, August 20, 2019, https://www.utilitydive.com/news/global-redox-flow-battery-market-set-to-reach-to-370m-by-2025-qy-research/561279/.

209. "Lazard's Levelized Cost of Storage Analysis—Version 6.0" (Lazard), accessed January 8, 2021, https://www.lazard.com/media/451418/lazards-levelized-cost-of-storage-version-60.pdf.

210. Kip Keen, "As Battery Costs Plummet, Lithium-Ion Innovation Hits Limits, Experts Say," As battery costs plummet, lithium-ion innovation hits limits, experts say | S&P Global Market Intelligence (S&P Global, May 14, 2020), https://www.spglobal.com/marketintelligence/en/news-insights/latest-news-headlines/as-battery-costs-plummet-lithium-ion-innovation-hits-limits-experts-say-58613238#:~:text=Lithium%2Dion%20battery%20packs%20cost,2019%2C%20according%20to%20BloombergNEF%20data.

211. "Did QuantumScape Just Solve a 40-Year-Old Battery Problem?" By Daniel Oberhaus. Wired. December 8, 2020. https://www.wired.com/story/quantumscape-solid-state-battery/. Accessed January 31, 2021.

212. Kalyan Dasgupta, et al., "Estimating Return on Investment for Grid Scale Storage within the Economic Dispatch Framework," 2015 IEEE Innovative Smart Grid Technologies - Asia (ISGT ASIA), November 2015, https://doi.org/10.1109/isgt-asia.2015.7387088.

213. Energy Storage Journal, "Belgium Li-Ion ESS Fire Cause Still Unknown Two Months Later," Energy Storage Journal (Energy Storage Journal, January 11, 2018), https://www.energystorage journal.com/belgiums-li-ion-ess-fire-cause-still-unknown-two-months-later/.

214. Garrett Hering, "Burning Concern: Energy Storage Industry Battles Battery Fires," Burning concern: Energy storage industry battles battery fires | S&P Global Market Intelligence, May 24, 2019, https://www.spglobal.com/marketintelligence/en/news-insights/latest-news-headlines /burning-concern-energy-storage-industry-battles-battery-fires-51900636.

215. Ibid.

216. Iea, "Energy Storage," IEA (International Energy Agency, June 1, 2020), https://www.iea .org/reports/energy-storage.

217. Garrett Hering, "Burning Concern: Energy Storage Industry Battles Battery Fires," Burning concern: Energy storage industry battles battery fires | S&P Global Market Intelligence, May 24, 2019, https://www.spglobal.com/marketintelligence/en/news-insights/latest-news-headlines /burning-concern-energy-storage-industry-battles-battery-fires-51900636.

218. "Just How Much of a 'Breakthrough' is Tesla's Tabless Battery Cell? " All About Circuits. By Steve Arar. October 07, 2020. https://www.allaboutcircuits.com/news/just-how-much -breakthrough-teslas-tabless-battery-cell/ Accessed January 31, 2021.

219. Inc. ESS, "ESS Inc. Partners With Munich Re to Launch Industry-First Insurance Coverage for Flow Batteries," GlobeNewswire News Room (GlobeNewswire, March 7, 2019), https://www .globenewswire.com/news-release/2019/03/07/1749856/0/en/ESS-Inc-Partners-With-Munich -Re-to-Launch-Industry-First-Insurance-Coverage-for-Flow-Batteries.html.

220. "Graphite Firms Integrate European Battery Supply Chain," Commodity & Energy Price Benchmarks (Argus Media Group, September 24, 2020), https://www.argusmedia.com/en /news/2144154-graphite-firms-integrate-european-battery-supply-chain.

221. Iea, "Energy Storage," IEA (International Energy Agency, June 1, 2020), https://www.iea. org/reports/energy-storage.

222. "Graphite: Natural Graphite Remains on EU Critical Raw Materials List, for Now," Roskill, September 11, 2020, https://roskill.com/news/graphite-natural-graphite-remains-on -eu-critical-raw-materials-list-for-now/.

223. Soumyajit Saha, "Australia's EcoGraf Inks Graphite Battery Material Supply Deal with Thyssenkrupp Unit," Reuters (Thomson Reuters, June 4, 2020), https://www.reuters.com /article/us-ecograf-deals-thyssenkrupp/australias-ecograf-inks-graphite-battery-material -supply-deal-with-thyssenkrupp-unit-idINKBN23B05E?edition-redirect=in.

224. "FREYR and 24M Sign Licensing and Services Agreement for Mass Production of Clean, Low-Cost and Safe Lithium-Ion Battery Cells," Business Wire (Business Wire, Inc., December 21, 2020), https://www.businesswire.com/news/home/20201221005094/en/FREYR-and-24M-Sign -Licensing-and-Services-Agreement-for-Mass-Production-of-Clean-Low-Cost-and-Safe -Lithium-ion-Battery-Cells.

225. Sybil Pan, "ABB, Talga Sign Agreement for Swedish Battery Anode Project," Fastmarkets (Fastmarkets, November 25, 2020), https://www.fastmarkets.com/article/3963814/abb-talga -sign-agreement-for-swedish-battery-anode-project.

226. Steve Dent, "Tesla and Panasonic Will No Longer Work Together on Solar Cells," Engadget, March 6, 2020, https://www.engadget.com/2020-02-26-tesla-and-panasonic-split-solar-cells -new-york.html.

227. ISA/IEC 62443 Cybersecurity Certificate Programs webpage. https://www.isa.org/training-and-certification/isa-certification/isa99iec-62443/isa99iec-62443-cybersecurity-certificate-programs

228. Mark Triplett, "How AI Is Changing Energy Storage O&M," Solar Power World, May 7, 2020, https://www.solarpowerworldonline.com/2020/04/how-ai-is-changing-energy-storage-om/.

229. "Battery Energy Storage System Uses Artificial Intelligence to Lower Energy Bills and Provide Utility Grid Services," PR Newswire: news distribution, targeting and monitoring, December 5, 2018, https://www.prnewswire.com/news-releases/battery-energy-storage-system-uses-artificial-intelligence-to-lower-energy-bills-and-provide-utility-grid-services-300760421.html.

230. "Comparing Lithium-Ion Battery Chemistries," Energy Sage (Energy Sage), accessed January 8, 2021, https://news.energysage.com/comparing-lithium-ion-battery-chemistries/#:~:text=Types%20of%20lithium%2Dion%20battery%20chemistries&text=Perhaps%20the%20most%20commonly%20seen,Resu%20and%20the%20Tesla%20Powerwall.

231. "OECD Due Diligence Guidance for Responsible Supply Chains of Minerals from Conflict-Affected and High-Risk Areas," OECD (Organization for Economic Co-operation and Development), accessed January 8, 2021, http://www.oecd.org/daf/inv/mne/mining.htm.

232. Andrew Fawthrop, "ICMM Raises Bar for Ethical Mining Standards with New Membership Framework," NS Energy (NS Energy, February 13, 2020), https://www.nsenergybusiness.com/news/icmm-mining-principles/.

233. "The Regulation Explained," European Commission (European Union), accessed January 8, 2021, https://ec.europa.eu/trade/policy/in-focus/conflict-minerals-regulation/regulation-explained/#regulation-what.

234. Garrett Fitzgerald et al., "The Economics of Battery Energy Storage" (Rocky Mountain Institute, October 2015), https://rmi.org/wp-content/uploads/2017/03/RMI-TheEconomicsOfBatteryEnergyStorage-FullReport-FINAL.pdf.

235. Kalyan Dasgupta, et al., "Estimating Return on Investment for Grid Scale Storage within the Economic Dispatch Framework," 2015 IEEE Innovative Smart Grid Technologies - Asia (ISGT ASIA), 2015, https://doi.org/10.1109/isgt-asia.2015.7387088.

236. Author's analysis of data from Lazard at "Lazard's Levelized Cost of Storage Analysis - Version 6.0" (Lazard), accessed January 8, 2021, https://www.lazard.com/media/451418/lazards-levelized-cost-of-storage-version-60.pdf., p. 11.

237. Rakesh Ranjan, "Levelized Cost of Storage for Standalone BESS Could Reach ₹4.12/KWh by 2030: Report," Mercom India (Mercom Capital Group, June 2, 2020), https://mercomindia.com/levelized-cost-storage-standalone-bess/.

238. Energy Transition Outlook 2019. DNV-GL (from the databases).

239. Susan Gourvenec, "This Is How We Can Make Floating Wind Farms the Future of Green Electricity," World Economic Forum (World Economic Forum, July 29, 2020), https://theconversation.com/floating-wind-farms-how-to-make-them-the-future-of-green-electricity-142847.

240. Ibid.

241. "Operational and Maintenance Costs for Wind Turbines," Wind Measurement International (Wind Measurement International), accessed January 13, 2021, http://windmeasurementinternational.com/wind-turbines.php.

242. "Setback for Japanese Offshore Wind Efforts," Windfair.net (Smart Dolphin GMBH, August 10, 2018), https://w3.windfair.net/wind-energy/news/29116-wind-turbine-floater-japan-coast-capacity-factor-yield-turbine-lower.

243. New Straits Times, "Japan Removes Two Last Wind Power Turbines," New Straits Times (Media Prima Group, December 17, 2020), https://www.nst.com.my/world/region/2020/12/650410/japan-removes-two-last-wind-power-turbines.

244. Samuel Roach et al., "Application of the New IEC International Design Standard for Offshore Wind Turbines to a Reference Site in the Massachusetts Offshore Wind Energy Area," IOP Science (Journal of Physics, 2020), https://iopscience.iop.org/article/10.1088/1742-6596/1452/1/012068/pdf.

245. "Decentralised Solutions: Large-Scale Battery Storage Is Happening," Vattenfall (Vattenfall AB), accessed January 13, 2021, https://group.vattenfall.com/what-we-do/roadmap-to-fossil-freedom/decentralised-solutions/energy-storage.

246. "World's Biggest Offshore Wind Producer Moves Into Hydrogen. Bloomberg Green. By Lars Paulsson and Will Mathis. January 20, 2021. https://www.bloomberg.com/news/articles/2021-01-20/world-s-biggest-offshore-wind-producer-moves-into-hydrogen Last accessed January 27, 2021.

247. David Foxwell, "Hedging Emerging as Potential Solution for Wind Energy Risk," Riviera (Riviera Maritime Media Ltd., November 29, 2017), https://www.rivieramm.com/news-content-hub/news-content-hub/hedging-emerging-as-potential-solution-for-wind-energy-risk-26391.

248. William Wilkes, Brian Parkin, and Jeremy Hodges, "Wind Turbine Factories Struggle For Parts in Virus Lockdown," BloombergQuint (Bloomberg Quint, April 21, 2020), https://www.bloombergquint.com/business/wind-turbine-factories-struggle-to-get-parts-in-virus-lockdown.

249. Eric Ng, "China Turbine Makers Winded as Ecuador Lockdown Leaves Them without Blades," South China Morning Post (South China Morning Post, April 16, 2020), https://www.scmp.com/business/article/3080227/china-turbine-makers-winded-after-ecuador-lockdown-leaves-them-without.

250. Michelle Froese, "Vestas Receives 15-Year Service Contract for Seven Wind Farms in Sweden," Windpower Engineering & Development (WTWH Media, LLC., May 15, 2017), https://www.windpowerengineering.com/vestas-receives-15-year-service-contract-seven-wind-farms-sweden/.

251. Paul Dvorak, "ENERCON Concludes Strategic Partnership with Lagerwey," Windpower Engineering & Development (WTWH Media, LLC., January 3, 2018), https://www.windpowerengineering.com/enercon-concludes-strategic-partnership-lagerwey/.

252. "Siemens Gamesa Strengthens Its Partnership with European Energy to Supply Wind Farms in Sweden and Poland," Siemens Gamesa (Siemens Gamesa Renewable Energy, S.A, December 2, 2020), https://www.siemensgamesa.com/en-int/newsroom/2020/12/201202-siemens-gamesa-press-release-european-energy-sweden-poland.

253. "North America Power Options," CME Group (CME Group Inc., January 12, 2021), https://www.cmegroup.com/trading/energy/north-america-power-options.html.

254. Michelle Froese, "TPI Composites Signs New Turbine Supply Agreement with Gamesa," Windpower Engineering & Development (WTWH Media, LLC., April 4, 2017), https://www.windpowerengineering.com/tpi-composites-signs-new-turbine-supply-agreement-gamesa/.

255. "TPI Composites Expands Mexican Blade Supply Deal with GE," Renewables Now (Renewables Now, August 8, 2018), https://renewablesnow.com/news/tpi-composites-expands-mexican-blade-supply-deal-with-ge-622854/.

256. "Senvion Secures Long-Term Service Contract Extension in Australia," Senvion (Senvion Wind Energy Solutions, September 19, 2018), https://www.senvion.com/fileadmin/Redakteur/Press_Media/Press_releases/2018/2018_09_19_Meridian_Energy_EN_final.pdf.

257. "Wind Cyber Defence Experts Urge Tighter Control of Supplier Access," Reuters Events | Renewables (Reuters News and Media, Ltd., October 24, 2018), https://www.reutersevents.com/renewables/wind-energy-update/wind-cyber-defence-experts-urge-tighter-control-supplier-access.

258. Ibid.

259. "Wind Cyber Defence Experts Urge Tighter Control of Supplier Access," Reuters Events | Renewables (Reuters News and Media, Ltd. , October 24, 2018), https://www.reutersevents.com/renewables/wind-energy-update/wind-cyber-defence-experts-urge-tighter-control-supplier-access.

260. "Cybersecurity Coming Soon to Invenergy's Wind Farms," Renewable Energy World (Clarion Events, May 24, 2017), https://www.renewableenergyworld.com/2017/05/24/cybersecurity-coming-soon-to-invenergy-s-wind-farms/.

261. " Optimized Bidding Solutions for Batteries And Renewable Generators," AMS (Advanced Microgrid Solutions), accessed January 13, 2021, https://www.advancedmicrogridsolutions.com/nem-energy-storage.

262. Katharina Günther, "Smart Wind - Power Systems Technology and Power Mechatronics," SmartWind (Bundesministerium für Wirtschaft und Energie), accessed January 13, 2021, http://www.enesys.ruhr-uni-bochum.de/news/news00034.html.en.

263. "2018 Annual Report: Focused on Execution," TPI Composites (TPI Composites, Inc., 2018), https://www.annualreports.com/HostedData/AnnualReportArchive/t/NASDAQ_TPIC_2018.pdf.

264. "MHI Vestas Signs Major Wind Power Supply Chain Contract in Taiwan," REVE News of the wind sector in Spain and in the world (Spanish Wind Energy Association, July 16, 2020), https://www.evwind.es/2020/07/16/mhi-vestas-signs-major-wind-power-supply-chain-contract-in-taiwan/75798.

265. "MHI Vestas Secures Footprint in Taiwan with Waterside Manufacturing Facility," MHI Vestas Offshore (MHI Vestas Offshore Wind, December 2, 2020), https://mhivestasoffshore.com/mhi-vestas-secures-footprint-in-taiwan-with-waterside-manufacturing-facility/.

266. Molly Lumsden, "Wind Power Costs Have Plummeted. How Can They Fall Even Further?," Into the Wind (American Wind Energy Association, July 8, 2019), https://www.aweablog.org/wind-power-costs-plummeted-can-fall-even/.

267. "Wind Farm Costs - Guide to an Offshore Wind Farm," Catapult Offshore Renewable Energy (BVG Associates, November 20, 2019), https://guidetoanoffshorewindfarm.com/wind-farm-costs.

268. Chris Martini, "Wind Turbine Blades Can't Be Recycled, so They're Piling up in Landfills," Los Angeles Times (Los Angeles Times, February 6, 2020), https://www.latimes.com/business/story/2020-02-06/wind-turbine-blades.

269. "GE Goes beyond Carbon Neutral to Cut Blade Waste ," Reuters Events: Renewables (Reuters News and Media Ltd., February 5, 2020), https://www.reutersevents.com/renewables/wind-energy-update/ge-goes-beyond-carbon-neutral-cut-blade-waste.

270. Chris Martini, "Wind Turbine Blades Can't Be Recycled, so They're Piling up in Landfills," Los Angeles Times (Los Angeles Times, February 6, 2020), https://www.latimes.com/business/story/2020-02-06/wind-turbine-blades.

271. NREL Q4 2019/Q1 2020 Solar Industry Update, p. 2

272. *PV Inverters Market.* Market Analysis. Boston Strategies International. June 2016.

273. "Battery Energy Storage," Trimark (Trimark Associates, Inc., November 13, 2019), https://trimarkassoc.com/trimark-products/battery-energy-storage/.

274. "Lazard's Levelized Cost of Storage Analysis - Version 6.0" (Lazard): 2-3, accessed January 8, 2021, https://www.lazard.com/media/451418/lazards-levelized-cost-of-storage-version-60.pdf.

275. "Solar Industry Research Data," SEIA (Solar Energy Industries Association), accessed January 8, 2021, https://www.seia.org/solar-industry-research-data.

276. "How AI Is Transforming The Solar Energy Industry," Utilities Middle East (ITP Media Group, January 2, 2021), https://www.utilities-me.com/news/16391-how-ai-is-transforming-the-solar-energy-industry .

277. "Which Solar Panel Type Is Best? Mono- Vs. Polycrystalline Vs. Thin Film?" MEP cell website. https://www.mepcell.com/which-solar-panel-type-is-best-mono-vs-polycrystalline-vs-thin-film/?lang=en. Last accessed January 17, 2021.

278. "Solar + Storage." Solar Energy Industries Associate website. https://www.seia.org/initiatives/solar-plus-storage. Last accessed January 17, 2021.

279. Pamela Largue, "TOTAL Solar DG Starts Building One of South-East Asia's Largest Microgrids," Power Engineering International (Power Engineering International, December 8, 2020), https://www.powerengineeringint.com/decentralized-energy/total-solar-dg-starts-building-one-of-southeast-asias-largest-microgrids/.

280. "EESL Plans to Set Up 1,500 MW Decentralised Solar Power Plants by 2021," ETEnergyworld.com (The Economic Times, February 26, 2020), https://energy.economictimes.indiatimes.com/news/renewable/eesl-plans-to-set-up-1500-mw-decentralised-solar-power-plants-by-2021/74310474.

281. "Concentrating Solar-Thermal Power." United States Department of Energy. https://www.energy.gov/eere/solar/concentrating-solar-thermal-power. Accessed March 2, 2021.

282. "DEWA CSP 700 MWe Project: What We Know so Far," MENA CSP KIP (Center for Mediterranean Integration, August 6, 2018), https://cmimarseille.org/menacspkip/dewa-csp-700-mwe-project-know-far/.

283. Reve, "China Makes Half of Global Newly-Built Solar Power Capacity in 2019," REVE News of the wind sector in Spain and in the world (REVE, January 15, 2020), https://www.evwind.es/2020/01/15/china-makes-half-of-global-newly-built-concentrated-solar-power-capacity-in-2019/73071.

284. Henry Brean, "Nevada Solar Plant Back Online after Eight-Month Outage," Las Vegas Review-Journal (Las Vegas Review-Journal, July 22, 2017), https://www.reviewjournal.com/business/energy/nevada-solar-plant-back-online-after-eight-month-outage/.

285. Nickie Louise, "The World's Biggest Solar Energy Project Failure: How This $1 Billion Boondoggle Solar Plant Project Became Obsolete before It Ever Went Online: Tech News: Startups News," Tech News | Startups News (Tech Startups, November 19, 2020), https://techstartups.com/2020/11/19/worlds-biggest-solar-energy-project-failure-1-billion-boondoggle-solar-plant-project-became-obsolete-ever-went-online/.

286. Ibid.

287. "Ivanpah," Energy.gov (U.S. Department of Energy), accessed January 8, 2021, https://www.energy.gov/lpo/ivanpah.

288. Nickie Louise, "The World's Biggest Solar Energy Project Failure: How This $1 Billion Boondoggle Solar Plant Project Became Obsolete before It Ever Went Online: Tech News: Startups News," Tech News | Startups News (Tech Startups, November 19, 2020), https://techstartups.com/2020/11/19/worlds-biggest-solar-energy-project-failure-1-billion-boondoggle-solar-plant-project-became-obsolete-ever-went-online/.

289. Doloresz Katanich, "Extraordinary Views from the World's First Mountainous Solar Farm," EuroNews (Living, January 14, 2021), https://www.euronews.com/living/2021/01/13/extraordinary-views-from-the-world-s-first-mountainous-solar-farm.

290. Darius Snieckus, "World's First Offshore Solar Array Rides out Storm Ciara off Netherlands," Recharge (Recharge, February 16, 2020), https://www.rechargenews.com/transition/worlds-first-offshore-solar-array-rides-out-storm-ciara-off-netherlands/2-1-757022.

291. Andrew Lee, "Fire Hits BP Venture's Flagship Floating Solar Plant in UK," Recharge (Recharge, September 17, 2020), https://www.rechargenews.com/transition/fire-hits-bp-ventures-flagship-floating-solar-plant-in-uk/2-1-877293.

292. LinkedIn post by David Steven Jacoby, November 18, 2020. https://www.linkedin.com/posts/david-steven-jacoby-8b9b918_products-and-solutions-activity-6734875169640243200--u-D

293. "Consolidation Underway in Solar EPC Market," Bridge to India (Bridge to India Energy Private Limited, December 4, 2019), https://bridgetoindia.com/consolidation-underway-in-solar-epc-market/.

294. Anu Bhambhani, "Cypress Creek Renewables To Eliminate EPC Division," TaiyangNews (Taiyang News, July 18, 2019), http://taiyangnews.info/business/cypress-creek-renewables-to-eliminate-epc-division/.

295. "First Solar Evolves U.S. EPC Delivery Approach with Third-Party Execution Model," First Solar (First Solar, Inc., September 19, 2019), https://investor.firstsolar.com/news/press-release-details/2019/First-Solar-Evolves-US-EPC-Delivery-Approach-with-Third-Party-Execution-Model/default.aspx.

296. "Solar developers warned of PV price rise as polysilicon supply troubles hit." ReNew Economy.com website. https://reneweconomy.com.au/solar-developers-warned-of-pv-price-rise-as-polysilicon-supply-troubles-hit-54043/

297. Liam Stoker, "Trina Adds to Supply Deal Run with Three-Year Daqo Polysilicon Contract," PV Tech (Solar Media Limited, November 30, 2020), https://www.pv-tech.org/news/trina-adds-to-supply-deal-run-with-three-year-daqo-polysilicon-contract.

298. Anand Gupta, "Glass Shortage Threatens Solar Panels Needed for Climate Fix," Energy News for the United States Oil & Gas Industry (EQ Mag Pro, November 5, 2020), https://energynow.com/2020/11/glass-shortage-threatens-solar-panels-needed-for-climate-fix/.

299. Mark Osborne, "Amtech Looking to Outsource Solar Equipment Production in China to Remain Competitive," PV Tech (Solar Media Limited, May 13, 2018), https://www.pv-tech.org/news/amtech-looking-to-outsource-solar-equipment-production-in-china-to-remain-c.

300. Lisa Wang, "URE Cuts Capacity, Turns to Outsourcing to Stay Competitive," Taipei Times (Taipei Times, October 7, 2019), http://www.taipeitimes.com/News/biz/archives/2019/10/08/2003723544.

301. Daniel Seeger, "Meyer Burger Outsources SWCT Production to Mondragon," PV Magazine (PV Magazine International, November 7, 2018), https://marketpublishers.com/report/company_reports/meyer-burger-technology-ag-mbtn-power-deals-n-alliances-profile.html.

302. Marija Maisch, "Comtec Ready to Swim against the Tide by Outsourcing Manufacturing," PV Magazine (PV Magazine International, September 6, 2019), https://www.pv-magazine.com/2019/09/06/comtec-ready-to-swim-against-the-tide-by-outsourcing-manufacturing/.

303. Tax Extenders Included in the Stimulus Bill Poised to Provide a Boost to Renewables. Pillsbury. https://www.pillsburylaw.com/en/news-and-insights/tax-extenders-stimulus-bill-renewables.html Accessed January 27, 2021.

304. Section 201 Solar Tariffs. Solar Energy Industries Association. https://www.seia.org/sites/default/files/2019-12/SEIA-Section-201-Factsheet-Dec2019.pdf Accessed January 28, 2021.

305. The Adverse Impact of Section 201 Tariffs. Solar Energy Industries Association. https://www.seia.org/research-resources/adverse-impact-section-201-tariffs#:~:text=Lost%20Jobs%2C%20Lost%20Deployment%20and,on%20the%20U.S.%20solar%20industry.&text=62%2C000%20fewer%20jobs%20from%202017,tons%20of%20carbon%20dioxide%20emissions Accessed January 28, 2021.

306. "Canadian Solar Signs 1,800 MW Module Supply Agreement with EDF Renewables North America—Largest Module Supply Agreement in the Company's History," Canadian Solar (Canadian Solar Inc., May 29, 2019), http://investors.canadiansolar.com/news-releases/news-release-details/canadian-solar-signs-1800-mw-module-supply-agreement-edf.

307. "ENGIE Signs 25-Year Power Purchase Agreement with Senegalese Government for Two Solar Photovoltaic Projects," PV Magazine (PV Magazine International, November 14, 2018), https://www.pv-magazine.com/press-releases/engie-signs-25-year-power-purchase-agreement-with-senegalese-government-for-two-solar-photovoltaic-projects/.

308. "Southern Nevada Water Authority Signs Power Purchase Agreement with SunEdison," PV Magazine (PV Magazine International, July 22, 2014), https://www.pv-magazine.com/press-releases/southern-nevada-water-authority-signs-power-purchase-agreement-with-sunedison_100015810/.

309. "EDF Renewables PPA Enables Storage Expansion on 100-MW California Solar Project," Saur Energy International (Saur Energy, September 28, 2020), https://www.solarpowerworldonline.com/2020/09/edf-renewables-ppa-enables-storage-expansion-on-california-solar-project/.

310. "Solarpack Wins a PPA Contract in India Increasing Its Backlog by 396 MW," Solar Pack (Solar Pack, June 30, 2020), https://www.solarpack.es/en/solarpack-wins-ppa-contract-in-india-increasing-its-backlog-by-396-mw/.

311. Catalin Cimpanu, "Cyber-Attack Hits Utah Wind and Solar Energy Provider," ZDNet (ZDNet, October 31, 2019), https://www.zdnet.com/article/cyber-attack-hits-utah-wind-and-solar-energy-provider/.

312. Sean Lyngaas, "Utah Renewables Company Was Hit by Rare Cyberattack in March," CyberScoop (Scoop News Group, October 31, 2019), https://www.cyberscoop.com/spower-power-grid-cyberattack-foia/.

313. "How AI Is Transforming The Solar Energy Industry," Utilities Middle East (ITP Media Group, January 2, 2021), https://www.utilities-me.com/news/16391-how-ai-is-transforming-the-solar-energy-industry.

314. Jordan Wirfs-Brock, "Learning How To Adapt To More Renewables As 'Duck Curve' Deepens," Inside Energy (Inside Energy, October 25, 2016), http://insideenergy.org/2016/10/25/learning-how-to-adapt-to-more-renewables-as-duck-curve-deepens/.

315. "Value of AI and UAV (Drone) Inspection in Solar O&M - A Case Study," Aerospec Technologies (Aerospec Technologies, April 26, 2018), https://aerospec.us/value-of-ai-and-uav-drone-inspection-in-solar-om-a-case-study/.

316. "Chinese Solar Panel Production Issues Are Mounting," IER (Institute For Energy Research, November 18, 2020), https://www.instituteforenergyresearch.org/renewable/solar/chinese-solar-panel-production-issues-are-mounting/.

317. "US Solar Maintenance Costs Plummet as Tech Gains Multiply," Reuters Events | Renewables (Reuters Events, February 9, 2019), https://www.reutersevents.com/renewables/pv-insider/us-solar-maintenance-costs-plummet-tech-gains-multiply.

318. Ala' K. Abu-Rumman, Iyad Muslih, and Mahmoud A. Barghash, "Life Cycle Costing of PV Generation System," Journal APRIE (Journal of Applied Research on Industrial Engineering, October 12, 2017), http://www.journal-aprie.com/article_54724_4e5a256ff89a93cd0a5b12c5116c96f3.pdf.

319. Emiliano Bellini, “South Korea to Introduce New Rules for PV Recycling,” PV Magazine (PV Magazine International, October 8, 2020), https://www.pv-magazine.com/2020/10/08/south-korea-to-introduce-new-rules-for-pv-recycling/.

320. Haley Rischar, “New California Rule Will Facilitate the Recycling of Solar Panels,” Waste Today Magazine (GIE Media, Inc., December 15, 2020), https://www.wastetodaymagazine.com/article/new-california-rule-will-facilitate-the-recycling-of-solar-panels/.

321. “Europe's First Solar Panel Recycling Plant Opens in France—ET EnergyWorld,” ETEnergyworld.com (The Economic Times, June 26, 2018), https://energy.economictimes.indiatimes.com/news/renewable/europes-first-solar-panel-recycling-plant-opens-in-france/64741561.

322. “Biomass for Power Generation,” IRENA (International Renewable Energy Agency, June 2012), https://www.irena.org/-/media/Files/IRENA/Agency/Publication/2012/RE_Technologies_Cost_Analysis-BIOMASS.pdf.

323. Ibid.

324. Ran Mei et al., “Evaluating Digestion Efficiency in Full-Scale Anaerobic Digesters by Identifying Active Microbial Populations through the Lens of Microbial Activity,” Scientific Reports (Springer Nature, September 26, 2016), https://www.nature.com/articles/srep34090#citeas.

325. According to the author’s analysis of data from “The European Commission's Knowledge Centre for Bioeconomy,” Europa.eu (European Union, 2019), https://publications.jrc.ec.europa.eu/repository/bitstream/JRC109354/biomass_4_energy_brief_online_1.pdf . and “DNV GL's Energy Transition Outlook 2020,” DNV GL (DNV GL AS, 2020), https://eto.dnvgl.com/2020/index.html#ETO2019-top.

326. “Wärtsilä & Vantaa Energy Partners for Carbon Neutral Synthetic Biogas Production in Finland,” Energetica India Magazine (Editorial Omnimedia, May 19, 2020), https://www.energetica-india.net/news/wrtsil--vantaa-energy-partners-for-carbon-neutral-synthetic-biogas-production-in-finland.

327. Lucas Morals, “Brazil's IBS Energy Secures EPC Contract for 80-MW Biomass Project,” Renewables Now (Renewables Now, March 17, 2020), https://renewablesnow.com/news/brazils-ibs-energy-secures-epc-contract-for-80-mw-biomass-project-691133/.

328. “Village Farms' Clean Energy Subsidiary Renews and Extends Vancouver Landfill Gas Contract to Transition to Renewable Natural Gas Model,” Yahoo! Finance (Yahoo!, November 10, 2020), https://finance.yahoo.com/news/village-farms-clean-energy-subsidiary-120000565.html.

329. “Blue Sphere Selects Anaergia for Dutch Biogas EPC Contract,” Bioenergy International (Bioenergy International, December 14, 2017), https://bioenergyinternational.com/biogas/blue-sphere-selects-anergia-dutch-biogas-epc-contract.

330. Ben Messenger, “£150m EPC Contract for 28 MW Waste to Energy Gasification Plant in Hull,” Waste Management World (WEKA Industrie Medien GmbH, November 23, 2015), https://waste-management-world.com/a/150m-epc-contract-for-28-mw-waste-to-energy-gasification-plant-in-hull.

331. “FLI Energy Breaks Ground on UK Biogas Project,” Renewable Energy Focus (Elsevier Ltd., May 12, 2014), http://www.renewableenergyfocus.com/view/38375/fli-energy-breaks-ground-on-uk-biogas-project/.

332. “PacBio Signs Long-Term Supply Contracts with Japanese Power Producers,” Canadian Biomass Magazine (Annex Business Media, January 21, 2019), https://www.canadianbiomassmagazine.ca/pacbio-signs-long-term-supply-contracts-with-japanese-power-producers-7215/.

333. Wayne Barber, “B&W Construction Subsidiary Sues Developer of New Hampshire Biomass Project,” Transmission Hub (Endeavor Business Media, April 14, 2014), https://

www.transmissionhub.com/articles/2014/04/bw-construction-subsidiary-sues-developer-of-new-hampshire-biomass-project.html.

334. "EQTEC Signs Agreement with Phoenix Energy," Renewable Energy Magazine (Renewable Energy Magazine, May 23, 2019), https://www.renewableenergymagazine.com/biomass/eqtec-signs-framework-agreement-with-phoenix-energy-20190523.

335. "GE Provides Steam Technology for Japanese Biomass Power Plant," Bioenergy Insight (Woodcote Media Ltd., February 11, 2020), https://www.bioenergy-news.com/news/ge-provides-steam-technology-for-japanese-biomass-power-plant/.

336. "Ichihara Yawatafuto Biomass Power Plant," NS Energy (NS Energy), accessed January 14, 2021, https://www.nsenergybusiness.com/projects/ichihara-yawatafuto-biomass-power-plant/.

337. "Albioma Commissions Vale Do Paraná Albioma, the Group's Fourth All-Bagasse Power Plant in Brazil," Albioma (Albioma, December 29, 2020), https://www.albioma.com/wp-content/uploads/2020/12/20201229_Albioma_CP_ENG.pdf.

338. Katie Fletcher, "Big Data Pays Big," Biomassmagazine.com (BBI International, January 2, 2017), http://biomassmagazine.com/articles/14057/big-data-pays-big.

339. Rajdeep Golecha, "Biomass Supply Chain Trade-Offs: Basis for Successful Bioenergy Businesses Biomassmagazine.com (BBI International, January 21, 2016), http://biomassmagazine.com/articles/12786/biomass-supply-chain-trade-offs-basis-for-successful-bioenergy-businesses

340. William L. Leffler and Martin Raymond, *Oil and Gas Production in Nontechnical Language*, Tulsa: PennWell, 2005.

341. *Mark Graham, Mark Cook and Frank Jahn, "Project and Contract Management," Hydrocarbon Exploration and Production* 55, 2001.

342. William L. Leffler and Martin Raymond, *Oil and Gas Production in Nontechnical Language,* Tulsa: PennWell, 2005.

343. Duncan Seddon, *Gas Usage and Value: The Technology and Economics of Natural Gas Use in the Process Industries,* Tulsa: PennWell, 2006.

344. Duncan Seddon, *Gas Usage and Value: The Technology and Economics of Natural Gas Use in the Process Industries,* p. 12. Tulsa: PennWell, 2006.

345. Bernard J. Duroc Danner, "The Role of Oil Field Services in the Energy Dynamic presentation at the National Oil Congress, London, June 2010.

346. Salah Hassan Wahbi, "Sudan's growing Exploration and Development presentation at the National Oil Conference, London, June 2010.

347. "Shell, Schlumberger in joint oil & gas multi-year research technology cooperation," *Tendersinfo News,* December 20, 2010.

348. "The First Cloud Enabled and Integrated Digital Oilfield for Kuwait Oil Company." By Shaikha Battal Al Qahtani, Kuwait Oil Company. 2010.

349. "Strategies to Sustainably Fulfill Energy Needs KOC presentation to National Oil Congress in London, June 22, 2010.

350. "Schlumberger Inaugurates Brazil Research and Geoengineering Center," Schlumberger Limited, accessed February 10, 2012, http://www.slb.com/news/press_releases/2010/2010_1116_brgz_slb.aspx.

351. "GE To Invest $500 Million In Brazil for Accelerated Growth GE press release, November 10, 2010, http://www.google.com/url?sa=t&rct=j&q=&esrc=s&source=web&cd=2&ved=0CDQQFjAB&url=http%3A%2F%2Fwww.genewscenter.com%2FPress-Releases%2FGE-TO-INVEST-500-MILLION-IN-BRAZIL-FOR-ACCELERATED-GROWTH-2cab

.aspx&ei=gC83T7vvGKPJ0AGOo6jPAg&usg=AFQjCNEvOkZGQWq3BoE2Jij15wdRhnqltA &sig2=InYId6__dTPlFbZnu-LEAQ.

352. Charlotte J. Wright and Rebecca A. Gallun, *Fundamentals of Oil & Gas Accounting,* Tulsa: PennWell, 2008.

353. Erik German, "Rent this oil rig for just 30 million bucks!" Global Post, February 3, 2010, accessed February 11, 2011. http://www.globalpost.com/dispatches/bric-yard/rent-oil -rig-just-30-million-bucks.

354. "The Most Challenging Oil and Gas Projects In The World." By James Stafford. OilPrice. com, March 25, 2015. https://oilprice.com/Energy/Energy-General/The-Most-Challenging-Oil -And-Gas-Projects-In-The-World.html Last accessed January 26, 2021.

355. "Setting the Standard," article on the BP website, http://www.bp.com/genericarticle .do?categoryId=9013611&contentId=7021424#

356. "Altus Oil & GAs Moves the World's Largest STP Buoy Case Study, " Altus, last accessed February 11, 2012, http://www.altuslogistics.com/i/Altus_case_FPSO.pdf.

357. David Steven Jacoby, "Balancing economic risks: Tips for a well-structured deal," *Middle East Energy,* September 2010, page 5.

358. Maximising the Value of Strategic Partnerships: NOCs & IOCs, S.N. Ebrahimi, presentation to the National Oil Companies Congress, June 21, 2010.

359. "Rebuilding Iraq: How Much Risk is Too Much? Smith, Gambrell & Russell, LLP, accessed June 30, 2009. http://www.sgrlaw.com/resources/trust_the_leaders/leaders_issues/ttl5/917/

360. Thomas E. Copeland, *Real Options: A Practitioner's Guide,* Texere, 2001.

361. Jason Margolis and Lisa Mullins, "The Energy Costs of Oil Production," The World from PRX (The World from PRX, November 12, 2012), https://www.pri.org/stories/2012-11-02 /energy-costs-oil-production.

362. Jason Margolis and Lisa Mullins, "The Energy Costs of Oil Production," The World from PRX (The World from PRX, November 12, 2012), https://www.pri.org/stories/2012-11-02 /energy-costs-oil-production.

363. Karl-Erik Stromsta, "Chevron to Build 500MW of Renewables to Power Oil and Gas Facilities - and It's Considering More," Greentech Media (Greentech Media, July 30, 2020), https:// www.greentechmedia.com/articles/read/chevron-to-build-500mw-of-renewables-globally-to -power-oil-and-gas-facilities.

364. "Chevron Greenhouse Gas Management," Chevron (Chevron, May 20, 2020), https://www .chevron.com/sustainability/environment/greenhouse-gas-management.

365. "Oil and Gas Companies Can Power Offshore Platforms With Renewables." Greentech Media. November 21, 2019. https://www.greentechmedia.com/articles/read/oil-and-gas -companies-can-power-offshore-platforms-with-renewables.

366. "Carbon Management," Aramco (Aramco), accessed January 22, 2021, https://www.aramco .com/en/creating-value/technology-development/globalresearchcenters/carbon-management.

367. Twesh Mishra, "Cairn Oil and Gas to Spend $1 Billion to Boost Production from Rajasthan Fields," Business Line (The Hindu Business Line, August 29, 2019), https://www.thehindu businessline.com/companies/cairn-oil-and-gas-to-spend-1-billion-to-boost-production-from -rajasthan-fields/article29286157.ece.

368. "Minimizing Gas Flaring in the Permian," Chevron (Chevron, July 21, 2020), https://www .chevron.com/stories/minimizing-gas-flaring-in-the-permian.

369. "Tackling Methane Emissions" White paper. Shell Oil. https://www.shell.com/energy -and-innovation/natural-gas/methane-emissions/_jcr_content/par/textimage_438437728

.stream/1587995196996/53beef2f8ba2e90560c074f56552e2acfe30582b/shell-methane-case-study.pdf

370. "FLIR Launches Its First Uncooled, Fixed-Mount, Connected Thermal Camera for Detecting Methane: GF77a," FLIR (FLIR Systems, Inc., January 30, 2020), https://www.flir.in/news-center/press-releases/flir-launches-its-first-uncooled-fixed-mount-connected-thermal-camera-for-detecting-methane-gf77a/ .

371. "Halliburton Claims Electric Frac First." Rigzone. Matthew Veazey. January 14, 2021. https://www.rigzone.com/news/halliburton_claims_electric_frac_first-14-jan-2021-164342-article/?rss=true

372. "ABB's Sustainable Energy Innovation Recognized by WEF," ABB (ABB Group, May 28, 2020), https://new.abb.com/news/detail/62869/abbs-sustainable-energy-innovation-recognized-by-wef.

373. "Vale bends to new quarterly iron ore pricing model," Mining.com, December 8, 2011, p. 1.

374. David McPhee, "New £31m Deal with Equinor 'Will Help Secure' More Business for Aberdeen Firm," Energy Voice (DC Thomson Media, October 15, 2019), https://www.energyvoice.com/oilandgas/north-sea/209765/new-31m-deal-with-equinor-will-help-secure-more-business-for-aberdeen-firm/.

375. Charles Kennedy, "Cesium—The Most Important Metal You've Never Heard Of," OilPrice.com (OilPrice.com, August 5, 2020), https://oilprice.com/Energy/Energy-General/Cesium-The-Most-Important-Metal-Youve- Never-Heard-Of.html.

376. "The Metal That Could Spark a New Resource War," Cision (Cision US Inc., May 11, 2020), https://www.prnewswire.com/news-releases/the-metal-that-could-spark-a-new-resource-war-301055968.html.

377. "Newkut Launches PDC Cutters for Oil and Gas Drill Bits," Newkut Industries, accessed February 10, 2012, http://www.newkut.com/polycrystalline-diamond-PDC-cutters-news.htm.

378. "Ruhrpumpen Foundry Fundemex," Ruhrpumpen, accessed February 10, 2012, http://www.ruhrpumpen.com/files/Brochures_Generales/Fundemex.pdf.

379. "BASF to build new methylamines plant in Geismar, Louisiana," BASF, last modified May 18, 2009, http://www.basf.com/group/pressrelease/P-09-241.

380. "China's CNOOC to buy Cinda AMC's Hubei Dayukou Chemical stake for 530 mln yuan Sulfuric Acid Today, December 17, 2007. http://www.h2so4today.com/publish/posts/1/chinas-cnooc-to-buy-cinda-amcs-hubei-dayukou-chemical-stake-for-530-mln-yuan.html.

381. Jim Redden, "Mud companies struggle with diminishing barite supplies *World Oil-Drilling Advances,* vol. 232, no. 12.

382. Jim Redden, "Mud companies struggle with diminishing barite supplies," World Oil Online, December, 2011, http://www.worldoil.com/December-2011-Drilling-advances.aspx.

383. "ONGC Inks Pacts with Schlumberger, Halliburton for Two Fields," Business Standard (Business Standard, December 7, 2016), https://www.business-standard.com/article/pti-stories/ongc-inks-pacts-with-schlumberger-halliburton-for-two-fields-116120701262_1.html.

384. "ESAB Cutting Systems: 1996–2008," ESAB, accessed January 30, 2012, http://www.esab-cutting.com/index.php?id=707.

385. "Commerce Finds Dumping and Subsidization of Drill Pipe from the People's Republic of China," Targeted News Service, January 4, 2011.

386. "U.S. pipe makers prevail Sharon Herald, September 9, 2009, http://sharonherald.com/local/x1081150087/U-S-pipe-makers-prevail.

387. "Alcan breaks ground for its new world-class facility in China," The Free Library, accessed February 10, 2012, http://www.thefreelibrary.com/Alcan+breaks+ground+for+its+new+world-class+facility+in+China.-a0167876858.

388. "General Cable Acquires Chinese Specialty Cable Facility." NEMA Industry News, February 23, 2007. http://www.nema.org/media/ind/20070223b.cfm.

389. "GE expands commitment to water technology in China with multi-million dollar investment." GE Water & Process Technologies, Last modified November 19, 2008. http://www.gewater.com/who_we_are/press_center/pr/11192008.jsp

390. "Toray, China BlueStar Establishes Water Treatment Joint Venture in China, to Hold Groundbreaking Ceremony," Toray Industries Inc., last modified August 25, 2009, http://www.toray.com/news/affil/nr090825.html

391. Santosh Mathilakath and Jonathan Rhoads, "Rigstore supply model aims to cut MRO costs," Offshore Magazine, April 1, 2008, http://www.offshore-mag.com/articles/print/volume-68/issue-4/drilling-technology-report/opened-for-business._printArticle.html.

392. "Program Changing Face of Procurement Aramco Expats, February 26, 2006, http://www.aramcoexpats.com/articles/2007/02/program-changing-face-of-procurement/.

393. ADG signs $25m deal with Traverse (Daily Business Alerts (Australia)) 28/05/2011, https://www.macquarie.com.au/mgl/au/personal/campaigns/trading/online-information/.

394. "Wärtsilä and CEVA Logistics expand their cooperation," Wärtsilä Corporation, last modified March 31, 2009, http://www.wartsila.com/en/press-releases/newsrelease302

395. "Technip and PETRONAS subsidiaries agree to establish a strategic business collaboration," Technip, last modified August 20, 2010, http://www.technip.com/en/press/technip-and-petronas-subsidiaries-agree-establish-strategic-business-collaboration.

396. "FMC Technologies: Transforming Planning and Scheduling Processes FMC company, June 24, 2010, http://supply-chain.org/civicrm/event/info?id=183&reset=1.

397. *Strategic Alliances Managing the Supply Chain,* Underhill, 1996.

398. Introduction GCR from Lincoln Electric, Lincoln Electric, http://www.lincolnelectric.com/en-us/company/custom-solutions/Pages/guaranteed-cost-reduction.aspx.

399. Ian Burdis, "Strategies and Practices to realize the maximum potential of NOC's and IOC's AGR Petroleum Services, presentation to NOC Congress, June 2010.

400. The National Oil Company Database Report. April 25, 2019. Natural Resource Governance Institute. https://resourcegovernance.org/analysis-tools/publications/national-oil-company-database. Last accessed January 25, 2021.

401. "Resource Nationalism Rises in 30 Countries," Verisk Maplecroft (Verisk, March 21, 2019), https://www.maplecroft.com/insights/analysis/resource-nationalism-rises-30-countries/.

402. David Steven Jacoby and Bruna Figueiredo, "The Art of High-Cost Sourcing Supply Chain Management Review, May/June 2008.

403. Mark Graham, Mark Cook and Frank Jahn, "Project and Contract Management," *Hydrocarbon Exploration and Production* 55, 2001, p. 61.

404. "ABB to work with BP on upstream electrical needs," Offshore Magazine (Endeavor Business Media, March 07, 2019), https://www.offshore-mag.com/field-development/article/16790320/abb-to-work-with-bp-on-upstream-electrical-needs.

405. "Schlumberger's Acquisition of Smith the Latest Evidence of a Takeover Trend Money Morning, February 22, 2010, http://moneymorning.com/2010/02/22/schlumberger-smith-takeover/.

406. Rima Ali Al Mashni, "Schlumberger opens Al-Khafji Oilfield Services base in Saudi Arabia AME Info., June 22, 2010, http://www.ameinfo.com/235993.html.

407. "Wood Group wins first supply chain services contract in Kazakhstan," Energyme.com, accessed February 10, 2012, http://www.energyme.com/energy/2006/en_06_0135.htm.

408. Santosh Mathilakath and Jonathan Rhoads, "Rigstore supply model aims to cut MRO costs," Offshore Magazine, April 1, 2008, http://www.offshore-mag.com/articles/print/volume-68/issue-4/drilling-technology-report/opened-for-business._printArticle.html.

409. "Ingersoll Rand Signs 4,000th Unit Under Its PackageCare Service Agreement," Ingersoll Rand website, January 12, 2012.

410. "KSB buys US pump Service Company," Water World, accessed February 10, 2012, http://www.waterworld.com/index/display/article-display/6656514988/articles/waterworld/drinking-water/distribution/2010/08/KSB-buys-US-pump-service-company.html.

411. 2010 Annual Report, SKF, p. 17 for inventory management contract, and p. 18 for PM contract.

412. "WTG, BTU and PI Intervention will now approach the market as one voice, with a new name and identity: Interwell," Interwell Norway AS, last modified March 30, 2011, http://www.interwell.com/news/becoming-interwell-article81-120.html.

413. "The rising fortunes of the Caspian BP website, accessed February 7, 2012, http://www.bp.com/sectiongenericarticle800.do?categoryId=9039889&contentId=7072885.

414. "Setting the Standard," British Petroleum, accessed February 8, 2012, http://www.bp.com/liveassets/bp_internet/globalbp/globalbp_uk_english/publications/frontiers/STAGING/local_assets/downloads/bpf14p22-31standardacg.pdf.

415. "Economies of Scale: How the oil and gas industry cuts costs through replication," Economic Intelligence Unit, accessed February 8, 2012, http://www.managementthinking.eiu.com/sites/default/files/downloads/Oil%20and%20Gas_%20Economies%20of%20Scale.pdf.

416. "Economies of Scale: How the oil and gas industry cuts costs through replication," Economic Intelligence Unit, accessed February 8, 2012, http://www.managementthinking.eiu.com/sites/default/files/downloads/Oil%20and%20Gas_%20Economies%20of%20Scale.pdf.

417. "Petronect: Reenergizing Latin America's Leading Oil And Gas Portal SAP case study, accessed February 11, 2012, http://www.google.com/url?sa=t&rct=j&q=&esrc=s&source=web&cd=1&ved=0CCUQFjAA&url=http%3A%2F%2Fdownload.sap.com%2Fdownload.epd%3Fcontext%3D27015EDBEAA93E8358A20DBDE58CE7F82C5FBA779F76883FBBF76112C012AD82B0DDF32C786801D3E997D5E8B6130DAEF1C1236A58CD9B26&ei=6Oc2T5CpFIPq2QXlj8WBAg&usg=AFQjCNHLzF433HlXaaZIue8hfvKBd-U5wg&sig2=Et9RvNNo-oif1uHVpUsw1Q.

418. "Tenaris awarded long-term agreement for tubulars and services by ADNOC." By Soumya Mutsuddi. IHS Markit. August 22, 2019. https://connect.ihsmarkit.com/upstream-insight/article/phoenix/2376655/tenaris-awarded-long-term-agreement-for-tubulars-and-services-by-adnoc.

419. Bermuda Hamilton, "Archer Secures Long Term Platform Drilling and Maintenance Contract from Aker BP for Offshore Norway," NewsWeb (Oslo Bors, November 24, 2020), https://newsweb.oslobors.no/message/518778.

420. "ADNOC Completes First Phase of AI Predictive Maintenance Project," Hydrocarbon Processing (Gulf Publishing Company LLC., October 10, 2020), https://www.hydrocarbonprocessing.com/news/2020/11/adnoc-completes-first-phase-of-ai-predictive-maintenance-project.

421. "ExxonMobil to Increase Permian Profitability through Digital Partnership with Microsoft," ExxonMobil (Exxon Mobil Corporation, February 22, 2019), https://corporate.exxonmobil.com/news/newsroom/news-releases/2019/0222_exxonmobil-to-increase-permian-profitability-through-digital-partnership-with-microsoft.

422. "Total to Develop Artificial Intelligence Solutions with Google Cloud," total.com (Total, April 24, 2018), https://www.total.com/media/news/press-releases/total-develop-artificial-intelligence-solutions-google-cloud.

423. Abdullah Al-Baiz, "How the 4IR Is Driving a New Wave of Energy Innovation," World Economic Forum (World Economic Forum, September 18, 2020), https://www.weforum.org/agenda/2020/09/how-the-4ir-is-driving-a-new-wave-of-energy-innovation/.

424. "Tackling Methane Emissions." Shell, n.d. https://www.shell.com/energy-and-innovation/natural-gas/methane-emissions/_jcr_content/par/textimage_438437728.stream/1587995196996/53beef2f8ba2e90560c074f56552e2acfe30582b/shell-methane-case-study.pdf

425. Mark Radka, "Methane Emissions," Shell Sustainability Report 2019 (Shell, 2019), https://shell.online-report.eu/sustainability-report/2019/sustainable-energy-future/managing-greenhouse-gas-emissions/methane-emissions.html.

426. "All About Optical Gas Imaging (OGI) - Part 1: Complying with Regulations," Opgal (Opgal, May 16, 2018), https://www.opgal.com/blog/gas-leak-detection/all-about-optical-gas-imaging-part-1-complying-with-regulations/.

427. "ExxonMobil Tests New Methane Monitoring Technologies," Globuc (Globuc, April 24, 2020), https://globuc.com/news/exxonmobil-is-leading-testing-for-the-most-promising-next-generation-methane-detection-technologies/.

428. Rebecca Addison, "Chevron Taps SecurityGate.io Risk Management Platform for Cybersecurity Efforts," Industrial Cyber (Expert Market Insight, September 9, 2020), https://industrialcyber.co/news/chevron-taps-securitygate-io-risk-management-platform-for-cybersecurity-efforts/.

429. "Leading Asia-Pacific Company Simulates Gas Separation to Increase Capacity, Drive Incremental Benefit of $60K/Day," AspenTech brochure, p. 1.

430. "Supply Chain: Does Manufacturing Have the Capacity to Meet Projected Demand," Flowserve, accessed February 10, 2012, http://www.flowserve.com/files/www/Collections/Spotlight/Industries%20Landing%20Page%20Spotlights/Power/EIM.pdf.

431. "Modular Block Bodies," ITT, accessed February 10, 2012, http://www.duhig.com/Images/ITT/pfmbb06.pdf.

432. "BP's Ties With Nalco Co.'s Corexit," TopNews, accessed February 10, 2012, http://topnews.co.uk/24532-bp-s-ties-nalco-co-s-corexit.

433. Hitachi Cable Annual report, 2006, p. 18.

434. "NSK Reduces Customers' Costs with Extended Bearing Life," NSK Corporation, accessed February 10, 2012, http://www.4e534b.com/PressRelease/Jan282009.html.

435. David Steven Jacoby, "RFID Finds a 'Sweet Spot,'" *Offshore Oil and Gas Drilling Logistics Digest,* January 2011, pp. 40–41.

436. David Steven Jacoby, "RFID Finds a 'Sweet Spot,'" *Offshore Oil and Gas Drilling Logistics Digest,* January 2011, pp. 40–41.

437. David Steven Jacoby, "RFID Finds a 'Sweet Spot,'" *Offshore Oil and Gas Drilling Logistics Digest, January 2011,* pp. 40–41.

438. Baker Hughes, Inc. "Baker Hughes 2008 Annual Report http://files.shareholder.com/downloads/BHI/1625622136x0x279889/ACAF6320-2A52-42B2-ACD9-B6A52C9A160A/baker_hughes_2008_annual_report.pdf.

439. "Air Products Installs GE's Revolutionary Wireless Monitoring." Bloomberg, May 05, 2009, http://www.bloomberg.com/apps/news?pid=newsarchive&sid=aXi45u6sq4UQ.

440. "Aker BP Contracts Optime Subsea Technology to Increase Safety, Reduce Cost and Risk," Drilling Contractor (Drilling Contractor, May 29, 2020), https://www.drillingcontractor.org/aker-bp-contracts-optime-subsea-technology-to-increase-safety-reduce-cost-and-risk-56546.

441. "Liquefied Natural Gas: Understanding the Facts." United States Department of Energy. 2005. p. 3.

442. Saeid Mokhatab, William A. Poe, and James G. Speight, *Handbook of Natural Gas Transmission and Processing,* Amsterdam: Gulf Professional Publishing, 2006.

443. Alternative Fuels Data Center—Fuel Properties Comparison. U.S Department of Energy Alternative Fuels Data Center. https://afdc.energy.gov/fuels/fuel_comparison_chart.pdf Last accessed January 26, 2021.

444. Brian Collins, "Midstream 2020 CapEx Cuts Drive Investments below Previous Oil Bust Levels," S&P Global (S&P Global, May 13, 2020), https://www.spglobal.com/marketintelligence/en/news-insights/research/midstream-2020-CapEx-cuts-drive-investments-below-previous-oil-bust-levels.

445. "No Business Sense: Caspian Pipelines Will Cost 5 Times Others And Suffer Loss Of 25% Revenue," BSI (Boston Strategies International, December 1, 2015), https://bostonstrategies.com/no-business-sense-caspian-pipelines-will-cost-5-times-others-and-suffer-loss-of-25-revenue/.

446. "Oil and Gas Pipelines," National Geographic (National Geographic Society, February 22, 2014), https://www.nationalgeographic.org/photo/europe-map/.

447. MarkWest Liberty Midstream & Resources and Sunoco Logistics Announce New Marcellus Ethane Pipeline and Marine Project, June 01, 2010.BusinessWire, https://www.businesswire.com/news/home/20110322005892/en/MarkWest-Liberty-Sunoco-Logistics-Announce-Expansion-Project

448. Midstream Operators Ramp Up in the Eagle Ford Shale," Meredith Cantrell. *Pipeline and Gas Technology Magazine,* November/December 2010, p. 14.

449. ISO 21809-5:2010, International Organization for Standardization, accessed February 11, 2012, http://www.iso.org/iso/iso_catalogue/catalogue_tc/catalogue_detail.htm?csnumber=40881

450. Vickey Du, Jeslyn Lerh, and Debiprasad Nayak, "Low Freight, Improving Storage Economics Spur More Short-Term Time Charters," S&P Global (S&P Global, September 11, 2020), https://www.spglobal.com/platts/en/market-insights/latest-news/oil/091120-low-freight-improving-storage-economics-spur-more-short-term-time-charters.

451. Saeid Mokhatab, William A. Poe, and James G. Speight, Handbook of Natural Gas Transmission and Processing, Amsterdam: Gulf Professional Publishing, 2006.

452. Yergin, Daniel. *The New Map: Energy, Climate, and the Clash of Nations.* Penguin Press, 2020. p. 36.

453. Yergin, Daniel. *The New Map: Energy, Climate, and the Clash of Nations,* Penguin Press, 2020, p. 111.

454. "Russia's Answer To The U.S. Shale Threat Is Finally Here." By Simon Watkins, February 24, 2021. Oilprice.com. https://oilprice.com/Energy/Energy-General/Russias-Answer-To-The-US-Shale-Threat-Is-Finally-Here.html. Last accessed March 2, 2021.

455. "Next U.S. LNG Project Could Begin Production as Soon as October." Bloomberg.com, https://www.bloomberg.com/news/articles/2021-02-25/next-u-s-lng-project-could-begin-production-as-soon-as-october. By Stephen Stapczynski. February 25, 2021. Last accessed March 2, 2021.

456. Greg Miller, "LNG Shipping Rates Just Hit $125,000 per Day," American Shipper (Freight Waves, Inc., November 2, 2020), https://www.freightwaves.com/news/lng-shipping-rates-just-hit-125000-per-day#:~:text=%E2%80%9CTime%2Dcharter%20rates%20in%20the,per%20day%20for%20stea m%20ships.

457. "LNG—market development and trends." Energy BrainBlog, by Simon Göß28. April 28, 2019. https://blog.energybrainpool.com/en/lng-market-development-and-trends/. Accessed January 30, 2021.

458. ISO 28460:2010, International Organization for Standardization, accessed February 11, 2012, http://www.iso.org/iso/iso_catalogue/catalogue_tc/catalogue_detail.htm?csnumber=44712.

459. Saeid Mokhatab, William A. Poe, and James G. Speight, *Handbook of Natural Gas Transmission and Processing,* Amsterdam: Gulf Professional Publishing, 2006.

460. Pipelines and the Exploitation of Gas Reserves in the Middle East. Dagobert L. Brito and Eytan Sheshinski, p. 6. https://scholarship.rice.edu/handle/1911/91620 Last accessed January 27, 2021.

461. Speight, Poe and Mokhatab, *Handbook of Natural Gas Transmission and Processing,* Gulf Professional Pub., 2006, p. 8.

462. Oil & Gas Pipelines brochure by Honeywell (Robert Ell), July 2016. https://www.honeywellprocess.com/library/marketing/presentations/one-honeywell-pipelines.pdf

463. Roger N. Anderson, *Computer-Aided Lean Management for the Energy Industry,* PennWell, 2008, p. 275.

464. The Real Financial Cost Of Nord Stream 2. Economic sensitivity analysis of the alternatives to the offshore pipeline. Piotr Przybyło. Casimir Pulaski Foundation. https://pulaski.pl/wp-content/uploads/2019/05/Raport_NordStream_TS-1.pdf Accessed January 27, 2021.

465. Nord Stream. "Pipeline Construction Progresses on Schedule last modified December 16, 2010. http://www.nord-stream.com/press-info/emagazine/pipeline-construction-progresses-on-schedule-24/.

466. Rogelio E. Garza, "Sacrificial anodes selected for Nord Stream system Offshore Magazine, May 1, 2011, http://www.offshore-mag.com/articles/print/volume-71/issue-5/flowlines-__pipelines/sacrificial-anodes-selected-for-nord-stream-system.html.

467. Steel Guru, "Kobe Steel acquires 44pct stake in Wuxi Compressor Co Ltd. May 18, 2011, http://www.steelguru.com/chinese_news/Kobe_Steel_acquires_44pct_stake_in_Wuxi_Compressor_Co_Ltd/205651.html.

468. Kobe Steel turns non-standard compressor affiliate in China into subsidiary. June 3, 2020. Kobelco website. https://www.kobelco.co.jp/english/releases/1204283_15581.html Accessed January 27, 2021.

469. Phil Reddin and Rick Hernandez, Egyptian LNG: The Value of Standardization, BG LNG Services, accessed February 11, 2012, http://lnglicensing.conocophillips.com/EN/publications/documents/GastechValueofStandardizationPaper.pdf.

470. Alexander Stevens, "Technology Arrives to Capture Stranded Gas," IER (Institute for Energy Research, June 10, 2020), https://www.instituteforenergyresearch.org/fossil-fuels/gas-and-oil/technology-arrives-to-capture-stranded-gas/.

471. "EDGE Pioneers New Way to Capture, Distribute and Monetize Stranded and Flared Gas," Blue Water Energy (Blue Water Energy, June 18, 2019), https://www.bluewaterenergy.com/news/edge-pioneers-new-way-to-capture-distribute-and-monetize-stranded-and-flared-gas.

472. "The Advantages of Outsourcing Gas Compression and Gathering Services." Audubon Companies website. https://auduboncompanies.com/the-advantages-of-outsourcing-gas-compression-and-gathering-services/

473. "Cheniere to outsource LNG Sabine Pass marketing." Reuters. April 16, 2008. https://www.reuters.com/article/energy-cheniere-lng/cheniere-to-outsource-lng-sabine-pass-marketing-idINN1632569220080416

474. "Energy Focus: LNG Brokers - do we need them?" ICIS Editorial, May 24, 2007. https://www.icis.com/explore/resources/news/2007/05/25/9296352/energy-focus-lng-brokers-do-we-need-them-/

475. Oil & Gas Pipelines brochure by Honeywell (Robert Ell), July 2016. https://www.honeywellprocess.com/library/marketing/presentations/one-honeywell-pipelines.pdf

476. "The Future of Long-term LNG Contracts." Peter R. Hartley, George & Cynthia Mitchell Professor of Economics and Rice Scholar in Energy Studies, James A. Baker III Institute for Public Policy, Rice University and BHP-Billiton Chair in Energy and Resource Economics, University of Western Australia. 2012.

477. Thomas O. Miesner, Oil and Gas Pipelines in Nontechnical Language, Tulsa: PennWell, 2006.

478. Philip Meyers, Aboveground Storage Tanks, New York: McGraw-Hill Professional, 1997, p.90.

479. "Cyberhawk Awarded Five-Year Agreement with Major LNG Producer in the Middle East," Cyberhawk (Cyberhawk, October 20, 2020), https://insights.thecyberhawk.com/news_and_blog/cyberhawk-awarded-five-year-agreement-with-major-lng-producer-in-the-middle-east .

480. Beth Bacheldor, "Manufacturer Tests RFID to Track Industrial-Size Containers of Liquid," RFID Journal, March 19, 2007, http://www.rfidjournal.com/article/view/3156.

481. "Top 10 RFID Security Concerns and Threats." SecurityWing. https://securitywing.com/top-10-rfid-security-concerns-threats/

482. United States GDP Growth Rate. TradingEconomics. https://tradingeconomics.com/united-states/gdp-growth Accessed January 27, 2021.

483. WTI Crude Oil Prices—10 Year Daily Chart. MacroTrends. https://www.macrotrends.net/2516/wti-crude-oil-prices-10-year-daily-chart Accessed January 27, 2021.

484. Crude oil stocks and days of supply. Energy Information Administration. https://www.eia.gov/petroleum/weekly/crude.php#tabs-stocks-supply Accessed January 27, 2021.

485. Despite the becoming a net petroleum exporter, most regions are still net importers. US Energy Information Administration. https://www.eia.gov/todayinenergy/detail.php?id=42735#:~:text=In%20September%202019%2C%20the%20United,net%20importer%20of%20crude%20oil. Accessed January 27, 2021.

486. Michael D. Tusiani, *LNG: A Nontechnical Guide,* Tulsa: PennWell, 2007.

487. Oil Supertankers Fetch Astronomical Rates with Vessels Scarce. Bloomberg. March 16, 2020. https://www.bloomberg.com/news/articles/2020-03-16/oil-supertankers-fetch-astronomical-rates-with-vessels-scarce Accessed January 27, 2021.

488. "Super Premium Efficiency (IE4) Induction Motors." Emerging Technologies. http://e3tnw.org/ItemDetail.aspx?id=638 Last accessed January 27, 2021.

489. James H. Gary and Glenn E. Handwerk, *Petroleum Refining: Technology and Economics,* New York: CRC Press, 2001, pp. 365–370.

490. James H. Gary and Glenn E. Handwerk, *Petroleum Refining: Technology and Economics,* New York: CRC Press, 2001, pp. 365–370.

491. Khalid Y. Al-Qahtani and Ali Elkamel, *Planning and Integration of Refinery and Petrochemical Operations,* Hoboken: John Wiley & Sons, 2010.

492. "ExxonMobil Chemical Completes Debottleneck Project Increasing Synesstic (TM) Alkylated Naphthalene (AN) Blendstocks Production by 40 Percent." OEM/Lube News, vol. 3 no. 26, June 25, 2007, http://www.imakenews.com/lubritec/e_article000845700.cfm?x=b11,0,w.

493. "PEMEX Selects Site For $10 Billion Refinery EBR, April 14, 2009, http://refiningand petrochemicals.energy-business-review.com/news/pemex_selects_site_for_10_billion _refinery_090414.

494. "Case Study: Chevron Employs APC Best Practices to Get Controllers Online Faster After Unit Turnarounds." AspenTech website. https://www.aspentech.com/en/resources /case-studies/chevron-employs-apc-best-practices-to-get-controllers-online-faster-after -unit-turnarounds

495. Chen Aizhu, "PetroChina's Jinxi Refinery Exports First Cargo of Low-Sulphur Marine Fuel," ed. Aditya Soni, Reuters (Thomson Reuters, March 13, 2020), https://in.reuters.com/article /china-petrochina-imo/petrochinas-jinxi-refinery-exports-first-cargo-of-low-sulphur -marine-fuel-idUSL4N2B62BE.

496. "Bharat Petroleum Confident of Meeting 10 per Cent Ethanol Blending Target by 2022," ETEnergyworld.com (ETEnergyworld.com, November 18, 2020), https://energy.economic-times.indiatimes.com/news/oil-and-gas/bharat-petroleum-confident-of-meeting-10-per-cent -ethanol-blending-target-by-2022/79274243.

497. "New Retail Fuels Channel Strategies for the Next Round of Downstream Growth and Competition." Oliver Wyman. January 2012. https://www.oliverwyman.com/our-expertise /insights/2012/jan/new-retail-fuels-channel-strategies-for-the-next-round-of-downst.html

498. Gateway Logistics Group acts as Logistics Provider for the first major equipment for the Abreu e Lima Refinery arrive in Pernambuco, article on Gateway Logistics Group's website, http://www.gateway-group.com/en/releases/printview.asp?59

499. ConocoPhillips brochure, publication # CSH-11-0318.

500. Martin Turk, "Decreasing OpEx in Oil & Gas Downstream Operations," Schneider Electric Blog (Schneider Electric, July 9, 2020), https://blog.se.com/oil-and-gas/2018/11/16 /higher-return-capital-employed-oil-gas-downstream-decreasing-OpEx/.

501. "Invensys implements InFusion-based enterprise control system at ExxonMobil Lubricants Plant" Reuters, January 9, 2008, http://www.reuters.com/article/2008/01/09 /idUS154667+09-Jan-2008+MW20080109.

502. "Intrepid Hackers Use Chinese Takeout Menu to Access a Major Oil Company." Gizmodo. By Ashley Feinberg, April 8, 2014. https://gizmodo.com/hackers-are-being-forced-to-target -chinese-takeout-menu-1560755886

503. Crude oil is defined by specific gravity (API rating), sulfur content (more sulfur makes oil "sour," and less sulfur makes them "sweet"), and Total Acid Number (TAN). See Refinery Technology and Economics.

504. Source: Alex Lidback (CMAI) Presentation to the ISM Chemicals Group, March 1, 2007.

505. Khalid Y. Al-Qahtani and Ali Elkamel, Planning and Integration of Refinery and Petrochemical Operations, Wiley-VCH, 2010, pages 4 and 85.

506. Khalid Y. Al-Qahtani and Ali Elkamel, Planning and Integration of Refinery and Petrochemical Operations, Wiley-VCH, 2010.

507. Jacoby, David Steven. "SPE-161035-PP: Major Accident Prevention and Lessons Learned in Oil & Gas." Presented at the Abu Dhabi International Petroleum Exhibition Conference (ADIPEC). November 11, 2012.

508. Manufacturing Energy Consumption. International Energy Agency.1998.

509. "Safety Drives Refinery Shutdowns, Restarts." Mark Green. American Petroleum Institute. September 13, 2017. https://www.api.org/news-policy-and-issues/blog/2017/09/13/safety-drives-refinery-shutdowns-restarts

510. "Informe De Costos De Tecnologías De Generación." Santiago: Comisión Nacional de Energía de Chile, March 2020.

511. "What Are the Advantages and Disadvantages of Geothermal Energy?," TWI Global (TWI Ltd.), accessed January 13, 2021, https://www.twi-global.com/technical-knowledge/faqs/geothermal-energy/pros-and-cons.

512. Martina Serdjuk et al., "Geothermal Investment Guide," GeoElec (GeoElec, November 2013), http://www.geoelec.eu/wp-content/uploads/2011/09/D3.4.pdf.

513. Hance, 2005; GTP, 2008, cited in IRENA 2017 p. 138.

514. Jonathan L. Mayuga, "EDC Eyeing to Build Binary Plants," Business Mirror (Business Mirror, February 28, 2020), https://businessmirror.com.ph/2020/02/28/edc-eyeing-to-build-binary-plants/.

515. Hiroshi Murakami, Yoshifumi Kato, and Nobuo Akutsu, "Construction of the Largest Geothermal Power Plant for Wayang Windu Project, Indonesia," GeothermalEnergy.org (Fuji Electric Co., June 10, 2000), https://www.geothermal-energy.org/pdf/IGAstandard/WGC/2000/R0187.PDF.

516. "Huge Gap between 'Demand' and Supply' of Iron Ore: Steel Minister Dharmendra Pradhan," The Hindu BusinessLine (THG Publishing PVT LTD. , November 19, 2020), https://www.thehindubusinessline.com/economy/huge-gap-between-demand-and-supply-of-iron-ore-steel-minister-dharmendra-pradhan/article33132248.ece.

517. "Rising Ore Prices, Floods in Australia to Impact Steel Companies' Q3 Numbers," The Economic Times (The India Times, February 2, 2011), https://economictimes.indiatimes.com/markets/stocks/earnings/rising-ore-prices-floods-in-australia-to-impact-steel-companies-q3-numbers/articleshow/7408426.cms?from=mdr .

518. Kevin Wallace, Marshall Ralph, and William Harvey, "Cooperative Models for Engineering Geothermal Power Plants," Geothermal-Energy.org, accessed January 13, 2021, https://www.geothermal-energy.org/pdf/IGAstandard/ARGeo/2008/Cooperative_Models_MashallRalph.pdf.

519. Alexander Richter, "Daldrup & Söhne Realigns Geothermal Business to Avoid Cluster Risk," Think GeoEnergy (ThinkGeoEnergy EHF, June 7, 2019), https://www.thinkgeoenergy.com/daldrup-sohne-realigns-geothermal-business-to-avoid-cluster-risk/.

520. "Toshiba Concludes Package Deal Agreement on Geothermal Power Generation Projects with Turkey's Zorlu Energy Group," Toshiba (Toshiba Corporation, September 19, 2017), https://www.toshiba.co.jp/about/press/2017_09/pr1901.htm.

521. Ibid.

522. Alexander Richter, "Nevis Geothermal Project Chooses Turbine Supplier and EPC Contractor," Think GeoEnergy (Think GeoEnergy EHF, March 18, 2016), https://www.thinkgeoenergy.com/nevis-geothermal-project-chooses-turbine-supplier-and-epc-contractor/.

523. "Life Cycle Analysis of a Geothermal Power Plant: Comparison of the Environmental Performance with Other Renewable Energy Systems. Riccardo Basosi et al. Sustainability 2020, 12, 2786. file:///C:/Users/dsj/Dropbox/My%20PC%20(DESKTOP-Q8DGU5H)/Downloads/sustainability-12-02786-v2.pdf.

524. "TransAlta's Geothermal Partnership CalEnergy Signs 24-Year Contract: Continued Investment in Renewable Energy," TransAlta (TransAlta Corporation, September 17, 2013), https://

www.transalta.com/newsroom/news-releases/transaltas-geothermal-partnership-calenergy-signs-24-year-contract-continued-investment-in-renewable-energy/.

525. Alexander Richter, "CalEnergy Executes 24 Year PPA for 86 MW with City of Riverside," Think GeoEnergy (Think GeoEnergy EHF, June 19, 2013), https://www.thinkgeoenergy.com/calenergy-executes-24-year-ppa-for-86-mw-with-city-of-riverside/.

526. "Large Geothermal Power Plant Improves Cybersecurity and Performance," Rockwell Automation (Rockwell Automation, Inc.), accessed January 13, 2021, https://www.rockwellautomation.com/en-in/company/news/magazines/large-geothermal-power-plant-improves-cybersecurity-and-performa.html.

527. Alexander Richter, "Start of Demonstration Project in Indonesia on Predictive Diagnostics for Geothermal Plants," Think GeoEnergy (Think GeoEnergy EHF, October 23, 2019), https://www.thinkgeoenergy.com/start-of-demonstration-project-in-indonesia-on-predictive-diagnostics-for-geothermal-plants/.

528. "Geothermal Technologies Could Push Energy Storage Beyond Batteries," NREL.gov (U.S. Department of Energy, September 17, 2018), https://www.nrel.gov/news/program/2018/geothermal-technologies-could-push-beyond-batteries.html.

529. Alexander Richter, "Toshiba Signs MOU on a Partnership in Geothermal Power Projects in Malawi, Africa," Think GeoEnergy (Think GeoEnergy EHF, May 4, 2018), https://www.thinkgeoenergy.com/toshiba-signs-mou-on-a-partnership-in-geothermal-power-projects-in-malawai-africa/.

530. U.S. Department of Energy Federal Energy Management Program (FEMP), ed., "Geothermal Electric Technology," Whole Building Design Guide (National Institute of Building Sciences, November 15, 2016), https://www.wbdg.org/resources/geothermal-electric-technology.

531. Thomas E. Copeland, Real Options: A Practitioner's Guide, Texere, 2001.

532. Matthew Wald, "Price of New Power Plants Rises Sharply" New York Times Online, July 10, 2007, http://www.nytimes.com/2007/07/10/business/worldbusiness/10iht-power.4.6593271.html

533. Tim Sharp, "SSE buys 15% stake in BiFab," The Herald Scotland, April 12, 2010, http://www.heraldscotland.com/business/corporate-sme/sse-buys-15-stake-in-bifab-1.1020...

534. Peter G. Hessler, Power Plant Construction Management: A Survival Guide, Tulsa: PennWell, 2005.

535. "Co-firing Biomass with Coal: a success story." IEA Bioenergy. https://www.ieabioenergy.com/blog/publications/co-firing-biomass-with-coal-a-success-story/

536. "Increasing the Efficiency of Existing Coal-Fired Power Plants." Congressional Research Service. Richard J. Campbell. December 20, 2013. https://fas.org/sgp/crs/misc/R43343.pdf

537. Department of Energy (DOE), Office of Fossil Energy (FE). Carbon Ore, Rare Earth and Critical Minerals (CORE-CM) Initiative for U.S. Basins. CFDA Number: 81.089.

538. Everett Woodruff and Herbert Lammers, Steam Plant Operation, McGraw-Hill, 2004.

539. Source: Boston Strategies International analysis, based on a subset of products in electrical distribution and control.

540. Peter G. Hessler, Power Plant Construction Management: A Survival Guide, Tulsa: PennWell, 2005, p. 121.

541. "CLP Power Procurement Principles & Practices," company brochure, undated.

542. "MHI Compressor Technology Licensee in China Begins Marketing." Mitsubishi Heavy Industries Compressor Corporation, last modified February 09, 2011, http://www.mhi.co.jp/en/news/story/1102091407.html.

543. "Black & Veatch Provides Technical Expertise for a Low-Emissions Natural Gas Plant in Thailand." Black & Veatch website. Accessed January 16, 2021. https://www.bv.com/projects/black-veatch-provides-technical-expertise-low-emissions-natural-gas-plant-thailand

544. "GE Signs 25-Year, US$330 Million Power Plant Agreement." Diesel & Gas Turbine Worldwide. Jack Burke. January 18, 2018. https://www.dieselgasturbine.com/7004988.article

545. "Supply Chain Security Guidelines on Provenance" findings of a working committee chaired by David Steven Jacoby for North American Electric Reliability Corporation (NERC). June, 2019. https://www.nerc.com/comm/CIPC_Security_Guidelines_DL/Security_Guideline-Provenance.pdf and https://www.nerc.com/comm/CIPC_Security_Guidelines_DL/Security_Guideline-Provenance_Presentation.pdf

546. Rex Kenyon, Process Plant Reliability and Maintenance for Pacesetter Performance, Tulsa: Pennwell, 2004.

547. "Monitran DSP monitors turbine blade conditions." Source the Engineer, September 7, 2009, http://source.theengineer.co.uk/measurement-quality-control-and-test/vision-sound-and-vibration-testing/vibration-sensors/monitran-dsp-monitors-turbine-blade-conditions/352130.article.

548. "GE Expands European Presence with Equity Share of Artesis Teknoloji Sistemleri," ARC Advisory Group, September 30, 2010, http://www.arcweb.com/asset-lifecycle-management/2010-09-30/ge-expands-european-presence-with-equity-share-of-artesis-teknoloji-sistemleri--1.aspx.

549. Allen J. Wood and Bruce F. Wollenberg, *Power Generation, Operation, and Control,* Wiley-Interscience, 1996.

550. "ACS 1000 variable speed drive replaces steam turbine in the petrochemical industry Case Study," Petrochemical industry, Repsol YPF, Argentina, Application: Blower, 3000 kW (4000 hp).

551. ABB Group, "ABB factory is voted best in Europe," last modified September 27, 2010, http://www.abb.com/cawp/seitp202/00cb18b719e98a1cc12577ab003ae7f2.aspx.

552. The Weir Group PLC. "Rapid Response Programme produces results." Weir Group Bulletin, March 2011, http://content.yudu.com/A1rcwe/wb-March-2011/resources/20.htm.

553. Dipankar Sen and Prosenjit Sen, RFID For Energy & Utility Industries, Tulsa: PennWell, 2005.

554. "Rolls-Royce wins £120M TotalCare® contract for industrial gas turbines from BP." Rolls-Royce plc., last modified December 18, 2007, http://www.rolls-royce.com/energy/news/2007/rr_wins_totalcare.jsp.

555. Form 10-Q, Capstone Turbine Corp, Yahoo Finance, Feb 7, 2011, http://biz.yahoo.com/e/110207/cpst10-q.html.

556. Andrea Prencipe, "Modular Design and Complex Product Systems: Facts, Promises and Questions," Complex Product Systems Innovation Centre, www.cops.ac.uk/pdf/cpn47.pdf.

557. "RB211 Gas Turbines: For power generation and mechanical drive," Rolls Royce, accessed February 8, 2012, http://www.rolls-royce.com/Images/rb211final_tcm92-21095.pdf.

558. Drew Robb, "Compressor Maintenance Trends: Modularity, Remote Monitoring and Outsourcing are key," Turbomachinery International Magazine, September/October 2011, accessed November 8, 2011, http://www.turbomachinerymag.com/sub/2011/SeptOct2011-CoverStory.pdf.

559. "Gas Turbines in Simple Cycle & Combined Cycle Applications," National Energy Technology Laboratory, accessed February 8, 2012, http://www.netl.doe.gov/technologies/coalpower/turbines/refshelf/handbook/1.1.pdf.

560. "Safety Management: Industry leading risk assessment and compliance support." Uniper Energy website. https://www.uniper.energy/services/solutions/compliance-risk-and-safety/safety-management Last accessed January 16, 2021.

561. David Ferris, "HYDROPOWER: Canada Has Too Much Clean Electricity. Anybody Want It?," E&E News (Politico, LLC, September 21, 2017), https://www.eenews.net/stories/1060061255#:~:text=Read%20the%20first%20here.,it%20is%20on%20the%20way.

562. Comparative Economics of Combined Cycle, Solar, Wind, Hydro, and Geothermal Power University of Calgary, Haskayne School of Business. Boston Strategies International. Calgary, Alberta, Canada, October 19, 2017.

563. "Hydropower Explained - Where Hydropower Is Generated," EIA.gov (U.S. Energy Information Administration (EIA), March 30, 2020), https://www.eia.gov/energyexplained/hydropower/where-hydropower-is-generated.php.

564. "2017 Hydropower Market Report," Office of Energy Efficiency and Renewable Energy (U.S. Department of Energy, 2017), https://www.energy.gov/eere/water/downloads/2017-hydropower-market-report.

565. "Types of Hydropower Plants," Energy.gov (U.S. Department of Energy), accessed January 20, 2021, https://www.energy.gov/eere/water/types-hydropower-plants.

566. "88-megawatt Canadian hydro-to-hydrogen plant to open in 2023." New Atlas. By Loz Blain, January 20, 2021. https://newatlas.com/energy/canada-hydro-hydrogen-electrolysis-plant/ Last accessed January 27, 2021.

567. Jesse Kathan, "Decline in Hydropower Hampered by Drought Will Impact Utility Costs," The Mercury News (Bay Area News Group, August 10, 2020), https://www.mercurynews.com/2020/08/09/decline-in-hydropower-hampered-by-drought-will-impact-utility-costs/.

568. Jiao Wang, Leah Schleifer, and Lijin Zhong, "No Water, No Power," World Resources Institute (World Resources Institute, September 26, 2018), https://www.wri.org/blog/2017/06/no-water-no-power.

569. Alvaro Lara, "The Pitfalls of Hydroelectric Power in Drought-Prone Africa," Climate & Environment at Imperial (Grantham Institute, August 21, 2018), https://granthaminstitute.com/2018/08/21/the-pitfalls-of-hydroelectric-power-in-drought-prone-africa/.

570. "The Thirst for Power: Hydroelectricity in a Water Crisis World," BRINK (Marsh & McLennan Advantage, October 19, 2016), https://www.brinknews.com/the-thirst-for-power-hydroelectricity-in-a-water-crisis-world/.

571. Phil Allan, "GCube Offers Renewables Firms Policy against Volatile Weather," Energy Voice (DC Thompson Media, September 21, 2015), https://www.energyvoice.com/otherenergy/88176/gcube-offers-renewables-firms-policy-against-volatile-weather/.

572. "CCI to Meet Cement Demand for New Hydropower Project," International Cement Review (Tradeship Publications Ltd., March 12, 2020), https://www.cemnet.com/News/story/168481/cci-to-meet-cement-demand-for-new-hydropower-project.html.

573. "GE Power India Sizzles as Consortium Bags Order for Hydro Power Plant," Business Standard (Business Standard, December 24, 2018), https://www.business-standard.com/article/news-cm/ge-power-india-sizzles-as-consortium-bags-order-for-hydro-power-plant-118122400264_1.html.

574. "Consortium Led by VINCI Signs Contract to Build Sambangalou Dam in Senegal," Hydro Review (Clarion Events, December 16, 2020), https://www.hydroreview.com/2020/12/16/consortium-led-by-vinci-signs-contract-to-build-sambangalou-dam-in-senegal/.

575. “Voith Awarded Supply Contract for 800-MW Mohmand Dam Hydropower in Pakistan,” Hydro Review (Clarion Events, October 19, 2020), https://www.hydroreview.com/2020/10/19/voith-awarded-supply-contract-for-800-mw-mohmand-dam-hydropower-in-pakistan/.

576. “Project Implementation Consultant Appointed for Bakaru I and II in Indonesia,” Hydropower & Dams International (Aqua Media International Ltd., July 30, 2020), https://www.hydropower-dams.com/news/project-implementation-consultant-appointed-for-bakaru-i-and-ii-in-indonesia/.

577. “2020-12-15 VH Voith Hydro Awarded the Contract for 174 MW Asahan 3 Hydropower Plant in Indonesia,” Voith (Voith GmbH & Co. , December 15, 2020), http://voith.com/es-en/news-room/press-releases/2020-12-15-vh-asahan-3-hydropower-plant-in-indonesia.html.

578. “Hydropower,” PPP LRC (World Bank Group), accessed January 11, 2021, https://ppp.worldbank.org/public-private-partnership/climate-smart/hydropower.

579. “Cybersecurity for Hydropower Generation,” Waterfall Security (Waterfall Security Solutions), accessed January 11, 2021, https://static.waterfall-security.com/WF_Hydro_Brochure_FINAL.pdf.

580. “Voith Utilizes Big Data and AI for Hydro Project,” Power Engineering International (Clarion Energy, May 8, 2018), https://www.powerengineeringint.com/digitalization/big-data/voith-utilizes-big-data-and-ai-for-hydro-project/.

581. “Surge of Hydropower Could Force Cutbacks of Solar, Wind,” USSD (United States Society on Dams), accessed January 11, 2021, https://www.ussdams.org/our-news/surge-of-hydropower-could-force-cutbacks-of-solar-wind/.

582. Peng Wei and Yang Liu, “The Integration of Wind-Solar-Hydropower Generation in Enabling Economic Robust Dispatch,” Mathematical Problems in Engineering (Hindawi, January 2, 2019), https://www.hindawi.com/journals/mpe/2019/4634131/.

583. Sarita Singh, “Norms Mandate Local Sourcing of Gear for New Power Projects,” The Economic Times (India Times, January 16, 2019), https://economictimes.indiatimes.com/industry/energy/norms-mandate-local-sourcing-of-gear-for-new-power-projects/articleshow/67563340.cms?from=mdr .

584. Adam Mazmanian, “Trump Fires TVA Chairman over IT Outsourcing,” FCW (1105 Media, Inc., August 4, 2020), https://fcw.com/articles/2020/08/04/mazmanian-tva-trump-outsource-layoffs.aspx.

585. David Appleyard, “Refurbishing Hydropower to Increase Project Life and Boost Capacity,” Renewable Energy World (Clarion Events, September 24, 2013), https://www.renewableenergyworld.com/2013/09/24/refurbishing-hydropower-to-increase-project-life-and-boost-capacity/.

586. “Evolugen and Gazifère announce hydrogen injection projects to be powered by hydro.” Renewable Energy World , March 2, 2021. https://www.renewableenergyworld.com/hydrogen/evolugen-and-gazifere-announce-hydrogen-injection-projects-to-be-powered-by-hydro/?f-bclid=IwAR38KmwDItfMsJbJNirqchl8VZdfg9MIyhQJ5zIowSPTC3aZie3Z9gpyC-w. Last accessed March 2, 2021.

587. Per author’s analysis of data from “DNV GL’s Energy Transition Outlook 2020,” DNV GL (DNV GL AS, 2020), https://eto.dnvgl.com/2020/index.html#ETO2019-top.

588. N. Sönnichsen, “Number of Nuclear Reactors Worldwide 1954–2019,” Statista (Statista, January 7, 2021), https://www.statista.com/statistics/263945/number-of-nuclear-power-plants-worldwide/.

589. “Investment in New Nuclear Declines to Five-Year Low,” World Nuclear News (World Nuclear Association, July 17, 2018), https://www.world-nuclear-news.org/NP-Investment-in-new-nuclear-declines-to-five-year-low-1707185.html.

590. "Plans for New Reactors Worldwide," World Nuclear Association (World Nuclear Association, January 2021), https://www.world-nuclear.org/information-library/current-and-future-generation/plans-for-new-reactors-worldwide.aspx#:~:text=Nuclear%20power%20capacity%20worldwide%20is,being%20created%20by%20plant%20upgrading.

591. "Nuclear Power Reactors," World Nuclear Association (World Nuclear Association, November 2020), https://www.world-nuclear.org/Information-Library.aspx.

592. "Nuclear Reactor Types." London: The Institution of Electrical Engineers, November 2005.

593. "NuScale and UAMPS to Facilitate Development of Carbon Free Power Project," Nuclear Engineering International (Progressive Engineering International, January 13, 2021), https://www.neimagazine.com/news/newsnuscale-and-uamps-to-facilitate-development-of-carbon-free-power-project-8453580/.

594. "Nuclear Power Reactors," World Nuclear Association (World Nuclear Association), accessed January 15, 2021, https://www.world-nuclear.org/information-library.aspx.

595. "Russia Relocates Construction of Floating Power Plant," World Nuclear News (World Nuclear Association, August 11, 2008), https://www.world-nuclear-news.org/NN-Russia_relocates_construction_of_floating_power_plant-1108084.html.

596. Charles Digges, "New Documents Show Cost of Russian Floating Nuclear Power Plant Skyrockets," Bellona.org (Bellona, May 25, 2015), https://bellona.org/news/nuclear-issues/2015-05-new-documents-show-cost-russian-nuclear-power-plant-skyrockets.

597. John Boyd, "Is the World Ready for Floating Nuclear Power Stations?," IEEE Spectrum: Technology, Engineering, and Science News (IEEE Spectrum, September 30, 2019), https://spectrum.ieee.org/energywise/energy/nuclear/is-the-world-ready-for-floating-nuclear-power-stations.

598. "Akademik Lomonosov," NTI.org (Nuclear Threat Initiative, September 24, 2019), https://www.nti.org/learn/facilities/942/.

599. "Unlocking Reductions in the Construction Costs of Nuclear," Nuclear Energy Agency (OECD, September 2, 2020), https://www.oecd-nea.org/jcms/pl_30653.

600. Samer Alsharif and Aslihan Karatas, "A Framework for Identifying the Causal Factors of Delay in Nuclear Power Plant Projects," Science Direct (Elsevier B.V., 2016), https://www.sciencedirect.com/science/article/pii/S1877705816301941 .

601. Edward W. Merrow, "Oil and Gas Industry Megaprojects: Our Recent Track Record," Spe.org (Society of Petroleum Engineers, April 2012), https://www.spe.org/media/filer_public/de/15/de15f740-fa58-4ca9-9383-ff54030f990f/153695.pdf .

602. Francois de Beaupuy, "EDF Cost Overrun at French Plant Piles Pressure on Nuclear Giant," BloombergQuint (Bloomberg L.P., October 10, 2019), https://www.bloombergquint.com/business/edf-lifts-cost-of-french-nuclear-reactor-by-14-to-13-6-billion.

603. "Transport of Radioactive Materials," World Nuclear Association (World Nuclear Association, July 2017), https://www.world-nuclear.org/information-library/nuclear-fuel-cycle/transport-of-nuclear-materials/transport-of-radioactive-materials.aspx.

604. Neil Hume, "Uranium Enters Bull Market after Covid-19 Hits Supply," Financial Times (Financial Times Ltd., April 8, 2020), https://www.ft.com/content/d5696032-e7d7-4945-a506-50898a15bccb.

605. Geert De Clercq, "French Nuclear Safety at Risk from Outsourcing: Report," Reuters (Thomson Reuters, July 5, 2018), https://www.reuters.com/article/us-france-nuclearpower-edf-idUSKBN1JV24Y.

606. "Industrial Way Forward Agreement Signed between the EDF Group and the Indian Energy Company NPCIL for the Implementation of 6 EPRs in Jaitapur," EDF France (EDF, March 10,

2018), https://www.edf.fr/en/edf/industrial-way-forward-agreement-signed-between-the-edf-group-and-the-indian-energy-company-npcil-for-the-implementation-of-6-eprs-in-jaitapur.

607. "Framatome Signs Long-Term Service Contract to Support Operation at Taishan EPRs in China," Framatome (Framatome, April 14, 2020), https://www.framatome.com/EN/businessnews-1892/framatome-signs-longterm-service-contract-to-support-operation-at-taishan-eprs-in-china.html.

608. "India and Kazakhstan to Renew Uranium Supply Contract for 2020-24," Business Standard (Business Standard, November 18, 2019), https://www.business-standard.com/article/pti-stories/india-kazakhstan-to-renew-uranium-supply-contract-for-2020-24-119111801605_1.html.

609. Kim Zetter, "An Unprecedented Look at Stuxnet, the World's First Digital Weapon," Wired (Conde Nast, November 3, 2014), https://www.wired.com/2014/11/countdown-to-zero-day-stuxnet/.

610. "SIMATIC WinCC / SIMATIC PCS 7: Information about Malware / Viruses / Trojan Horses," Siemens (Siemens AG, April 1, 2011), https://support.industry.siemens.com/cs/document/43876783/simatic-wincc-simatic-pcs-7%3A-information-about-malware-viruses-trojan-horses?dti=0&lc=en-WW.

611. Sonal Patel, "Advanced Nuclear Reactor Designs to Get Digital Twins," POWER Magazine (Access Intelligence, May 14, 2020), https://www.powermag.com/advanced-nuclear-reactor-designs-to-get-digital-twins/.

612. "GEMINA - Generating Electricity Managed by Intelligent Nuclear Assets," Advanced Research Projects Agency - Energy (U.S. Department of Energy), accessed January 15, 2021, https://arpa-e.energy.gov/sites/default/files/documents/files/GEMINA_Project_Descriptions_FINAL.pdf.

613. "Project Launched to Digitally Clone All French Reactors," World Nuclear News (World Nuclear Association, September 23, 2020), https://www.world-nuclear-news.org/Articles/Project-launched-to-digitally-clone-all-French-rea.

614. "Nuclear Power Reactors," World Nuclear Association (World Nuclear Association, November 2020), https://www.world-nuclear.org/information-library/nuclear-fuel-cycle/nuclear-power-reactors.aspx.

615. Ibid.

616. "Belgian Nuclear Operator to Cut O&M Costs by 5% by 2018," Reuters Events (Reuters, June 28, 2016), https://www.reutersevents.com/nuclear/belgian-nuclear-operator-cut-om-costs-5-2018.

617. "Huntsman Completes First Phase of REACH Registration Huntsman press release, December 16, 2010, http://www.huntsman.com/eng/News/News/Huntsman_Completes_First_Phase_of_REACH_Registration/index.cfm?PageID=8583&News_ID=7971&style=72.

618. David Steven Jacoby, "Global Trade Restrictions and Related Compliance Issues Pertaining to Oil and Gas Production Chemicals OTC paper reference OTC 22005-PP, May 10, 2011.

619. David Steven Jacoby, "Global Trade Restrictions and Related Compliance Issues Pertaining to Oil and Gas Production Chemicals OTC paper reference OTC 22005-PP, May 10, 2011.

Index

A

B

C

D

E

J

K

L

N

O

P

Q

R

S

T

U

V

W

Z

About the Authors

David Steven Jacoby is the President of Boston Strategies International, a global strategy consultancy, and Managing Director of BSI Energy Ventures. He serves on the advisory boards of New York Energy Week, WP Advisory (an M&A firm), and the International Supply Chain Educational Association. Previously he held international positions at Kearney and Oliver Wyman in the United States, France, Brazil, and Hong Kong. David taught Operations Management at Boston University's Questrom Graduate School of Business and earned his MBA and a Master of Arts from the Wharton School at the University of Pennsylvania. He also studied petroleum engineering and holds six supply chain certifications. His academic and professional contributions to the field of supply chain management span more than 300 publications and media events.

Alok Raj Gupta is an expert in energy and sustainability strategies focused on economic gains. He is the Founder & CEO of Envecologic, an energy & sustainability think tank, advisory, and training firm, and a Partner at Boston Strategies International. Alok holds a B.A. Honors in Economics from University of Delhi and MS degree in Economics (Resource & Environment Economics) from TERI University, India. He has been practicing as an energy and sustainability economist with a host of global corporations, multilateral agencies, and governments. Alok is considered to be one of Asia's emerging thought leaders, with over a hundred articles, essays and white papers in publication. He has delivered keynote talks and conducted training/workshops at numerous international platforms on addressing the most pressing energy and sustainability challenges of our times.

Other Supply Chain Books by David Steven Jacoby

The Guide to Supply Chain Management
ISBN: 1576603458

Optimal Supply Chain Management in Oil, Gas & Power Generation
ISBN: 1593702922

The High Cost of Low Prices: A Roadmap to Sustainable Prosperity
ISBN: 1631578278

Cover Legend

Solar

Gas / coal fired power

Power transmission

Hydrogen

Nuclear

Hydropower

Oil & gas

Wind

Utility storage

Geothermal

Biomass